620.1

19.99

The Properties of
Engineering Materials

Second Edition

The Properties of Engineering Materials

Second Edition

Raymond A. Higgins BSc (Birm), CEng, FIM

Formerly Senior Lecturer in Materials Science, West Bromwich College of Commerce and Technology; former Chief Metallurgist, Messrs Aston Chain and Hook Co. Ltd., Birmingham; and Examiner in Metallurgy to the Institution of Production Engineers, The City and Guilds of London Institute, The Union of Lancashire and Cheshire Institutes and The Union of Educational Institutes.

A member of the Hodder Headline Group
LONDON • SYDNEY • AUCKLAND

First published in Great Britain 1977
Second edition 1994
Third impression 1997 by Arnold,
a member of the Hodder Headline Group
338 Euston Road, London NW1 3BH

British Library Cataloguing in Publication Data
Higgins, Raymond A.
  Properties of Engineering Materials.—
  2Rev.ed
  I. Title
  620.11

ISBN 0 340 60033 0

Typeset in 10/11 Times by Wearset, Boldon, Tyne and Wear
Printed and bound in Great Britain by St Edmundsbury Press,
Bury St Edmunds, Suffolk and J W Arrowsmith Ltd, Bristol

PREFACE
To the First Edition

This book was written primarily as an introduction to Materials Science for those engineering students who become involved with the subject at advanced technical college level or, as first-year undergraduates, feel the need of an introduction to its basic principles. It may also prove useful to those students of metallurgy who find it necessary to extend their knowledge to include the properties of non-metallic materials.

The treatment of materials science in this book is of a descriptive and qualitative nature. The author has little regard for those writers who, shirking the task of setting down a lucid explanation of some principle or dogma, proclaim that '. . . This is governed by the following mathematical expression . . .'. Similarly the temptation to write a textbook which claims to be an 'introduction' to the subject yet lambastes the luckless reader with differential equations on page 2, has also been resisted.

Some years ago the author was lecturing on Fick's Law of Diffusion to a class of engineering students and, having given a description and explanation of the principles involved, remarked that for the benefit of devotees of the differential calculus, these Laws could be expressed mathematically as follows. . . . At this point one particular student visage became illuminated with the light of understanding and, in response to a doubtlessly acrid remark from the lecturer, its owner admitted: 'I've just realised what calculus is all about.' The student in question had been exposed to the calculus in a presumably sterile pure-mathematical form for at least three years previously.

When the author began to teach metallurgy to engineers he quickly discovered that their conception of the nature of solids was one in which minute billiard balls—called 'atoms'—were stuck together with some mysterious 'celestial glue'. We now appreciate that even a mildly academic study of the mechanical properties of engineering materials is meaningless if it does not seek to relate the strengths of these materials to the forces which operate between the fundamental particles contained in them. Most modern syllabuses have therefore been written along these lines. Nevertheless many students still enter engineering courses without having offered chemistry even as an 'O-level' subject. With this problem in mind the relationship between atomic structure and chemical bonding is discussed in simple terms in the first three chapters. Similarly an elementary knowledge of van der Waals forces has been deemed necessary in order that the reader can compare and contrast the mechanical properties of metals with those of plastics materials. It is hoped that the treatment of these topics under the headings of 'The Atom', 'The Molecule' and 'The Crystal' will help the reader to clarify his ideas sufficiently so that he can embark on the study of materials science which follows.

No doubt in an effort to 'get into the market' quickly many authors of otherwise excellent textbooks of metallurgy have, in recent years, added to their books an extra chapter on 'Plastics' and have then considered that they were justified in describing their works as dealing with 'Materials Science'. Such a course of action is rarely fruitful in that the non-metallics chapter remains totally divorced from the rest and this author resisted the temptation to proceed in this way with his 'Engineering Metallurgy' which will continue as a separate publication dealing solely with metals.

Finally the author wishes to thank his friend and colleague of many years standing, E. A. Boyce Esq., F.R.I.C., of the West Bromwich College, for reading the first three chapters of this book and for making helpful suggestions relating to the basic chemistry they contain.

R. A. Higgins
Division of Materials Technology,
West Bromwich College of Commerce and Technology,
Wednesbury, West Midlands.

*PREFACE*_____
To the Second Edition

The Swedish scientist Svante August Arrhenius died at about the time I was receiving my first school lessons in physics but it is only quite recently that his 'Equation' has found its way into college syllabuses for Materials Science. Other fundamental scientific 'laws', 'hypotheses' and 'relationships' have been similarly assimilated. The science of materials has ceased to deal with substances on a mainly descriptive basis and now follows a more 'quantitative' approach so that the student, in addition to retaining a basic knowledge of physics and chemistry, must be prepared to use expertise acquired in his accompanying maths course.

This change in the approach to the study of Materials Science during recent decades has enabled the examiner to pose questions requiring a numerical answer and, in turn, this has helped him to appease examinations assessors, referees, committees and the like who were continually clamouring for more questions of a 'searching nature' whilst rejecting those which, they claimed, merely called for 'a regurgitation of class notes'. Consequently in this edition I have introduced a few typical examination questions, along with suggested answers, on such topics as are suitable. Nevertheless the reader should regard any practical application of some of these problems with caution. Thus, whilst the differential equations associated with Fick's Laws indicate to us that diffusion in metallic solid solutions is not a simple 'linear' process and also explain to us why coring is never completely removed even after prolonged annealing, these Laws are of limited use in practical calculations (see footnote to section 9.9.1).

Since this book was first written there has been considerable progress in some branches of Materials Science. Whilst the properties of metals and alloys had been fairly extensively explored during the first half of the twentieth century, ceramics and composite materials had received only limited attention. The closing decades of the twentieth century, however, saw both ceramics and a wide range of composites becoming firmly established as engineering materials. Consequently chapters dealing with these topics have been almost completely rewritten.

My thanks are due to Dr Russell W. H. Parry, Senior Editor with Messrs

Edward Arnold, who, as a result of his travels around Britain, is able to keep me informed of many of the developments taking place in university and college course syllabuses.

Mention at the beginning of this preface of early science lessons I received as a boy brought to mind one introducing us to 'light'. This phenomenon was explained as being propagated by 'wavemotion travelling through the æther (or ether), a *fluid* which pervades the whole Universe'. Similarly one often heard the phrase, 'I've just heard it over the æther!' (a reference to some momentous BBC announcement such as the marriage— or divorce—of one of their star 'crooners'). Much later in my school days the neutron was discovered—until then it had disguised itself as a proton clutching to its bosom a captive electron. Ah well, things have moved on since those far-off days and it's been quite a task keeping up with it all. In the words of the Preacher some three thousand years ago—

> '. . . of making many books there is no end; and much study is a weariness of the flesh.'
>
> (Ecclesiastes, XII, 12)

R. A. HIGGINS
Walsall,
West Midlands.

CONTENTS _____

To
THE WOMEN OF MY LIFE—
My Wife, Helen and our
Daughter, Alice.

'Whatever happens, happens as it should. Observe carefully and you will find this to be true.'

Marcus Aurelius (AD 121–180), 'Meditations' (IV, 10)

or, put another way

'When the tree falls, how can the shadow stand?'

Mary Lavin, 'In the Middle of the Fields'.

1

The Atom

1.1   Until a few hundred years ago—a mere blink of the eye in the time scale of the Universe—educated Man believed that our Earth was the fixed hub about which the whole of Creation revolved. The modern astronomer, however, visualises this planet as but a speck of dust in a galaxy, one of billions of other galaxies, that go to make up an ever-expanding Universe, the visible edge of which is some 10^{23} km distant. Meanwhile the atomic scientist concerns himself with the architecture of the atom itself and in so doing is dealing with minute particles of the order of 10^{-31} kg in mass. Somewhere between these two extremes in size ordinary Man gazes out into the apparently limitless void and tries to make sense of it all.

The culture of the Ancient Greeks was literary rather than scientific, philosophical rather than technological. Certainly they built impressive temples but the building blocks of these they fixed together with iron clamps in order to save themselves the trouble of inventing cement. Although we still remember the intellectual observations of Archimedes, experimental research was not really the fashion in those days and so we find Aristotle proclaiming that the Universe consisted of four 'elements' and four 'essences' (Fig. 1.1). Unfortunately this line of thought persisted even as late as the Renaissance, often with Man trying to 'bend' what facts he had accumulated in order to fit them into some preconceived romantic idea. The mathematician, Johannes Kepler, for example, tried to calculate the planetary orbits so that they were based on some harmonic relationship thus providing the totally imaginary 'music of the spheres'. Though, on the credit side, one must admit that this was an indication that thinking men were using their knowledge of geometry to produce order out of chaos.

1.1.1   Possibly the most important hypothesis left to us by the Greeks was Democritus' idea of the atom. He suggested that all matter was composed of small but indivisible particles which he calls 'atoms'. Our modern word is in fact derived from the Greek *atomos*, meaning 'indivisible', and though we now know that these atoms can be subdivided into a series of even

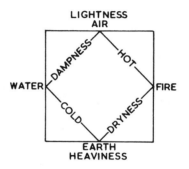

Fig. 1.1 Aristotle's idea of the nature of the Universe. Even much later men were still obsessed with ideas of simple geometrical patterns as governing natural laws. Students of the occult still play around with 'pentacles' and the like.

smaller particles, by this fundamental concept Man had laid the corner-stone of modern chemistry.

The ideas of Democritus and his friends appear to have been neglected both during the Dark Ages and the Renaissance which followed and it was not until 1803 that the English chemist John Dalton enunciated his Atomic Theory. From then onwards progress in the field of pure chemistry was rapid and by 1868 the Russian chemist Dimitri Mendeléef produced his first great classification of the known elements, a system by which he was able to predict the properties of other elements at that time still unknown.

Science deals largely with systems of classification. Thus, the biologist classifies living things first as either animal or vegetable. The former he subdivides into vertebrate and invertebrate—and so on. Chemistry is, above all, a science of classification. Both elements and their compounds are placed into groups depending upon their chemical and physical properties. In this book we shall be classifying substances mainly because, as engineers, we are interested in their mechanical properties, but first it will be necessary to seek reasons which explain why some materials are strong whilst others are weak, and, in particular, what sort of sub-atomic 'glue' makes atoms 'stick' together at all.

The Nature of Fundamental Particles

1.2 As long ago as 1833 Michael Faraday reached the conclusion that a flow of electricity was due to the movement of separate particles through a conductor whilst, later, Sir William Crooke showed that cathode rays could be bent by an electromagnetic field thus suggesting that these rays consisted of electrically-charged particles of some kind. These particles ultimately became known as 'electrons' and were in fact the first of the sub-atomic particles to be discovered. Today we make considerable use of the electromagnetic deflection of electrons, as for example, in the cathode-ray tube of a television set.

1.2.1 The Electron   is a very small particle carrying a charge of what we arbitrarily refer to as 'negative electricity', it was examined quantitatively in 1897 by Sir J. J. Thomson. He assigned the first value for the ratio 'charge/mass' (e/m) to the electron by observing the deflections of cathode rays in magnetic and electrical fields of known strengths. His early results were inaccurate—as are those of any explorer with few 'stars to guide'—and the currently accepted values are:

Charge: 1.6×10^{-19} coulombs (negative electricity)
Mass: 9.106×10^{-31} kg (0.000 548 8 a.m.u.)

(Since the masses of these particles of atomic and sub-atomic size are inconveniently small when related to the kilogram, the *atomic mass scale* is generally used to express the masses of fundamental particles and atoms. The atom of the carbon isotope $^{12}_{6}C$ is taken as the standard at 12.000 00 atomic mass units (a.m.u.).

During the course of its history the electron has acquired other names such as the 'negatron' and the 'β-particle' and is represented by various symbols such as β, e^- and $_{-1}^{0}e$.

The electron has another special property which influences its effect on atomic properties. Besides having mass and electrical charge it behaves as though it were spinning on an axis. This 'spin' causes the electron to act as though it were a minute magnet and so established the close relationship between magnetism and electricity. It also means that the electron possesses angular momentum and this must be taken into account in spectrographic measurements.

1.2.2 The Proton   In 1886 Eugene Goldstein noticed luminous rays emerging from holes in a perforated cathode which he had used in a cathode-ray tube. These rays were travelling in a direction *opposite* to that of the cathode rays. Later it was shown that they were deflected by both magnetic and electrical fields but in the opposite direction to that of a stream of electrons (cathode rays). Consequently it was realised that the particles of which these new rays consisted were positively charged. Initially they were known as 'canal rays' since they passed through channels in the cathode but in 1907 J. J. Thomson suggested the name 'positive rays' and when a determination of the value charge/mass (e/m) was made it showed them to consist of particles which were much heavier than electrons. Ultimately it was found that the mass of the lightest of these new particles was roughly that of the ordinary hydrogen atom (1.3.2) when stripped of its lone electron. The name *proton*, derived from the Greek word meaning 'first', was suggested by Rutherford in 1920.

The proton carries an equal but opposite (hence positive) charge to that of the electron. Its mass is 1.6725×10^{-27} kg (1.007 263 a.m.u.) making it some 1836 times heavier than the electron. It is represented by a number of symbols such as p, ^1H and ^1_1H; the latter being the most commonly used and indicating that it is the nucleus of the hydrogen atom with a mass of one and a positive charge of one.

1.2.3 The Neutron During my school days chemists got along quite well with just two particles with which to explain the properties of atoms—the negative electron and the positive proton. Philosophically the concept of these two particles seemed to fit in with the existence of two sexes in the biological world! It was realised, however, that all atoms with the exception of hydrogen were more massive than they should be when taking into account the numbers of positively charged protons in their nuclei. The difficulty was overcome by assuming that also present in the nucleus were a number of protons which were 'paired' with electrons on a one for one basis thus rendering the composite unit electrically neutral. Thus we classified electrons as being either 'nuclear electrons' or 'planetary electrons'. So N. V. Sidgwick's *The Electronic Theory of Valency*, published in 1927, was an erudite treatise on chemical combination, the bulk of which is still valid today, yet it makes no mention of the *neutron* since this particle was not discovered until 1932 by Sir James Chadwick. (Nevertheless during the year 1934/35 when I studied inorganic chemistry at university the lecturer still disregarded the existence of the neutron and in fact our recommended text book was Sidgwick—I still have my copy!)

The neutron is so called because it has no resultant electrical charge. It has a mass of 1.675×10^{-27} kg (1.008665 a.m.u.) making it very slightly more massive than the proton so that earlier chemists can be forgiven for assuming that these units of 'dead mass' in the atomic nucleus were protons 'neutralised' by the presence of comparatively light electrons.

Our knowledge of the rôle of the neutron in nuclear science is still far from complete. It is a relatively stable particle within the nucleus of most atoms but in a number of atoms with very large nuclei it tends to be unstable giving rise to radioactivity and other forms of induced nuclear reaction. Generally the presence of the neutron has little effect on ordinary chemical and physical properties (other than atomic mass). Chemical properties are governed by the disposition of electrons present in the atom.

1.2.4 Since Chadwick's discovery of the neutron the number of elementary particles has proliferated. These can be very roughly classified into three main groups:

(i) *baryons*   protons and other particles of mass greater than that of the proton;

(ii) *mesons*   any of a group of particles having a rest mass between those of the electron and the proton and having an integral 'spin';

(iii) *leptons*   which include electron, muon, positron (positive electron) and neutrino (which possesses neither mass nor charge—but only spin!).

The term 'quark' has been adopted to describe any one of a number of hypothetical elementary particles with charges of $\frac{1}{3}e^-$ or $-\frac{1}{3}e^-$ (where e^- is the charge on the electron) and considered to be the fundamental units of all baryons and mesons. Incidentally the word 'quark' was coined by James Joyce in *Finnegan's Wake*! However, the reader need not be discouraged by the complexity of atomic structure. Only three particles—the electron, the proton and the neutron—have any direct effect on the distinctive properties of each element so no consideration of other baryons, mesons, leptons—or quarks—need be given and will not be mentioned further in this book. For, in truth, it seems that:

'We are such stuff as dreams are made of.'

Before we leave the intriguing subject of fundamental particles (or should we refer to them as 'entities'?) perhaps a brief mention should be made of the apparent dual nature of the electron. In a study of particle physics the electron can be equally well considered as a discrete particle or as a form of wave motion in which the electron constitutes a *wave packet*. (Light which is usually described as a form of electromagnetic radiation may also be considered as a beam of 'particles' or 'photons'.) However, in studying the structures of atoms it will be more convenient to regard the electron as a particle even though this gives a restricted vision of the subject.

Table 1.1   *Summary of Properties of the Principal Fundamental Particles*

Particle	Approximate mass (relative to the proton)	Actual mass (a.m.u.)	Resultant electrical charge	
Electron	1/1836	0.0005488	Negative	
Proton	1	1.007263	Positive	} Equal but opposite
Neutron	1	1.008665	Zero	

The Structure of the Atom

1.3   Everything is built up of atoms. They are very tiny particles and the copper-nickel alloy British 5p coin in your pocket contains some 30 700 000 000 000 000 000 000 atoms—give or take a few trillions whilst, in America, the dime will contain roughly 21 900 000 000 000 000 000 000 atoms—give or take a few quintillions.

Each atom has at its centre a nucleus consisting of a group of protons and neutrons whilst 'in orbit' around the nucleus are electrons. An individual atom consists almost entirely of empty space. Although most of the mass is concentrated in the nucleus both the nucleus and the individual electrons in 'orbitals' around it are of approximately the same diameter. If we imagine the nucleus of an atom as consisting of a sphere the size of a golf ball then the outermost electron will be represented by a golf ball in an orbit of some 100 metres radius from the nucleus.

We must be careful, however, not to accept too literally the ideas which diagrams representing atoms present. These diagrams (Fig. 1.2) are merely a convenient way of summarising the general make-up of an atom rather than giving an actual picture of what the atom would look like were it possible for it to be made visible. Thus, such diagrams bear the same relationship to the actual nature of an atom as does a paper street guide to

Fig. 1.2 A 'particle' diagram of the oxygen atom.

the pictorial view of a town—it shows the locality of the 'Cock and Magpie' but gives no hint of the actual appearance of that excellent hostelry.

1.3.1 Although the original picture of the atom as visualised by the Danish physicist Niels Bohr in 1913 is now known to give an incomplete picture of the structure of the atom it is still of considerable use in showing, subject to the sort of limitations just mentioned, the general particle make-up of the atom.

Briefly, Bohr suggested that electrons orbited the nucleus in fixed 'shells' but that when an electron received energy it moved from a lower to a higher level within the shell. Conversely when it gave up that energy it fell back again to its former level (Fig. 1.3). Thus Bohr applied a quantum theory to electron orbits in roughly the way in which Max Planck and Albert Einstein had quantised light.

Knowledge of the grouping of electrons was derived originally from spectromagnetic studies whereby the wavelength (λ) of light emitted was measured. From these studies it was concluded that a small 'packet' or *quantum* of energy is required to make an electron move from one energy level to the next-higher energy level. Conversely a quantum of energy, known as a *photon*, was emitted when an electron fell back from a higher

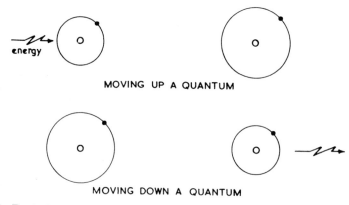

Fig. 1.3 The hydrogen atom contains one proton in the nucleus and one electron in orbit around it. The orbit in which the electron exists depends upon the energy state of the electron.

to the next-lower level. The energy E of the photon can be calculated from its wavelength λ using the equation:

$$E = \frac{h.c}{\lambda}$$

where h is Planck's constant, c is the velocity of light and c/λ becomes the wave frequency. In practical terms Planck's constant is 6.62×10^{-34} joule.seconds.

1.3.2 The somewhat tenuous nature of the electron has already been touched upon. In fact physicists no longer think of an electron orbit as consisting of a specific path. The definite 'orbit' has been replaced by the term *orbital* which refers to a mathematical function representing the distribution of the electron within the space occupied by the atom. The electron can be visualised as a sort of 'mist' of electricity rather than as a single particle; the orbital prescribing the density of that 'mist' at any point within the atom. Alternatively the electron may be regarded almost in the nature of a 'resultant particle' (here the term 'resultant' is used in the mechanical sense) and the orbital interpreted as denoting the statistical probability of finding that particle at any given point around the nucleus. Consequently it must be stressed again that pictures showing an atom as consisting of a central nucleus about which a number of electrons are moving in definite orbits must not be interpreted too literally. They are purely symbolic and offer a simple diagrammatic means of showing the electron complement of the various quantum shells. Such diagrams will be used in the pages which follow in order to explain in simple terms the methods of chemical bonding.

The modern idea of the structure of the atom, as applied to that of hydrogen, is indicated in Fig. 1.4. Instead of an electron occupying a single fixed orbit the probability of it existing at any location is indicated by the graph (ii) and resultant diagram (iii).

Mendeléef had observed the periodicity of chemical properties of the then-known elements and this knowledge ultimately gave rise to the Periodic Table as we know it today (Table 1.3). Not only is there a

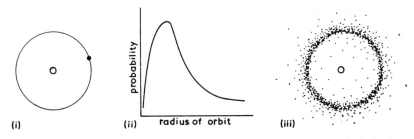

Fig. 1.4   A pictorial interpretation of the possible position of the lone electron in the hydrogen atom.
(i) The 'Bohr atom'. (ii) A graph indicating the probability of the orbit being of a particular radius.
(iii) The resultant idea of the position occupied by the electron.

repeating pattern with regard to chemical properties of the elements in the same group but also periodicity in connection with their electrical, magnetic and mechanical properties. Modern atomic theory successfully explains why the arrangement of elements within the Periodic Table is governed by the manner in which additional electrons are situated in elements of ascending 'Z number'.

1.3.3   Electrons surrounding the atomic nucleus are not all at the same energy level. Consequently it is convenient to group electrons into shells with different characteristics. The first *quantum shell* contains only two electrons, whilst the second shell contains a maximum of eight, the third eighteen, and the fourth thirty-two. The geatest number of electrons in a given shell is $2n^2$ where n is known as the *quantum number* of the shell. As already suggested this electron shell concept is somewhat over-simplified. Amongst other things it suggests that all electrons within a shell are equal whereas in practice they are not. The application of what is known as *Pauli's exclusion principle* is of help here. This principle states that there are definite rules governing the energy levels and probable positions of electrons which are in orbit around a nucleus. Thus the single electron of a hydrogen atom is normally at the lowest possible energy level so that the most probable position of this electron is that indicated in Fig. 1.4(ii). No more than two interacting electrons can have the same orbital quantum numbers. Even these two are not entirely identical since they exhibit inverted magnetic behaviours or 'spins'.

Figure 1.5 shows the electron complement—depicted in the manner of the Bohr atom—for the first eleven elements of ascending atomic number (Z). An atom of the first element, hydrogen, contains a single proton in the nucleus and a single electron in orbit around it. The atom is of course electrically neutral. Helium contains two protons in the nucleus and two electrons in the first shell. These two electrons fill the first shell making helium a very stable element which does not react with other elements. Lithium has three protons ($Z = 3$), and since the first quantum shell is

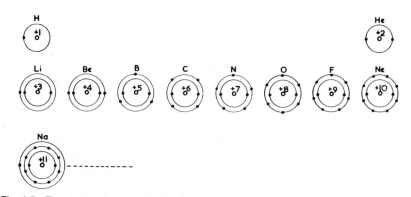

Fig. 1.5   The electron 'make-up' of the first eleven elements of the periodic classification. The atoms are depicted in the Bohr manner.

already filled (with the helium atom) the third electron in the lithium atom must go into the second quantum shell. In atoms of each of the elements which follow—beryllium, boron, carbon, nitrogen, oxygen, fluorine and neon—one more electron is added to the second quantum shell. However, as indicated by *Pauli's exclusion principle* no more than two electrons can have the same energy levels and, hence, the same probability of position. Additional location probabilities are therefore introduced and *subshells* established in the manner indicated in Fig. 1.2 showing a simple 'particle diagram' of the oxygen atom.

The fact that spectrometric methods were used in early work on atomic structures led to the adoption of spectrographic nomenclature. Thus, since the most definite spectral lines are obtained from those electrons which fall to the lowest energy level within a given shell, the notation s has been used for those electrons in each quantum shell which have the lowest energy levels. Hence $1s^2$ indicates that two electrons (of opposite magnetic spins) are present in the low-energy first quantum shell (formerly referred to as the 'K shell'). Similarly $2s^2$ indicates that two electrons are present in the lowest energy level of the second quantum shell (formerly the 'L shell'). Since two is the maximum number of electrons which can exist in this s-subshell, further subshells are created to hold the other six electrons which are members of this second principal shell.

Each of the succeeding shells contains two or more subshells which are designated p, d and f. The greatest number of electrons in these subshells is six, ten and fourteen respectively. We will consider again the example of oxygen (Fig. 1.2) which has six electrons in its second quantum (or L) shell. It has the electron notation $1s^2$; $2s^2$, $2p^4$ which indicates that two electrons are in the first (or K) shell and six electrons in the second (or L) shell (two in the lower subshell and four in the next-higher subshell). The complete electron notation system for the chemical elements in the first four periods of the periodic table is given in Table 1.2.

1.3.4 A study of Table 1.2 will show that in several instances quantum shells of higher number begin to accept electrons in their levels before the preceding shells of lower quantum number have been filled. This is due fundamentally to overlapping of energy levels of succeeding subshells and also of succeeding principal shells. Since electrons are most stable when they are in a position of least energy they naturally seek out suitable spaces in shells or subshells where vacancies of lowest energy level occur. For these reasons there are groups of 'transition' elements in the periodic classification in which an outer—or valency—subshell begins to fill before a full quota of electrons is present in the previous subshell. For example in the scandium–nickel series the $4s$ subshell begins to fill before the $3d$ subshell has received its full complement of ten electrons. Similar situations occur with other transition series such as yttrium-palladium and in the lanthanide ('rare earth') and actinide (trans-uranic) elements.

Table 1.2 *Electron notation of the elements in the first four periods of the periodic table*

Element	Atomic Number (Z)	1s	2s	2p	3s	3p	3d	4s	4p	4d	4f
H	1	1									
He	2	2									
Li	3	2	1								
Be	4	2	2								
B	5	2	2	1							
C	6	2	2	2							
N	7	2	2	3							
O	8	2	2	4							
F	9	2	2	5							
Ne	10	2	2	6							
Na	11	2	2	6	1						
Mg	12	2	2	6	2						
Al	13	2	2	6	2	1					
Si	14	2	2	6	2	2					
P	15	2	2	6	2	3					
S	16	2	2	6	2	4					
Cl	17	2	2	6	2	5					
Ar	18	2	2	6	2	6					
K	19	2	2	6	2	6		1			
Ca	20	2	2	6	2	6		2			
Sc	21	2	2	6	2	6	1	2			
Ti	22	2	2	6	2	6	2	2			
V	23	2	2	6	2	6	3	2			
Cr	24	2	2	6	2	6	5	1			
Mn	25	2	2	6	2	6	5	2			
Fe	26	2	2	6	2	6	6	2			
Co	27	2	2	6	2	6	7	2			
Ni	28	2	2	6	2	6	8	2			
Cu	29	2	2	6	2	6	10	1			
Zn	30	2	2	6	2	6	10	2			
Ga	31	2	2	6	2	6	10	2	1		
Ge	32	2	2	6	2	6	10	2	2		
As	33	2	2	6	2	6	10	2	3		
Se	34	2	2	6	2	6	10	2	4		
Br	35	2	2	6	2	6	10	2	5		
Kr	36	2	2	6	2	6	10	2	6		
etc.											

Shell 1 — 1s

Shell 2 — 2s, 2p ← sub-shells

Shell 3 — 3s, 3p, 3d ← sub-shells

Shell 4 — 4s, 4p, 4d, 4f sub-shells

4s begins to fill before 3d

4f is not filled until element no.71 (Lu) by which time 6s has already been filled.

The Atomic Nucleus

1.4 The nucleus of an atom consists of an association of protons and neutrons. Since these are particles some 1836 times heavier than the electron it follows that the mass of an atom tends to be concentrated in its nucleus. The number of protons in the nucleus of a stable atom is always equal to the total number of electrons in the various shells around the nucleus. Since the charge on a proton is equal but opposite to that on an electron it follows that a stable atom is electrically neutral. The number of protons in the nucleus of an atom is called the Atomic Number (Z) of the element.

Table 1.3 *The periodic classification of the elements. All elements with atomic numbers above 92 are 'artificial' – the products of the nuclear scientist. Since this classification was last revised numbers 107 (Uns), 108 (Uno) and 109 (Une) have been reported. Fortunately the fashion for assigning 'patriotic' names to these relatively unimportant metals is now past, and IUPAC (the International Union of Pure and Applied Chemistry) allocates to each new element a name which states its atomic number in 'Dog Latin'. Thus 'Unq' (104) is 'Unniliquadum', ie Un-nil-quadum or 1-0-4; whilst 'Uun' (110) is 'Ununnilium', ie Un-un-nilium or 1-1-0.*

Fig. 1.6 Diagrammatic representation of two isotopes of chlorine.

The neutrons, also present in the nucleus, have little effect on the *chemical* properties of the element since they carry no resultant electrical charge. They behave as a form of 'nuclear ballast' increasing the total mass of the atom. The number of neutrons present in the nucleus may vary even from one atom to another *of the same element*. Consequently when we measure the *relative atomic mass* of an element we find that it is rarely a whole number since we are in fact measuring the average mass of a large number of atoms of several different masses. Originally the relative atomic mass* of an element was the average specific mass of an atom of that element as compared with the mass of an atom of hydrogen. For various reasons a carbon atom of atomic mass 12.0000 is now used as the standard. However, in an understanding of the properties of the elements the atomic number (Z) is of much greater significance than the relative atomic mass.

1.4.1 Isotopes Whilst the number of protons in the nucleus of any atom of a single element is fixed, the number of neutrons can vary from one atom to another. Thus the nucleus of any chlorine atom will always contain 17 protons, but about 75% of all chlorine atoms contain 18 neutrons whilst roughly 25% contain 20 neutrons. These two different chlorine atoms are known as *isotopes* and chlorine is said to be *isotopic*. The relative atomic masses of these two chlorine isotopes will be 35 and 37 respectively (Fig. 1.6) but since the lighter isotope is the more plentiful this explains why the average relative atomic mass of chlorine is 35.46. It is important to remember that these two isotopes will have exactly similar chemical properties since both have the same atomic number and consequently the same electron configuration.

Hydrogen too is isotopic and occurs as three isotopes (Fig. 1.7). The 'ordinary' hydrogen atom (sometimes called *protium*)† contains a single proton in its nucleus, whilst the nucleus of isotope 2 (generally called *deuterium*)† contains a neutron in addition to the proton. The nucleus of isotope 3 (known as *tritium*)† consists of two neutrons in addition to the

---

* Previously known as 'atomic weight'.

† This naming of isotopes is to be discouraged—it may give the impression that separate *elements* are being described rather than isotopes of a single element, in this case hydrogen.

mass number = A [A = Z + N] protons = Z ⊢H	STRUCTURAL DIAGRAM	NUCLEUS		ELECTRONS
		protons (Z)	neutrons (N)	
$^1_1$H HYDROGEN		O	none	●
$^2_1$H DEUTERIUM		O	◉	●
$^3_1$H TRITIUM		O	◉◉	●
$^3_2$He HELIUM		OO	◉	● ●

CHEMICALLY IDENTICAL

Fig. 1.7 The structural make-up of the three isotopes of hydrogen. The isotope of helium, $^3_2$He, has a similar atomic mass to tritium but is of a very different nature because of its 'filled' electron shell.

proton. Although the relative atomic masses of the three isotopes are 1, 2 and 3 respectively the average relative atomic mass of hydrogen is no more than 1.00797 because isotopes 2 and 3 are extremely rare.

1.4.2 Nuclear Binding Forces The nucleus of an atom consists of a collection of protons and neutrons. Now like-charged bodies repel each other with a force which is inversely proportional to the square of their distance apart (the Coulomb Law). Since protons are positively charged particles why does not the nucleus disintegrate due to repulsion between these protons? The answer to this question is that forces which bind together the particles of the nucleus are not in fact of an electrostatic nature. Neither are they magnetic or gravitational in character but appear to be separate forces entirely and about which little is yet known. These forces act not only between the protons but also between the neutrons and are known as nuclear binding forces.

An atomic nucleus may contain a large number of *nucleons* (i.e. protons and neutrons) bound together as a coherent unit. Nucleons in the core are subjected to their special forces of attraction by all neighbouring nucleons but those on the surface are attracted inwards towards the core by the bulk of the nucleons situated there. This produces something in the nature of a

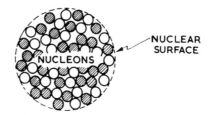

Fig. 1.8 Nucleons (i.e. protons ○ and neutrons ⦿) at the surface are attracted inwards by those at the core.

'surface tension' effect at the outer limits of the nucleus so that nuclei can be regarded as being constituted after the fashion of droplets of moisture.

Whilst much remains to be discovered about these nuclear binding forces it is clear that they are effective over an extremely limited range, often less than the radius of the nucleus. Figure 1.9 is a graph in which the potential energy possessed by a proton, due to forces acting between it and a nearby nucleus, is plotted as a function of the distance separating them. When the proton is at A simple electrostatic repulsion tends to push it away from the nucleus which has its centre at O. Thus the proton can be considered to possess potential energy in the same way that a compressed spring possesses potential energy and if released it would recoil from the nucleus. As the proton gets closer to O the electrostatic force—and hence the potential energy of the proton—increase predictably according to the Coulomb Law, and potential energy would be expected to go on increasing in the direction of X until contact was achieved between proton and

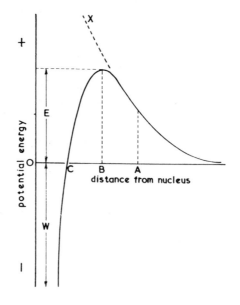

Fig. 1.9

nucleus. At B, however, the curve flattens off at some potential energy value, E, because the proton now comes close enough to be under the influence of the nuclear binding force. Beyond C the nuclear binding force is much greater than the electrostatic force of repulsion and so the proton is now attracted to the nucleus instead of being repelled by it. Relative to its previous condition it can now be considered to possess 'negative potential energy'. The captured proton is said to be in a 'potential well' of depth $E + W$ and this energy would be required to remove it from the nucleus.

1.4.3 An important point to appreciate is that whilst *electrostatic repulsion acts over fairly large distances, the nuclear binding force operates over very small distances*. Therefore a small number of nucleons can hold together because it is geometrically possible for the nuclear binding force to overcome the electrostatic repulsion between protons, but there is a limit to this number. A very large nucleus is certain to possess many nucleons which are outside the influence of the short-range nuclear binding force but any protons will still be influenced by the mutual repulsion between each other's long-range electrostatic fields. Therefore as a nucleus becomes large the total electrostatic repulsion between protons increases whilst beyond a certain size the total nuclear binding force between nucleons does not.

Thus in a heavy nucleus the total binding force is not strong enough to hold the nucleus together against the total repulsion forces between all of the protons. This is one reason why 'extra' neutrons are present in these heavy nuclei of uranium and radium—to provide extra binding force to the total whilst adding nothing to the electrostatic forces of repulsion, since neutrons have no charge. However, these extra neutrons make the nucleus more massive still and the total binding force still weaker so that in spite of its extra neutrons a massive nucleus tends to break up.

For a nucleus to be stable it needs to be less than a certain maximum size, determined by the relative magnitude of the opposing forces of attraction and repulsion and containing an optimum ratio of charged particles (protons) to uncharged particles (neutrons). All naturally occurring nuclei (with the exception of hydrogen ^1_1H) are a mixture of protons and neutrons and the ratio of neutrons to protons increases from $1:1$ for the lightest nuclei to roughly $1.5:1$ for the most massive nuclei. Those nuclei with atomic mass numbers greater than 209 are unstable to some degree and no atom with a mass number greater than 238 occurs naturally. Those massive nuclei therefore undergo radioactive decay.

1.4.4 Nuclear Binding Energy By using an apparatus known as the mass spectrograph it is possible to determine atomic masses with a precision of about 1 in 100 000. In this way the mass of the proton has been determined as 1.007 263 a.m.u. and the mass of the neutron as 1.008 665 a.m.u. Since the numbers of protons and neutrons in any atomic nucleus are known and since the masses of the elementary constituents have been accurately determined it would seem possible to claculate the mass of a nucleus with precision from these values. For example the helium nucleus $^4_2$He contains two protons and two neutrons. Hence its mass should be:

$$2 \text{ protons} = 2 \times 1.007\,263 = 2.014\,526 \text{ a.m.u.}$$
$$2 \text{ neutrons} = 2 \times 1.008\,665 = 2.017\,330 \text{ a.m.u.}$$
$$\text{Total} = 4.031\,856 \text{ a.m.u.}$$

But in fact the mass of the helium nucleus is 4.002\,764 a.m.u. The difference between the expected mass and the actual mass is:

$$4.031\,856 - 4.002\,764 \text{ a.m.u.} = 0.029\,092 \text{ a.m.u.}$$

This difference is known as the *mass defect*. It follows that in the formation of the helium nucleus from its constituent protons and neutrons a portion of matter amounting to 0.029\,092 a.m.u. is converted into energy. This energy is present in the form of the binding force which holds the particles together as suggested in the previous section. The amount of energy released by the conversion of this amount of matter may be derived from Einstein's mass-energy equation:

$$E = mc^2$$

where c is the velocity of light $(3 \times 10^8 \text{ ms}^{-1})$; m is the mass involved and E the energy produced. Hence the substitution of above values in this equation indicates the energy released:

$$E = 0.029\,092 \times 1.661 \times 10^{-27} \times (3 \times 10^8)^2 \text{ J}$$
$$= 4.35 \times 10^{-12} \text{ J}$$

(where 1 atomic mass unit $= 1.661 \times 10^{-27}$ kg).

States of Matter

1.5   All elements can exist as either gases, liquids or solids. The *state* in which an element exists depends upon the combination of temperature and pressure which prevails at that time.

Fig. 1.10   The nucleus of the helium isotope $^4_2$He weighs less than the four nucleons (2 protons and 2 neutrons) taken separately.

Fig. 1.11 Forces operating between single atoms.
(i) Here the atoms are so far apart that only weak forces of attraction operate between them. (ii) Now the atoms are close together and strong forces tend to pull them closer together. (iii) When the electron orbitals are close or overlapping a strong force of repulsion builds up.

Forces of attraction and repulsion operate between atoms and the total force involved may be the resultant of a number of forces. First, forces of repulsion will operate between the nuclei of two atoms since both nuclei are positively charged. Similarly the electrons of one atom will repel those of another atom since electrons are negatively charged. Opposing these forces of repulsion are forces of attraction in which the positively charged nucleus of one atom is attracted by the negatively charged electrons of the other atom. At the same time there will be a gravitational force of attraction operating between the massive nuclei of both atoms. Whether the resultant force between the two atoms is of attraction or repulsion depends upon the distance apart of their centres.

If the two atoms are an infinite distance apart then their *potential energy of interaction* can be assumed to be zero. As the atoms move closer together a force of attraction between them increases and potential energy *falls* because work is being done by the atoms as they move closer together under the force of attraction. As they get very close a short-range force of repulsion builds up and ultimately tends to balance the force of attraction. At this stage the potential energy rises because the atoms are now being pulled together. This state of affairs is indicated in Fig. 1.12 where the potential energy due to forces of attraction (curve a), repulsion (curve b) and the resultant (curve c) is shown in relation to the interatomic distance between the atom centres. Potential energy is at a minimum (v_0) at the equilibrium position (r_0) and v_0 represents the work required to separate the atoms completely.

In a gas at relatively high temperatures the average energy of the atoms is sufficient for the forces of attraction between them to be rendered negligible. They approach each other on a collision path and only at short distances will strong forces of repulsion lead to separation. As the temperature of the gas is decreased so is the kinetic energy of each atom.

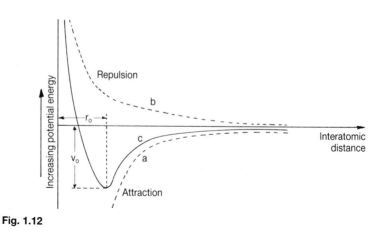

Fig. 1.12

Consequently forces of attraction between atoms become more significant. At some temperature a stage is reached where large groups of atoms are held together since their thermal activation is insufficient to pull them apart, i.e. the gas condenses to form a liquid. At a lower temperature still the force of attraction predominates to the extent that atoms fall into a fixed pattern and so form a solid.

Avogadro's Constant

1.6 In ordinary chemical reactions, i.e. where the release of nuclear energy is not involved, the *Law of Conservation of Matter* applies. This states that during such a chemical reaction matter is neither created nor destroyed. Thus our early school chemistry taught us how to 'balance' chemical equations and, by reference to the appropriate atomic mass numbers of the elements involved, calculate reacting masses of chemical reagents, thus:

C	$+ O_2$	$\rightarrow CO_2$
12	$2 \times 16^* = 32$	$12 + (2 \times 16) = 44$
'parts by weight'	'parts by weight'	'parts by weight'

In the above equation for 'parts by weight' (or more properly 'mass') we can substitute any unit of mass—grams, kilograms, tonnes or even the old Imperial hundredweights—if we wish. The equation then tells us that by burning 12 tonnes of carbon (in the form of coal or coke) we are using 32 tonnes of atmospheric oxygen and producing no less than 44 tonnes of carbon dioxide—which goes towards aggravating the Greenhouse Effect. If the coal also contains 1% sulphur our 12 tonnes of carbon will involve the burning also of 0.12 tonnes of sulphur. This forms the gas sulphur dioxide, SO_2, which enters the atmosphere and is ultimately washed down

---

* Reference to Appendix II will show the accurate value for the atomic mass of the oxygen atom as 15.9994 but for most purposes approximate values are used.

as 'acid rain' (dilute sulphuric acid) on our northern forests and those of our luckless Scandinavian neighbours:

$$
\begin{array}{llll}
 & S & + O_2 & \rightarrow SO_2 \\
 & 32 & 2 \times 16 = 32 & 32 + (2 \times 16) = 64 \\
\text{or} & 0.12\,t & 0.12\,t & 0.24\,t
\end{array}
$$

This 0.24 tonnes of SO_2 produces 0.3675 tonnes of sulphuric acid.

Although there may be a moral involved here our object is to show how simple 'chemical arithmetic' can be used to calculate reacting quantities of materials on an industrial scale. In materials science, however, we often wish to study substances at atomic level. Then we need to know the actual masses and dimensions of these tiny particles.

1.6.1 The 19th century Italian physicist, Amedeo Avogadro, suggested that equal volumes of all gases at the same temperature and pressure contain equal numbers of molecules. During my school days this was known as Avogadro's *Hypothesis* but in more recent years has generally been accepted as his *Law*. Obviously the molecules in question may vary in mass from one gas to another. For example one molecule of carbon dioxide $(CO_2 = 12 + (2 \times 16) = 44)$ is 22 times more massive than one molecule of hydrogen $(H_2 = 2 \times 1 = 2)$. From the Hypothesis was developed the useful information that:

The gram-molecular mass of any gas occupies a volume of 22.4 litres (dm^3) at standard temperature and pressure.

Here the term 'gram-molecule' refers to the relative molecular mass of a gas expressed in grams. Thus a gram-molecule of carbon dioxide contains 44 g of the gas. This quantity the 'gram-molecular mass' (often still called the 'gram-molecular weight'—a hangover from pre-SI days) is important in many chemical calculations and is now known as a 'mole' (or 'mol') of the substance being considered.

1.6.2 Avogadro's original Hypothesis dealt with the relationship between numbers of molecules and the volume they occupied in common gases. Since all elements (and many compounds) can exist in the gaseous state it follows that a mole of any pure substance will contain the same number of elementary particles in which form the substance normally exists. Thus the mole is defined in SI terms as *the amount of substance that contains as many elementary entities as there are atoms in 0.012 kg of carbon* $^{12}_{6}C$. The 'entity' may be an atom, a molecule, an ion, a radical, etc. This number of entities is known as Avogadro's Constant or Number, N_A and $N_A = 6.022\,52 \times 10^{23}$ atoms (etc.)/mole.

Q Calculate the number of atoms in 1 mm³ of copper given that the relative density of copper is 8.96 and its atomic mass is 63.546. ($N_A = 6.022\,52 \times 10^{23}$ atoms/mole)

A 12 g (the gram-atomic mass) of carbon contains $6.022\,52 \times 10^{23}$ carbon atoms. Therefore 63.546 g (the gram-atomic mass) of copper also contains $6.022\,52 \times 10^{23}$ copper atoms. 1 mm³ of copper weighs $8.96 \div 1000\,\text{g} = 0.008\,96$ g. Hence 0.008 96 g of copper contains

$$\frac{0.008\,96}{63.546} \times 6.022\,52 \times 10^{23} = \underline{8.491 \times 10^{19} \text{ atoms}}$$

2

The Molecule

2.1   In the previous chapter we considered some of the properties of the atom as a single free particle. Nevertheless, apart from those of the 'noble gases', atoms rarely occur as single particles but are generally attached to other atoms forming small or large groups. Which is just as well otherwise our Earth—if it existed at all—would presumably consist of a mass of rarefied gas.

What is the nature of the force of cohesion which apparently binds atoms together? In this chapter and the next we shall attempt a simplified answer to this question.

We have seen (1.3.3) that in the second and third periods of the Periodic Table the outer electron shell of an atom is generally complete when the number of electrons in it reaches eight.* Consequently as the result of a chemical reaction between two atoms in these periods there is a general tendency for the outer shells of electrons in each atom to finish with eight electrons.

Most metallic atoms contain one, two or three electrons in their outermost shells (or subshells in the case of metals in period four onwards) whilst in the outer shells of atoms of non-metallic elements the number of electrons is nearer to eight. Thus, when one atom is strongly metallic (the element occupies a position on the left-hand side of the Periodic Table) and the other strongly non-metallic (the element is situated on the extreme right of the Table), chemical bonding is achieved by the metallic atom *donating* its outer-shell electrons to fill the outer shell of the non-metallic atom. This produces what is variously termed an *ionic, polar* or *electrovalent* bond. Its formation gives rise to the production of *ions*, the building blocks of some crystal structures which will be dealt with in the next chapter.

---

* In the first period, i.e. H→He, the number of electrons involved is of course only two, whilst in the case of periods four and onwards the outer shells increase in size, i.e. 18, 18, 32, etc., each terminating with the appropriate 'noble gas' structure.

If the atoms concerned are of elements near to each other in the centre of the Periodic Table they can combine most easily by *sharing* electrons from their outer shells forming common orbitals which are the basis of the *covalent* bond.

2.1.1 By whatever means chemical combination takes place it follows that as we are dealing with small numbers of 'active' electrons, atoms will combine in simple ratios, one type with the other. The *valency* (or *valence*) of an element is usually described as 'the number of atoms of hydrogen with which an atom of that element can combine'. Since one atom of oxygen is able to combine with two atoms of hydrogen (Fig. 2.4) the valency of oxygen is said to be two.

The Covalent Bond

2.2 Two atoms of the gas fluorine will combine by sharing a pair of electrons (Fig. 2.1). In this way the outer shell of each atom now has a complete 'neon octet' and the two atoms are held together by the electrical

Fig. 2.1 The covalent bond in a molecule of fluorine formed by the sharing of a pair of electrons.

forces involved by sharing electrons belonging to the outer orbitals of both atoms. For this reason the covalent bond is a strong and stable chemical bond.

Two hydrogen atoms combine in a similar way except that the 'helium duplet' of each is completed by sharing the pair of electrons thus brought together (Fig. 2.2).

Fig. 2.2 The covalent bond in a molecule of hydrogen—in this case sharing of the electrons completes the 'helium duplet' of each atom.

Often more than two electrons are involved in the formation of the covalent bond. Thus, the outer shell of the nitrogen atom contains five electrons and two nitrogen atoms will combine by sharing six electrons— three donated to the bond by each atom (Fig. 2.3).

In each case the particle formed by the combination of two atoms in this

Fig. 2.3 The covalent bond in a molecule of nitrogen formed by sharing six electrons.

manner is called a *molecule*. Molecules can contain more than two atoms—in fact some molecules of the more complex carbon compounds contain many thousands of atoms. Most of the common non-metallic gases, however, exist as molecules which are diatomic—that is, the molecule contains two atoms. The 'noble gases' (helium, neon, argon, etc.) are exceptions to this general rule. Since the outer electron shells of their atoms are already complete they will not normally combine with other atoms, either of their own type or of other elements. Hence, they are monatomic, that is they exist as single atoms and do not form molecules.

2.2.1 Atoms of unlike elements combine to form molecules in a similar manner. Thus two hydrogen atoms are covalently bonded to an oxygen atom to produce a molecule of the compound we call water. A pair of electrons is shared between the oxygen atom and each of the two hydrogen

Fig. 2.4 The molecular structure of water, H_2O.

atoms (Fig. 2.4) so that the neon octet of the oxygen atom is completed, as are the helium duplets of each of the hydrogen atoms.

The structure of the water molecule is such that the *bond angle* between the two hydrogen atoms is 105° (Fig. 2.5). As we shall see later (2.4.4), this

Fig. 2.5 A 'model' representation of the water molecule.

particular shape of the water molecule offers an explanation of many of the properties of water.

2.2.2 Molecular Mass   The molecular mass—still commonly known as the 'molecular weight'—of a compound which exists in the form of stable molecules is virtually the sum of the atomic masses of all atoms within the molecule. Thus:

$$\text{Molecular mass of water } (H_2O) = (2 \times 1.008) + 16$$
$$= 18.016$$

where the atomic masses of hydrogen and oxygen are 1.008 and 16.000 respectively. Molecular mass (or more properly *relative* molecular mass) has no units since, like atomic mass, it is relative to the mass of the carbon atom $^{12}_{6}C$ (12.0000) (1.4).

Carbon and its Compounds

2.3   The element carbon occupies a position in the centre of the Periodic Table. Thus it may be expected to form covalently bonded compounds more readily than most elements. This it does, particularly with hydrogen, oxygen, nitrogen and the halogens and so produces a vast range of thousands of substances we call 'organic compounds'. These include many complex substances found in living organisms, both animal and vegetable, as well as many materials extracted from coal and petroleum (which are derived from living organisms). The groups of organic compounds of particular interest to the materials scientist are those known as 'super-polymers' or, more commonly, as 'plastics materials'.

2.3.1   Possibly the simplest of the organic compounds is the gas methane, the main constituent of 'natural gas'. A molecule of methane consists of an atom of carbon covalently bonded to four hydrogen atoms. Figure 2.6 indicates diagrammatically the electron distribution within the molecule. The true three-dimensional shape of the molecule will be something like that shown in Fig. 2.7. Methane is in fact the simplest compound in a series of compounds known collectively as alkanes (formerly the 'paraffins'). This

Fig. 2.6   The methane (CH_4) molecule represented diagrammatically.

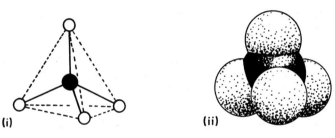

Fig. 2.7 The methane molecule.
(i) Indicates the spatial positions of the hydrogen atoms at the points of a tetrahedron at the centre of which is the carbon atom. (ii) A 'model' of the molecule.

is an example of what is called a *homologous series*. As will be seen from Fig. 2.8 the length of the carbon 'chain' increases by the unit:

from one compound to the next. As molecular size increases so do forces of intermolecular attraction (2.4). Hence, whilst the compounds methane, ethane and propane are gases at normal temperatures and pressures, butane is a gas which is readily compressed to form a liquid ('Camping Gaz'). Successive members of the series are volatile liquids followed by less volatile liquids such as paraffin oil ('kerosene' in the USA); whilst at the upper end of the series are solids such as paraffin wax with a formula of the order $C_{50}H_{102}$ (Table 2.1). Incidentally all of these compounds in the series are derived from crude petroleum.

Fig. 2.8 The *alkane* series of organic compounds—previously known as the 'paraffins'.

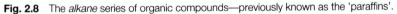

Table 2.1 The Alkane Series

Number of carbon atoms	Name	Formula (C_nH_{2n+2})	Molecular mass	Relative density as liquid	State at 0°C	M.pt. (°C)	B.pt. (°C)	Product obtained from petroleum	Uses
1	Methane	CH_4	16	0.4240	gas	−183	−162	'Natural gas'	Heating and lighting
2	Ethane	C_2H_6	30	0.5462		−172	−87		
3	Propane	C_3H_8	44	0.5826		−187	−42	Calor gas 'Camping Gaz'	
4	n-Butane	C_4H_{10}	58	0.5788		−135	−1		
5	n-Pentane	C_5H_{12}	72	0.6264		−130	36	'Benzine'	Dry cleaning
6	n-Hexane	C_6H_{14}	86	0.6594		−94	69		
7	n-Heptane	C_7H_{16}	100	0.6837	liquid	−90	98		
8	n-Octane	C_8H_{18}	114	0.7028		−57	126	Petrol (Gasoline)	Automobile fuel
9	n-Nonane	C_9H_{20}	128	0.7179		−54	151		
10	n-Decane	$C_{10}H_{22}$	142	0.7298		−30	174		
16	n-Hexadecane	$C_{16}H_{34}$	226	0.7794		18	280	Paraffin oil Kerosene	Heating and lighting
24	n-Tetracosane	$C_{24}H_{50}$	338	0.7786	solid	51	De-compose before b.pt.	'Mineral oils' Greases	Lubricating oils 'Vaseline'
50	n-Pentacontane	$C_{50}H_{102}$	702	0.7940		92		Paraffin wax (hard) / Residue	Candles, waxed paper / Asphaltic bitumen

n-BUTANE   or   iso-BUTANE (methyl propane)

Fig. 2.9   The two isomers of butane.

2.3.2   There is only one basic pattern in which the carbon atoms can be arranged in the first three compounds in the series, namely, methane, ethane and propane. With the compound butane, however, carbon atoms can be arranged in two different patterns as indicated in Fig. 2.9. In fact two different substances do exist with different physical properties (Table 2.2) even though a molecule of each compound contains the same numbers of atoms of carbon and hydrogen. These two substances, each having the same molecular formulae, yet different molecular *structures* are known as *isomers*. In the case of pentane C_5H_{12} there will be *three* isomers; the number of isomers increasing as the molecular formula increases. In fact the molecule Tetracontane, $C_{40}H_{82}$, will be represented by no less than 62 491 178 805 831 possible isomers.

Table 2.2   *The Isomers of Butane*

Isomer	Melting point (°C)	Boiling point (°C)
n-Butane	−138	−0.5
iso-Butane	−159	−10.2

2.3.3   The alkane series described above consists of hydrocarbons which are said to be 'saturated'. Briefly, this means that carbon atoms are joined to their neighbours by means of single bonds. Another series exists in which the compounds are 'unsaturated', that is, double valency bonds connect certain of the carbon atoms. These compounds are of the general formula C_nH_{2n} and the first member of the series is ethylene (Fig. 2.10). Collectively known as the alkenes (formerly the 'olefins'), these hydrocar-

represented by ——   or   or C_2H_4

A MODEL OF THE ETHYLENE MOLECULE

Fig. 2.10   The structural formula of ethylene, C_2H_4.

bons tend to be less chemically stable than do the alkanes and the double bond can be 'broken' relatively easily by chemical means.

Ethylene is a by-product of the petroleum refinery since, during the 'cracking' of some of the heavier oils to produce light and more volatile hydrocarbons (such as octane), an unsaturated hydrocarbon group is left over—in this case ethylene, C_2H_4. If ethylene is suitably treated the double bond between the carbon atoms is broken and simultaneously large numbers of these units are formed. These immediately link up—or *polymerise*—to form long chain-like molecules of the substance 'polyethylene', or polythene as it is commonly known. The sequence in this process is indicated in Fig. 2.11.

Fig. 2.11   The polymerisation of ethylene to form poly(ethylene) or 'polythene'.

Polythene is one of a large number of these super-polymers (commonly called 'plastics') which are finding increasing use as engineering materials. They will be discussed in more detail in Chapter Twelve.

Intermolecular Forces

2.4   The *atoms* in a molecule are, as we have seen, held together by covalent bonds. These are strong primary bonds which are dependent upon relatively powerful electrostatic forces. At the same time weaker electrostatic forces of a secondary nature give rise to attractions between any one *molecule* and its neighbours. Only in this way can we explain the fact that a gas condenses to form a liquid and that the liquid subsequently crystallises to form a solid. The secondary electrostatic forces operating between molecules influences such important properties as melting point, solubility and, in plastics, tensile strength. Powerful modern adhesives of the 'impact' variety rely on these secondary intermolecular forces and illustrate the collective effect of large numbers of such secondary bonds operating simultaneously.

As early as 1873 Johannes Diderik van der Waals a Dutch physicist, sought to explain deviations from the Gas Equation $(PV = RT)$. As readers may know, this relationship is true only for a *perfect* gas. No gas is 'perfect' in this sense and van der Waals introduced corrections into the Gas Equation such that:

$$\left(P + \frac{a}{V^2}\right)(V - b) = RT$$

where b is a constant related to the actual volume of the molecules of the gas; a is also a constant and a/V^2 represents the loss of pressure due to *forces of attraction between the molecules of the gas*.

A stable molecule or a stable atom (e.g. an atom of a noble gas) is, taken as a whole, electrically neutral since the number of protons present is balanced by an equal number of electrons. However, the resultant 'centre' of the electropositive charges does not necessarily coincide with the centre of the negative charges. Then the particle—whether molecule or single uncombined atom—can be assumed to possess a resultant 'dipole moment'

(i)   (ii)

Fig. 2.12 In (i) the centres of positive and negative charges coincide, but in (ii) they are separated so that the molecule (or single atom in the case of the noble gases) will have a resultant 'dipole moment'.

(Fig. 2.12). These intermolecular forces, often referred to as van der Waals forces, are of four main types:

2.4.1 Attraction Between Permanent Dipoles A molecule of the gas hydrogen chloride contains one atom of hydrogen and one of chlorine, covalently bonded (H—Cl). However, the electrons tend to be concentrated rather more densely in the region of the chlorine nucleus (which contains a larger positive charge) than around the hydrogen nucleus (which contains only a single positive charge due to the presence of one proton). The uneven distribution of electrons in the molecular orbital is equivalent to a separation of charges. For this reason the hydrogen nucleus tends to be rather 'exposed' so that that part of the molecule has a resultant small positive charge (δ^+). Conversely the part of the molecule containing the

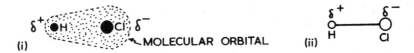

(i)   MOLECULAR ORBITAL   (ii)

Fig. 2.13 The molecular orbital of hydrogen chloride (i); showing, (ii) the formation of a dipole moment.

chlorine nucleus will carry an equal negative charge (δ^-) (Fig. 2.13). Thus the hydrogen chloride molecule will possess a resultant dipole moment.

Such a molecule is said to be *polar* whereas a hydrogen molecule, in which the charges were equally distributed, would be *non-polar* since it possessed no resultant di-pole moment, because of its relative symmetry.

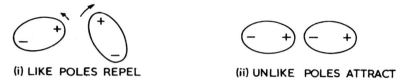

(i) LIKE POLES REPEL   **(ii) UNLIKE POLES ATTRACT**

Fig. 2.14 Attraction and repulsion between permanent dipoles follows the well-known laws of electrostatic attraction and repulsion.

Polar molecules will repel each other or attract each other according to the way in which they are orientated (Fig. 2.14).

The degree of mutual dipole alignment will govern the extent of intermolecular attraction. Since this alignment is opposed by thermal motion, dipole forces are dependent largely upon temperature. Thus, as temperature rises these forces decrease. For example, the plasticity of thermoplastic polymers increases with the temperature.

2.4.2 Attraction Between Permanent Dipoles and Induced Dipoles In some cases the orbital of a non-polar atom or molecule may be affected by the close proximity of a strongly polar molecule. In Fig. 2.15 it is assumed

Fig. 2.15 The formation of an induced dipole.

that the 'positive end' of a permanent dipole has approached a non-polar atom or molecule in which the electron cloud is large, diffused and capable of easy distortion. The electron cloud of the non-polar particle will be attracted towards the positive end of the permanent dipole. This distorts the electron cloud of the non-polar particle to the extent that a temporary dipole is formed. Such a distortion of an electron cloud is referred to as *polarisation*. The positive pole of the permanent dipole attracts the electrons of the non-polar atom or molecule and so builds up a negative charge facing it. Hence the permanent dipole and the induced dipole are in a position in which they will attract each other.

2.4.3 Dispersion Forces Although there are many molecules which appear to be non-polar, that is the centres of positive and negative charges would seem to be coincident, all molecules—and the single atoms of noble gases—have time-varying dipole moments the nature of which will depend upon the position of the electrons at any particular instant.

Consider the case of two atoms of the noble gas, argon, in close proximity to each other (Fig. 2.16). If the electron clouds are evenly distributed as in (i) then there will be no resultant dipole moment in either

Fig. 2.16 The formation of dispersion forces between two atoms of argon.

atom and they will neither attract nor repel each other. If, however, an uneven distribution of the electron clouds is present as in (ii) then the atoms will repel each other; but if the electron clouds are distributed as in (iii) then the electron cloud of atom A will be attracted by the 'unprotected' nucleus of atom B.

The displacement of electrical charges necessary to produce a dipole is temporary and random so that a molecule, as a whole, over a period of time has no resultant dipole. However, if it has a dipole moment at any one instant it will induce another in a near neighbour and a force of attraction will be set up between the two. It is thought that such interactions must be largely responsible for van der Waals bonding between non-polar atoms and molecules. The functioning of these dispersion forces was postulated by London as long ago as 1930 and, except in a few cases where very strong dipole moments exist, these forces exceed in magnitude all other intermolecular forces, when considered collectively.

The dispersion effect is evident when considering the relationship between molecular mass and boiling point. Massive molecules contain more electrons than do light ones and so the attraction due to dispersion effects is greater in the massive molecules. Heat of evaporation is a measure of the energy required to overcome forces of attraction between molecules so that they can separate. Since the boiling point is proportional to the heat of evaporation it is a convenient criterion in assessing these intermolecular forces. Thus normal covalently bonded compounds with massive molecules have higher boiling points than those with small molecules. An excellent example of this principle is given by the homologous series, the alkanes, mentioned earlier in this chapter. The lower members are gases at ordinary temperatures and pressures, then, as the series is ascended, liquids of increasing boiling point are followed by higher members which are waxy solids (Table 2.1).

Dispersion forces, along with permanent dipoles figure largely in the intermolecular forces which prevail between the chain-like molecules in super-polymers.

2.4.4 The Hydrogen Bond   When a hydrogen atom is covalently bonded to a relatively large atom such as nitrogen, oxygen or fluorine, a powerful permanent dipole is set up. This is because the electron cloud tends to become concentrated around that part of the molecule containing the nitrogen, oxygen or fluorine nucleus, thus leaving the positively charged hydrogen nucleus relatively 'unprotected'. In this way particularly strong dipole–dipole forces of attraction can be set up. These are fundamentally similar in nature to the permanent dipole–dipole forces described earlier but are of much greater magnitude.

Fig. 2.17   The basis of the 'hydrogen bond' (or 'bridge') in the water molecule.

This effect is apparent in the case of the water molecule. Here the oxygen part of the molecule, carrying the un-paired electrons, will be negatively charged whilst the hydrogen portions will be correspondingly positively charged. The hydrogen atoms form 'bridges' (shown thus ----- in Fig. 2.18) between themselves and oxygen atoms of other molecules thus linking the molecules together. It must be remembered that Fig. 2.18 shows a two-dimensional diagram whereas in practice a three-dimensional spatial pattern exists.

Fig. 2.18   The formation of hydrogen bonds between molecules in water.

Approximately two-thirds of the heat of vaporisation of water at its boiling point is used up in disrupting these hydrogen bonds or 'bridges'. Sufficient hydrogen bonds are formed to account for the abnormally high boiling point of water. For example, since the sulphur atom is more massive than that of oxygen the compound hydrogen sulphide, H_2S, might be expected to have a higher boiling point than that of water, H_2O. In fact the boiling point of hydrogen sulphide is $-60°C$, some $160°C$ *lower* than that of water. This is apparently due to the relatively *large* size of the sulphur atom causing a lesser concentration of the electrons so that a weaker dipole moment is produced in the molecule. In the same way the molecule of hydrogen chloride, HCl, does not exhibit a strong hydrogen bond because of the large size of the chlorine atom, whereas the hydrogen fluoride, HF, molecule does possess a powerful hydrogen bond because of the smaller size of the fluorine atom and also its very strong electronegative character.

The hydrogen 'bridge' associated with the hydroxyl (—OH) and other groups in many organic polymers contributes considerably to their overall cohesion. Whilst it is much stronger than other van der Waals forces it is nevertheless weaker than the ordinary covalent bond. In terms of bond energy a covalent bond may represent as much as 1700 kJ/mole whilst the hydrogen bond would be equivalent to some 45 kJ/mole. Bond energies of the weaker van der Waals forces vary between 0.1 kJ and 20 kJ/mole (3.2.6).

3

The Crystal

3.1 When used in a non-scientific sense the term 'crystal' conveys an impression of a material which is geometrically regular in shape and which is both lustrous and transparent—as the expression 'as clear as crystal' suggests. In fact crystals of many materials have an irregular external shape though the particles within these crystals are arranged in some regular pattern. Many crystalline materials are opaque and do not have a lustrous appearance. In general terms a substance can be described as crystalline when the particles from which it is built are arranged in a definite three-dimensional pattern which repeats over a long range. Metals and indeed all truly solid substances are crystalline in structure. Non-crystalline 'solids' such as glass, pitch and many polymers can often more convenient-ly be considered as extremely viscous liquids in so far as their physical and mechanical properties are concerned.

Again the question arises—what is the nature of the forces which bind together the units from which a crystal is built? In some types of crystal the binding forces consist of covalent bonds, in others hydrogen 'bridges' (2.4.4); whilst in mineral crystals and those of inorganic salts generally, the force of attraction is due to what is described variously as an ionic, polar or electrovalent bond.

The Electrovalent Bond

3.2 In the previous chapter we saw that a covalent bond is formed when electrons from the outer shell of two (or more) atoms are *shared* in such a way that the appropriate noble-gas structures are attained in the outer electron shells concerned. However, chemical combination can also take place by the *transfer* of electrons from one atom to another such that a noble-gas structure is attained in the valency shells of each resultant particle. Thus, a metallic atom which contains only one or two valency electrons tends to give up these electrons to a non-metallic atom which already has a valency shell of six or seven electrons. In this way the outer electron shell of each resultant particle attains a noble-gas structure.

3.2.1   The metal sodium reacts vigorously with the poisonous gas chlorine to produce a very stable substance, sodium chloride (common salt). In this instance the sodium atom which carries a single electron in its outer shell donates this electron to a chlorine atom which previously had seven electrons in its outer shell. By means of this transfer the sodium particle is left with a complete outer shell (the 'neon shell') and the chlorine particle also has a completed shell (the 'argon shell'). Since an electron transfer has taken place from one atom to another we can no longer refer to the resultant particles as 'atoms'. The sodium particle is obviously something less than a sodium atom whilst the chlorine particle is something more than a chlorine atom. Particles produced by electron transfer in this manner are known as *ions*.

Prior to the chemical reaction during which the electron transfer took place, the atoms were electrically neutral because in each case the number of positively charged protons in the nucleus was balanced by an equal number of negatively charged electrons in orbitals around the nucleus. During the reaction the sodium atom *lost* an electron and the ion so formed must therefore carry a resultant positive charge. Similarly, since the chlorine atom *received* this electron the chlorine ion produced must carry a resultant negative charge (Fig. 3.1).

Since the two ions formed as a result of this chemical reaction carry equal but opposite charges they will attract each other and it is this force of coulombic attraction which constitutes the electrovalent bond. Although generally weaker than a covalent bond, this electrovalent bond is much stronger than any of the relatively feeble van der Waals forces.

3.2.2   When sodium and chlorine react with each other even on the laboratory scale many millions of atoms are involved and an equally large

Na	+	Cl	Na+	Cl⁻
SODIUM ATOM		**CHLORINE ATOM**	**SODIUM ION**	**CHLORINE ION**
11 protons		17 protons	11 protons	17 protons
11 electrons		17 electrons	10 electrons	18 electrons
Therefore		Therefore	Therefore	Therefore
electrically		electrically	positively	negatively
neutral.		neutral.	charged.	charged.

Fig. 3.1   The formation of an electrovalent—or ionic—bond between a sodium atom and a chlorine atom.

(i)   (ii)

Fig. 3.2 Each Na$^+$ ion is surrounded by six Cl$^-$ ions.

number of ions is produced. Whilst there is mutual attraction of sodium ions for chlorine ions, forces of repulsion operate between all like ions. That is, each sodium ion is repelled by all other sodium ions whilst each chlorine ion is repelled by all other chlorine ions. As a result the system attains equilibrium when each chlorine ion is surrounded by six sodium ions and, conversely, when each ion of sodium is at the same time surrounded by six chlorine ions (Fig. 3.2). In this instance the relative sizes of the chlorine ions and sodium ions are such that a cubic type of pattern is formed. Since each sodium ion is surrounded by six chlorine ions it is said to have a co-ordination number of six. For a similar reason chlorine also has a co-ordination number of six.

Diagrams representing crystal structures are somewhat clearer if the centre of each ion is represented by a small circle (Fig. 3.2(ii) and Fig. 3.3(ii)). In this cubic type of structure the three axes of symmetry are at right angles to each other. The chlorine ions take up positions as far as possible from other chlorine ions and since they are equally charged they are also equally spaced. Each chlorine ion has six sodium ions equally spaced around it and in contact with it. Since there is an electrostatic field extending into the space outside this unit other sodium ions and chlorine ions will be attracted so that the structure depicted in Fig. 3.3 builds up and extends as long as ions are available. The framework on which the units of a crystal arrange themselves according to some fixed pattern is known as a *space lattice*.

3.2.3 The ions arrange themselves in positions dictated principally by the forces of attraction between the positive sodium ions and the negative chlorine ions. In addition to this force of attraction between oppositely charged ions other forces operate between ions. Thus there is the gravitational force of attraction which will always act between adjacent massive nuclei. Also when ions come into close proximity to each other their outer completely filled electron shells begin to *overlap* and a force of electrostatic repulsion is set up between the like negative charges.

At some specific separation distance these forces of attraction and

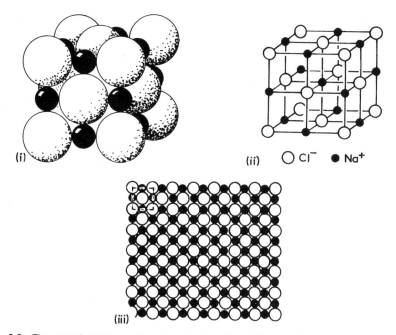

Fig. 3.3 The crystal structure of sodium chloride. (i) and (ii) show the unit cell of sodium chloride, whilst (iii) indicates the continuous structure of the crystal such that each face of a unit cell is shared by the next cell. In fact we have two interpenetrating *face-centred cubic* structures, one of Na^+ and one of Cl^- (3.6.2).

repulsion will be in equilibrium. In this condition the potential energy of the system will be at a minimum and of negative sign since an input of energy would be required to displace ions from their equilibrium positions. The potential energies associated with attraction and repulsion are inversely proportional to some power of the distance of separation, r, between nuclei. The energy due to attraction is of negative sign and is equal to $-A/r^m$ whilst the potential energy (PE) due to repulsion is equal to B/r^n. A and B are constants dependent on the natures of the ions, i.e. such factors as nuclear mass and magnitude of charge. m and n are powers (to be determined) where n is greater than m.

The resultant potential energy, V, is given by:

$$V = \frac{-A}{r^m} + \frac{B}{r^n}$$

The relationship between potential energy and interatomic distance is indicated in Fig. 1.12 where curve b represents the PE associated with forces of repulsion, curve a the PE associated with forces of attraction whilst curve c represents the PE associated with the resultant force, each as a function of the distance between nuclei. The potential-energy 'well' is deepest at the equilibrium spacing, r_0, and its depth, V_0, the energy required to separate ions completely.

3.2.4 The distances between the centre of any chlorine ion and the centres of the six nearest sodium ions is less than the distances between the centre of that same chlorine ion and the centres of the six nearest chlorine ions. Consequently the forces of attraction between any ion and its neighbours are greater than the forces of repulsion, and the crystal is therefore a stable structure.

The unit structure in the case of the substance sodium chloride is the crystal. Sodium chloride does not exist in the form of molecules* and it is therefore not appropriate to write the formula 'NaCl' since this would imply the existence of one molecule of sodium chloride. Nevertheless busy chemists are in the habit of representing sodium chloride by the empirical formula, NaCl, as a type of 'chemical shorthand'.

3.2.5 Sodium chloride will dissolve readily in water. This is due to two important properties of water:

(i) it forms a polar molecule (2.4.4);
(ii) it has a high dielectric constant.

The hydrogen 'end' of the water molecule is attracted to the negatively charged chlorine ions and the oxygen 'end' of the molecule to the positively

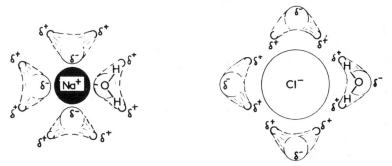

Fig. 3.4 The attraction between ions and water molecules.

charged sodium ions. This causes water molecules to cluster round ions at the surface of a sodium chloride crystal until their forces of attraction are sufficient to overcome the electrovalent bond between sodium ions and chlorine ions. The ions then mix freely with water molecules in solution. This process of separation of sodium ions from chlorine ions on solution is termed *dissociation*. It is sometimes incorrectly called ionisation. The latter process in fact took place during the original reaction between sodium and chlorine when atoms of each element became ions.

3.2.6 *Bond strength* is most easily measured in terms of the energy required to break a chemical bond, that is the quantity of heat which is necessary to vaporise the solid phase and so completely separate the

*Though I know of at least one text book on materials science in which the author describes a "sodium chloride molecule"!

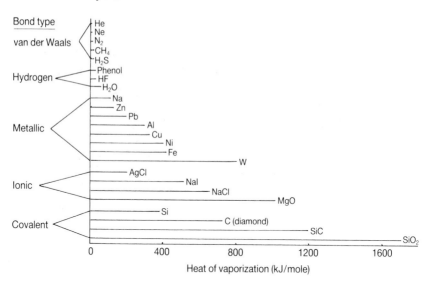

Fig. 3.5 Bond strength is conveniently measured by the energy required to break it. This diagram indicates the approximate bond energies in terms of the heat energy (including latent heat) required to change one mole of the solid material to vapour in each case. The general relationship between bond energy and the nature of the bond is clear to see.

constituent atoms. In each case the heat of vaporisation of one gram-mole is measured so that we shall always be dealing with the same number of bonds, i.e. N_A Avogadro's Number.

As would be expected van der Waals bonds are weakest and require between 0.1 and 20.0 kJ/mole whilst the covalent bond requires 400–1700 kJ/mole depending upon the material (Fig. 3.5). Between these two extremes come metallic and ionic bonds, those operating in sodium chloride being an example of the latter.

In any solid at temperatures above 0 K the atoms are in a state of constant vibration and when a solid is heated these vibrations increase until a point is reached where the bonds joining these atoms are ruptured. Atoms then move apart, that is, melting occurs. Thus there is a relationship between bond strength and melting point. If we compare a series of 'like elements' e.g. the Alkali Metals of Group I, we note that melting point (as a function of bond strength) decreases as the atomic number, Z, increases:

Element	Atomic number, Z	Relative atomic mass	Melting point, °C
Lithium	3	6.941	180.5
Sodium	11	22.989	97.8
Potassium	19	39.102	63.6
Rubidium	37	85.468	38.8
Caesium	55	132.905	29.7

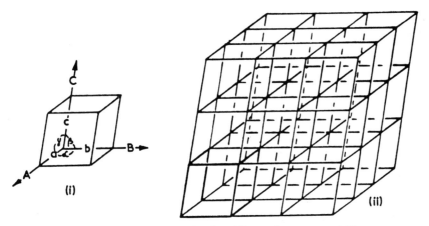

Fig. 3.6 The relationship between (i) a unit cell and (ii) a continuous space lattice.

Therefore one can assume that electrons of higher principal quantum number form weaker bonds so that smaller vibrational energies at lower temperatures are sufficient to disrupt these bonds.

Space Lattices

3.3 A space lattice can be defined as a distribution of particles in three dimensions such that every particle has identical surroundings. It is generally convenient to imagine a space lattice as continuing to infinity in all directions, for although crystals themselves are of finite size the units from which they are built are exceedingly small. Thus a small crystal of copper some 0.1 mm in diameter will nevertheless contain roughly 3×10^{16} atoms of copper.

A section of a space lattice of general form is illustrated in Fig. 3.6. Since the points at which particles are considered to exist are regularly distributed, the fundamental geometry of the space lattice can be described by three lattice vectors A, B and C. The vectors define the unit cell and the geometry of the space lattice is specified satisfactorily by the lattice constants (vector lengths) a, b and c and the inter-axial angles α, β and γ.

3.3.1 There are only fourteen different patterns in which points can be arranged in order to satisfy the definition of a space lattice. These are the *Bravais space lattices* (Fig. 3.7). Of these not more than six are met with in metallic structures and in fact only three are really common.

3.3.2 Frequently in a study of crystals it becomes necessary to identify a particular plane of atoms running through the structure. Such planes and directions are best described by a set of numbers known as *Miller indices*.

40   The Crystal

Fig. 3.7 The fourteen Bravais space lattices. (P = primitive; C = centred on 'ab' face; I = body centred and F = face centred).

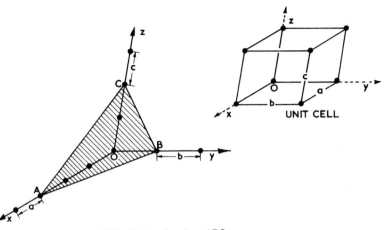

Fig. 3.8 The derivation of Miller indices for plane ABC.

These are widely used in X-ray analyses and other methods used in deriving crystal structures.

Miller indices are proportional to the reciprocals of the intercepts which the plane makes with the three principal axes (x, y and z) of the system. They are derived as follows:

(i)   determine the intercepts which the plane makes with the three crystal axes. These intercepts are expressed as the number of axial lengths (a, b or c) from the origin. In Fig. 3.8 these intercepts are:

$$x = 3; y = 1; z = 2$$

(ii)   take reciprocals of these intercept values, i.e. $\frac{1}{3}$; 1; $\frac{1}{2}$;

(iii)   convert these reciprocals to the smallest integers which are in the same ratio, i.e. 2; 6; 3. This is written in the form (263). Thus plane ABC in Fig. 3.8 is represented by the Miller indices (263).

These indices represent not only the plane ABC but all planes parallel to it. For example a plane with the intercepts 6, 2, 4 will be parallel to plane ABC and will be represented by the same Miller indices.

A plane on the opposite side of the origin would have negative intercepts on the axes and would be denoted ($\bar{2}\,\bar{6}\,3$).

The above derivation of Miller indices represents the general case (a triclinic structure in which $a \neq b \neq c$ and $\alpha \neq \beta \neq \gamma$). We will now consider some typical planes for a simple cubic system. In Fig. 3.9(i) the plane indicated has the following intercepts:

$$x = 1; y = \infty; z = \infty$$

Consequently the Miller indices will be (100)—the reciprocals of 1, ∞, ∞. The plane indicated in Fig. 3.9(v) has intercepts:

$$x = 1; y = 1; z = \infty$$

and hence Miller indices (110).

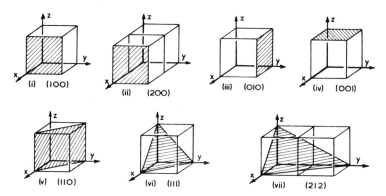

Fig. 3.9 Miller indices for some typical planes in a simple cubic system.

Figure 3.9(vi) having equal intercepts on all axes is represented by Miller indices (111).

In the case of the simple cubic structure, planes which form parallel groups are enclosed in brackets, e.g. {hkl}. Thus the following are equivalent parallel planes:

$$111 = (111); (\bar{1}11); (1\bar{1}1); (11\bar{1}); (\bar{1}\bar{1}1); (\bar{1}1\bar{1}); (1\bar{1}\bar{1}); (\bar{1}\bar{1}\bar{1})$$

3.3.3 Directions through a crystal can be specified in terms of the indices which give the integral co-ordinates of a point on a line drawn between the point and the origin. Direction indices are written in square brackets [ ] to distinguish them from Miller indices writtten in parentheses ( ) which represent a plane. In the case of the simple cubic cell the *direction* normal to the *plane* (hkl) is [hkl]. Figure 3.10 indicates some of the important directions for a simple cubic cell. A general *family* of directions is written ⟨hkl⟩.

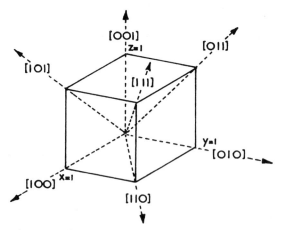

Fig. 3.10 Some important directions in a simple cubic cell.

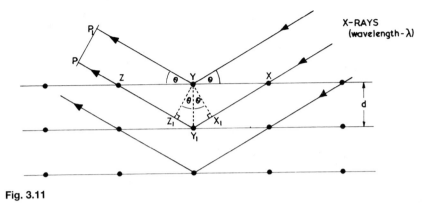

Fig. 3.11

Experimental Derivation of Crystallographic Planes

3.4 Interplanar distances for crystalline materials are generally determined by X-ray methods. The X-rays used are of short wavelength and hence great penetrating power. They are also as nearly as possible 'monochromatic', that is, of a single wavelength. The method used depends upon the fact that in a crystal, planes occur which are occupied by atoms according to a regular pattern and that X-rays will be either reflected or transmitted at these planes.

Figure 3.11 represents diagrammatically two layers of atoms in a crystal structure. The layers are distance d apart. It is assumed that two incident waves strike the crystal at X and Y. The wave which strikes the first layer at X will either be reflected or transmitted to a point Y_1 in the second layer. Here the process is repeated. The wave reflected from Y interferes constructively at PP_1 with the wave which is transmitted from X to Y_1 and subsequently reflected providing that $X_1 Y_1 Z_1$ is an integral multiple of the wavelength λ of the incident X-rays. Using simple geometry it can be seen that:

$$X_1 Y_1 = Y_1 Z_1 = d . \sin \theta$$

3.4.1 Bragg's Law   For constructive interference of X-rays, Bragg's Law states that:

$$n\lambda = 2d . \sin \theta$$

where θ is the incident angle of X-radiation. For the principal diffraction line, $n = 1$ and so d can be calculated:

$$d = \frac{\lambda}{2 \sin \theta}$$

Interplanar spacings are very small and since the maximum value of $\sin \theta$

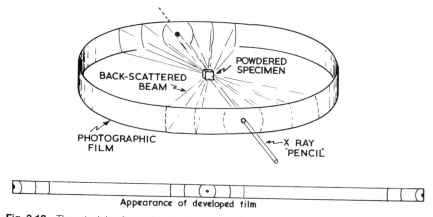

Fig. 3.12   The principle of a method for X-ray crystallography (the Debye-Scherrer Method).

is one, short wavelength monochromatic X-rays must be used so that λ is of the same order as the interplanar spacings, i.e. approximately 10^{-10} m.

In practice if a beam of X-rays falls on a crystal reflection from a given plane can only take place if that plane is inclined to the 'pencil' of X-rays at the necessary angle, θ. This condition is unlikely to be satisfied and a powdered sample of the crystalline material is used so that, inevitably, a few particles will be suitably orientated to make reflection possible. The powdered sample is placed at the geometric centre of a circular strip of photographic film and subjected to a pencil of X-rays (Fig. 3.12). Diffraction occurs from planes that are more or less randomly orientated with respect to the source. Hence the diffraction pattern will be spherically distributed such that the circular arrangement of the film intercepts also the back-diffraction patterns.

This is a very sensitive method for the determination of interplanar spacings, giving errors of no more than 0.05% in measurements of the order of 2×10^{-10} m.

The Metallic Bond

3.5   About three-quarters of the elements are metals, only about one-eighth are non-metals, whilst the remainder, the metalloids, have properties which are intermediate between metals and non-metals. The non-metals are characterised by relatively high electronegativities, by being poor conductors of heat and electricity and by having melting points and boiling points which are generally low. Metals, however, have relatively high melting points and boiling points and are good conductors of heat and electricity. Moreover they possess plasticity, the property which permits permanent deformation without accompanying fracture. All of these properties are related to the nature of the metallic bond, and are connected with the fact that metallic atoms have only one or two electrons in their

POSITIVELY-CHARGED IONS

NEGATIVELY-CHARGED
ELECTRON 'CLOUD'

Fig. 3.13   A very much simplified picture of the nature of the metallic bond.

valency subshells. These electrons are bound to the atom much less strongly than are those in the atoms of non-metals.

The atoms of a metal are arranged such that their ions conform to some regular crystal pattern (Fig. 3.13) whilst their valency electrons behave as though they are mobile, forming a common 'cloud' surrounding the ions. These electrons can be described as *delocalised* and no particular electron 'belongs' to any particular atom. Although the individual electrons are bound to metallic atoms much less strongly than are those in non-metallic elements, the shared electrons bind metallic ions very tightly into the lattice since there is attraction between the positively charged nuclear protons and the delocalised electrons of the common 'cloud'. This results in the high melting points and boiling points of metals, whilst the mobility of the electrons confers high conductivity of both heat and electricity.

3.5.1   It must be admitted that the above interpretation of the metallic bond is oversimplified and fails under more detailed quantitative examination. Nevertheless this simplified picture is useful at this stage in considering some of the mechanical properties of metals, and a slightly more detailed examination of the nature of the metallic bond will be presented later in this book (18.3.1).

3.5.2 Electrical Properties   Electrons in the electron 'cloud' are not bound to fixed orbitals but are able to move freely throughout the body of metal, that is their movement is random. If a relatively small electrical potential difference is applied to a piece of metal the random electron movements are converted into an orderly electron stream towards the higher potential. Thus metals are excellent conductors of electricity and this flow of electrons produced by the application of a potential difference is what we call a 'current' of electricity.

In non-metallic elements the outer-shell electrons of atoms are confined in covalent bonds. They are therefore unaffected by the applications of an electrical potential difference unless it is very great. Consequently most non-metallic elements and compounds are bad conductors, particularly those of the organic polymer type where covalent bonding exists throughout large molecules (2.3.3). Many such substances are in fact used as insulators.

Some elements, often referred to as 'metalloids', possess properties which are intermediate between those of metals and non-metals. This is reflected in their electrical properties. In silicon and germanium for example the energy necessary to detach electrons from their covalent bonds and move them through the structure of the solid is much less than with most non-metals. These elements are termed semi-conductors.

Crystal Structures in Metals

3.6 As we have seen, the forces binding the atoms in a metal are non-directional, and since in a pure metal all atoms are of the same kind and size, they will arrange themselves in the closest possible packing patterns which are associated with positions of minimum potential energy. Most of the metals of engineering importance crystallise in one of three principal types of structure in all of which the atoms are held together solely by the metallic bond. The spheres used in the illustrations which follow represent atomic cores between which the electron cloud will be distributed.

3.6.1 The Close-packed Hexagonal Structure (CPH)   When we place the snooker balls in the triangular frame at the beginning of a game we are in fact packing them in the closest possible arrangement in a single plane. The arrangement is that indicated in Fig. 3.14(i) and it will be seen that any sphere, X, is touched by six immediate neighbours. Lines joining the centres of these spheres form a regular hexagon. If we consider any three spheres, lines joining their centres form an equilateral triangle and at the centre of this triangle is a small space between the spheres. Now consider a sphere put on top of these first three so that it rests in the hollow formed by them. The space so enclosed by the four spheres is termed a *tetrahedral site* since it is enclosed by four spheres whose centres lie on the points of a regular tetrahedron. This type of site is of great importance in a study of crystallography.

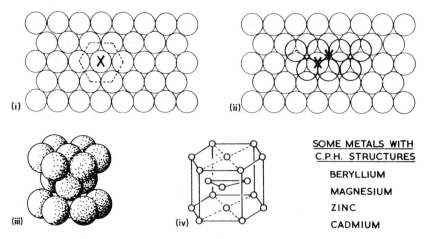

SOME METALS WITH
C.P.H. STRUCTURES

BERYLLIUM

MAGNESIUM

ZINC

CADMIUM

Fig. 3.14 The close-packed hexagonal (CPH) structure.

If a second layer of spheres is now placed on top of the first layer and allowed to find its own level, each sphere in this second layer will come to rest in one of the hollows formed by the first layer as indicated in Fig. 3.14(ii).

There are *two* possible positions in which spheres of a third layer can be placed—either in those positions marked X or in those marked Y. If we place them in positions marked X, the centre of each sphere in the third layer will be vertically above the centre of a sphere in the first layer. This is the close-packed hexagonal (CPH) structure and if the arrangement in the first layer is represented by A and that in the second layer by B, the building order involved here is ABABABAB . . . etc. Any atom in the system is touched by six atoms in its own plane (Fig. 3.14(i)) and by three atoms in the plane beneath. The CPH structure is so called because it has an axis of sixfold symmetry perpendicular to the layers of closest packing, that is, perpendicular to the plane of the paper in Fig. 3.14(i). A unit cell of this structure is shown in Figs 3.14(iii) and (iv); Fig. 3.14(iii) indicating the atomic cores in contact and Fig. 3.14(iv) the space lattice.

Plastic deformation in metals takes place by a process known as 'slip', that is by layers of atoms gliding over each other. This occurs most frequently by the close-packed planes slipping in close-packed directions. In the CPH structure such slip is limited since it will occur mainly along basal planes of the hexagon. Thus, most metals with a CPH structure are much less ductile than those with a face-centred cubic structure.

3.6.2 The Face-centred Cubic Structure (FCC)   In this structure the basic positions of spheres in the *first two* layers are the same as those indicated in Fig. 3.14(ii). For the *third* layer, however, the spheres are placed in the hollows marked Y (Fig. 3.14(ii)). Thus in this third layer the spheres are in different positions from those in the first layer (thus differing from the CPH structure). If we represent this third layer by C then the building order will be ABCABCABCABC . . . etc. This is the closest possible packing in a cubic form (Fig. 3.15(i)).

If the unit cell of this structure is examined it will be found that there is a sphere in the centre of each of six faces of the cube. The structure is therefore known as *face-centred cubic*. It will also be noted that the spheres forming the faces of the unit cell are not closely packed, and if the building up of the crystal is continued (Fig. 3.15(iii)) it will be observed that the spaces in this crystal face are bounded in each case by four spheres. If a sphere is now placed at the front of one of these spaces and another at the rear, the gap so enclosed is called an *octahedral* site since it is bounded by six touching spheres with their centres at the points of a regular octahedron. Octahedral sites, like tetrahedral sites, are important in a study of crystallography.

If the sphere marked S in Fig. 3.15(i) is now removed the exposed layer, which is along the (111) planes, has six spheres (Fig. 3.15(iv)) which reveal the close-packed arrangement formed in the building up of the structure. These close-packed layers are in (111) planes which are at right angles to

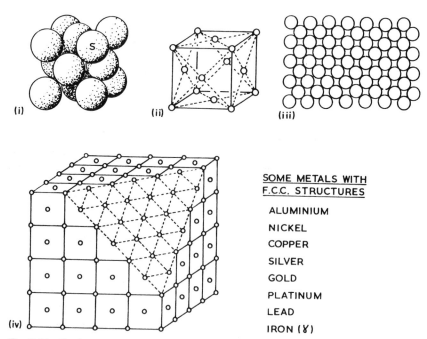

SOME METALS WITH
F.C.C. STRUCTURES

ALUMINIUM

NICKEL

COPPER

SILVER

GOLD

PLATINUM

LEAD

IRON (γ)

Fig. 3.15   The face-centred cubic (FCC) structure. (i) and (ii) depict the unit cell whilst (iii) shows the continuous structure of an 'end face'. (iv) Illustrates the arrangement of atoms on the (111) close-packed plane.

the body-diagonal of the cube and the sequence of layers along that diagonal is ABCABCABCABC . . . etc.

Figure 3.15(iv) reveals that the (111) plane is one of close-packing. The direction of closest packed rows of atoms is [110]. Altogether there are four sets of closely packed (111) planes in the face-centred cubic structure and twelve sets of close-packed directions. As already mentioned, plastic deformation of metallic crystals generally occurs by the slipping of close-packed planes over each other and in the close-packed directions. For this reason the face-centred cubic structure would be expected to show a considerable capacity for plastic deformation. This is indeed supported by the fact that all of the more ductile and malleable metals have face-centred cubic structures. They include gold, silver, copper, aluminium, platinum, lead and nickel.

3.6.3 The Body-centred Cubic Structure (BCC)   The unit cell for this structure consists of a sphere at the centre of a cube with eight other spheres at the corners of the cube and touching the central sphere (Fig. 3.16). Thus packing is not so close as in the other two systems. So when iron, which is an allotropic (3.6.4) metal, changes from a BCC structure to one which is FCC on heating to 910°C, a measurable contraction occurs spontaneously.

Fig. 3.16 The body-centred cubic (BCC) structure.

The most closely packed planes are (110) of which there are six. The closest packed direction is that of (111) of which there are four. Since the closest packed planes are still more open than those of FCC and there are less close-packed directions, it follows that the BCC structure is less ductile than the FCC structure. For example, iron in its FCC form is more malleable than in its BCC form.

3.6.4 Allotropy is defined as the ability of a single substance to exist in more than one physical form. These allotropes are generally stable over different temperature ranges. Thus iron (α-iron) is body-centred cubic at ambient temperatures but on heating it loses its ferromagnetism at 780°C (the *Curie point*) though retaining its BCC structure. This phase is judged to be a different allotrope—β-iron. Further heating to 910°C leads to recrystallisation to a face-centred cubic structure (γ-iron) which at 1400°C reverts to a body-centred cubic form (termed δ-iron to distinguish it from the BCC form existing below 910°C). These allotropic changes are reversible:

$$
\begin{array}{ccccc}
 & 780°C & & 910°C & & 1400°C \\
\text{α-iron} & \longleftarrow & \text{β-iron} & \longleftarrow & \text{γ-iron} & \longleftarrow & \text{δ-iron} \\
\text{(BCC)} & \longrightarrow & \text{(BCC)} & \longrightarrow & \text{(FCC)} & \longrightarrow & \text{(BCC)}
\end{array}
$$

The allotropy of solids which relies solely on difference in crystal structure is known as *polymorphism*. Many metals are polymorphic and sometimes two polymorphs will be face-centred cubic and close-packed hexagonal (CPH) respectively. This is feasible since the change FCC⇆CPH takes place with relative ease.

3.6.5 Coordination Number and Packing Fraction An examination of Fig. 3.14 shows that the central atom X in the upper plane of the CPH structure has six other atoms on the same plane as nearest neighbours. (When represented by spheres in these diagrams atoms are shown as touching the central atom.) This central atom is also touched by three atoms on the plane below and similarly by three atoms on the plane directly above (not shown in the diagram). Thus, in all, twelve atoms are in contact with the central atom as nearest neighbours. This total, 12, is the *coordination number* of the CPH structure.

In the FCC structure (Fig. 3.15) the central atom in the upper plane has four atoms touching it as nearest neighbours, four atoms in the plane below and similarly four atoms in the plane above (not shown). Again the total, 12, is the coordination number of the FCC structure and is characteristic of

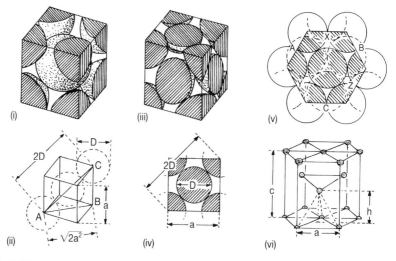

Fig. 3.17

all metallic structures with the closest possible packing of atoms. In the BCC structure, however, any atom—for example the central atom in the single unit shown in Fig. 3.16(i)—is touched by only eight other atoms, that is, four in the plane above and four in the plane below, giving a coordination number of eight.

Only one metal, radioactive polonium, is known to crystallise in simple cubic form, that is with one atom at each corner of a cube. Here any atom is touched by six nearest neighbours so that this very loose packing of atoms has a coordination number of six. Thus it will be seen that the coordination number is generally an indication of how closely atoms (or ions) are packed in a crystal structure.

We can calculate the density of packing more precisely in terms of a *packing fraction* which is the proportion of space in a unit cell actually occupied by the atoms.

In the BCC system (Fig. 3.17(i)) a unit cell contains one complete atom at its centre and one-eighth of an atom at each of the eight corners of the unit cube giving a total of two atoms per unit cell. Hence the total volume of these atoms in terms of their diameter D is:

$$2 \times \left[\frac{4}{3}\pi \left(\frac{D}{2} \right)^3 \right] = \frac{\pi D^3}{3}$$

The internal diagonal AC (Fig. 3.17(ii)) of the unit cell of parameter, a, is of length equal to two diameters of the atoms concerned.

The diagonal AB of a face of the cube is equal to $\sqrt{a^2 + a^2}$. Hence the internal diagonal AC is given by

$$\sqrt{a^2 + 2a^2} = \sqrt{3a^2}$$

$$= 2D$$

or
$$a = \frac{2D}{\sqrt{3}}$$

and the volume of the unit cell by $a^3 = \left(\frac{2D}{\sqrt{3}}\right)^3$

$$\text{Packing fraction} = \frac{\text{Volume of 2 atoms}}{\text{Volume of unit cell}}$$

$$= \frac{\pi D^3}{3} \div \left(\frac{2D}{\sqrt{3}}\right)^3$$

$$= \frac{\pi\sqrt{3}}{8}$$

$$= 0.680$$

In the FCC system (Fig. 3.17(iii)) each of the eight corners of a unit cell contains one-eighth of an atom, whilst each of the six faces of the unit cell contains half an atom, giving a total of four atoms per unit cell. The length of a diagonal of a face of the unit cell (Fig. 3.17(iv)) is equal to two atom diameters ($2D$) so that:

$$2D = \sqrt{a^2 + a^2}$$

or $\quad a = D\sqrt{2}$

Hence:

$$\text{Packing fraction} = \frac{\text{Volume of 4 atoms}}{\text{Volume of unit cell}}$$

$$= 4 \times \left[\frac{4}{3}\pi\left(\frac{D}{2}\right)^3\right] \div (D\sqrt{2})^3$$

$$= \frac{2\pi}{3} \div 2\sqrt{2}$$

$$= 0.740$$

This illustrates the magnitude of the contraction which occurs when BCC (α) iron transforms to FCC (γ) iron at 910°C and the packing fraction increases from 0.68 to 0.74.

In a unit cell of the CPH system (Fig. 3.17(v)) half an atom at the centre of the upper face is surrounded by six, one-sixth atoms giving a total for that plane of one-and-a-half atoms. Similarly a further one-and-a-half atoms are contributed by the bottom plane of the unit cell. Of the three atoms in the central plane of the unit cell segments A, B and C (Fig. 3.17(v)) are 'lost' to neighbouring cells but those 'losses' are balanced by 'gains' of similar segments X, Y and Z from atoms external to the cell. Hence three atoms are contributed by this plane making a total of six atoms per unit cell. The volume of those six atoms is:

$$6 \times \frac{4\pi}{3}\left(\frac{D}{2}\right)^3 = \pi D^3$$

As is indicated in Fig. 3.17(v) and (vi) side a of the regular hexagon is equivalent to one atom diameter D. The height c of the hexagonal prism is equal to $2h$ where h is the height of the regular tetrahedron of edge, a. But $h = a\sqrt{(2/3)}$ so that $c = 2h = 2a\sqrt{(2/3)}$. The area of the hexagonal base of the unit cell (of side a) is $[(3\sqrt{3})/2]a^2$ so that the volume of the hexagonal prism (height $2a\sqrt{(2/3)} = 3a^3\sqrt{2}$ which, since $a = D$, is $3D^3\sqrt{2}$. Hence:

$$\text{Packing fraction} = \frac{\text{Volume of 6 atoms}}{\text{Volume of unit cell}}$$

$$= \frac{\pi D^3}{3D^3\sqrt{2}}$$

$$= \frac{\pi}{3\sqrt{2}}$$

$$= 0.740$$

In the simple cubic system a unit cell contains only one atom in total (one-eighth of an atom at each of the eight corners of the cube). The volume of the unit cell is D^3 since each side is equal in length to one atom diameter. Hence:

$$\text{Packing fraction} = \frac{\frac{4}{3}\pi\left(\frac{D}{2}\right)^3}{D^3}$$

$$= \frac{\pi}{6}$$

$$= 0.523$$

This confirms the very open packing of atoms in this system.

Q | The atomic radius of aluminium is 1.43×10^{-10} m and its atomic mass number is 26.98. Given that Avogadro's Number, N_A, is 6.022×10^{23} k/mole calculate the density of aluminium.

A | The atomic radius of aluminium is 1.43×10^{-10} m. Hence the diameter is given by:

$$D = 2.86 \times 10^{-10}\,\text{m}$$

For the FCC structure (aluminium) the length of one side of the unit cell is equal to:

$$D\sqrt{2} = \sqrt{2} \times 2.86 \times 10^{-10}\,\text{m}$$

Hence the volume of a unit cell of aluminium is equal to:

$$(\sqrt{2} \times 2.86 \times 10^{-10})^3 = 6.617 \times 10^{-29}\,\text{m}^3$$

A unit cell of FCC aluminium contains four atoms. 1 k/mole of aluminium (atomic mass 26.98) contains N_A atoms. Therefore, mass of four aluminium atoms is:

$$\frac{4 \times 26.98}{N_A} = \frac{4 \times 26.98}{6.022 \times 10^{26}}$$

Hence, density of aluminium is:

$$\frac{\text{Mass of 4 aluminium atoms}}{\text{Volume occupied by 4 aluminium atoms}} = \frac{4 \times 26.98}{6.022 \times 10^{26} \times 6.617 \times 10^{-29}}$$

$$= 2.708 \times 10^3 \text{ kg/m}^3$$

i.e. relative density $\quad\quad\quad = \underline{2.708}$

This value is slightly higher than that determined experimentally (2.699) due to the presence of various small cavities in laboratory (or factory) produced metals. This will be discussed later in this and subsequent chapters (3.7.2; 5.3).

The Crystallisation of Metals

3.7   All pure elements exist either as gases, liquids or solids and the *state* in which an element* exists depends upon the prevailing combination of temperature and pressure. Of the metallic elements, only mercury is a liquid at ordinary temperatures, though a number of metals melt below red heat. A few, like zinc, cadmium and mercury, boil at relatively low temperatures and can be purified by distillation.

In 1827 botanist Robert Brown was using a microscope to examine some minute pollen particles floating on the surface of water. He was surprised to observe that these particles were in a state of continuous motion which was agitated and irregular. At the time he was completely baffled by the phenomenon but later it was realised that the motion produced in the particles was the result of impact by moving invisible water molecules at the surface of the water. The same kind of motion, known as 'Brownian motion', can be seen in the solid particles comprising smoke. When these particles, suspended in air, are viewed through a microscope they are seen to pursue irregular jerky paths due to bombardment by fast-moving oxygen and nitrogen molecules. Observations of this kind showed that molecules in gases and liquids were in a state of continuous motion and this led ultimately to the formulation of the *kinetic theory of matter*.

3.7.1   Metals form monatomic gases, that is, the vapour consists of single atoms of the metal. These atoms are in constant motion and are colliding

---

*The same is not necessarily true of compounds since many of these decompose at a temperature below their boiling point, and some below their melting point.

continually with the walls of their container. The collisions give rise to the pressure exerted by the gas. If temperature is increased atoms move more quickly as heat energy is transformed into kinetic energy. Hence collisions with the sides of the container will increase and so pressure will increase, assuming that the volume of the gas is kept constant by restraining expansion.

If the temperature is allowed to fall the average kinetic energy of the atoms decreases and is lost as heat energy (specific heat). Eventually an average energy content is reached where van der Waals forces acting between atoms are able to overcome the kinetic movement of many of the atoms. Hence those with less than the average kinetic energy are attracted to each other so that they stick to each other, that is they condense to form droplets of liquid which fall under gravity. At this point the remainder of the kinetic energy is given up as heat (latent heat of vaporisation). No orderly arrangement of atoms exists in the liquid metal and atoms are still free to move with respect to each other, that is, a liquid possesses *mobility*.

At a lower temperature still the liquid begins to solidify. This is a process where atoms change from a disordered or *amorphous* state to an ordered *crystalline* state. As this occurs potential energy is lost (Fig. 3.18)—this time as latent heat of fusion—and since the atoms arrange themselves according to some orderly pattern, shrinkage generally accompanies solidification. For this reason the upper portion of a metal ingot is 'piped' (Fig. 3.26).

3.7.2 Pure metals, like other pure substances, solidify at a fixed single temperature. Ideally this will occur in accordance with the type of cooling curve shown in Fig. 3.19(i). However, in most cases there is some degree of

Fig. 3.18 Energy and states of matter for a pure metal.

Fig. 3.19 Typical cooling curves for crystalline and amorphous solids. In (iii) there was insufficient molten metal to provide latent heat which would otherwise have caused a return to equilibrium as in (ii).

undercooling of the liquid before the onset of crystallisation. This is due to a lack of nucleation of the system (Fig. 3.19(ii) and (iii)). Once a crystal nucleus forms it provides a solid/liquid interface where crystallisation can proceed (Fig. 3.20). Under industrial conditions nucleation may occur on a particle of slag or dross, but with very pure liquid metal some degree of undercooling will occur. When nucleation ultimately takes place the greater the extent of undercooling the greater will be the rate of crystallisa-

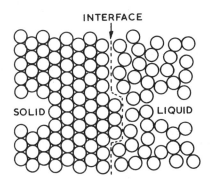

Fig. 3.20   Schematic representation of a solid/liquid interface.

tion and, consequently, the rate at which latent heat is liberated. As a result the temperature of the melt will rise to reach equilibrium at the freezing point of the metal (Fig. 3.19(ii)). If only small amounts of molten metal are involved insufficient latent heat may be available to allow the freezing point to be regained (Fig. 3.19(iii)).

The nuclei which form will be simple units of the crystal pattern in which the metal solidifies, generally either CPH, FCC or BCC. Crystal growth will tend to occur in the opposite direction to that in which heat is being conducted from the melt. Thus a 'spike' begins to grow from the nucleus into a region of under-cooled liquid, but as this occurs, latent heat is liberated and this warms up the liquid in front of the growing spike. Hence growth of the spike is retarded and as a result secondary spikes begin to

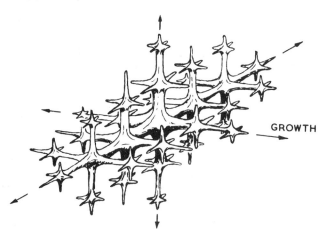

Fig. 3.21 Early stages in the growth of a metallic dendrite showing primary, secondary and tertiary arms.

grow from the primary one and these are followed by tertiary spikes (Fig. 3.21). These primary, secondary and tertiary branches follow crystallographic planes and give rise to the general regularity of the structure. The crystal skeleton which develops is called a *dendrite*, a reference to its tree-like growth (Gk. *dendron*, 'a tree'), in which the primary, secondary and tertiary arms resemble the trunk, branches and twigs of a tree.

The main branches of the dendrite continue to grow until the outer fringes make contact with those of neighbouring dendrites (Fig. 3.23). Since they are thus restrained from further outward growth these existing arms thicken as heat flows outwards from the region. Each dendrite develops independently so that the outer branches of neighbouring dendrites make contact at irregular angles. Moreover heat flow is not necessarily uniform and so the final crystals are of irregular shape though

(Courtesy of Dr. J. Moore, W.B.C.C.T.)

Fig. 3.22 Dendrite of zinc. This was grown from an aqueous solution of zinc sulphate by electrolysis. ×15.

(i)	(ii)	(iii)	(iv)

Fig. 3.23 The dendritic solidification of a metal. (i) Dendrites begin to develop from newly formed nuclei by putting out primary and secondary arms. (ii) Tertiary arms form and meet others growing in the opposite direction. (iii) Dendrites continue to grow until their outer arms touch those from neighbouring dendrites. Existing arms then thicken. (iv) When the metal is completely solid there is little evidence of the dendritic method of growth since in a pure metal all atoms are similar. Then only grain boundaries are visible.

the atoms within any one crystal are regularly spaced with respect to each other in a crystal lattice.

If the metal under consideration is pure only crystal boundaries will be visible when a section of the metal is examined under the micrscope since all atoms present in the structure are similar. The presence of impurities, however, may reveal the dendritic pattern to some extent as these impurities tend to congregate in that metal which solidifies last, that is, between the dendrite branches and particularly towards the crystal boundaries. Shrinkage cavities, formed due to lack of liquid as solidification contraction occurs, may also outline the dendrite shape to a limited extent (Fig. 3.24).

SHRINKAGE CAVITY

RELATIVELY PURE METAL

IMPURE METAL NEAR CRYSTAL BOUNDARIES

Fig. 3.24 The dendritic structure may be revealed by the presence of impurities and/or shrinkage cavities.

3.7.3   Although the dendrites in Fig. 3.23 are shown as being geometrically perfect in respect of the directionality of their branches, in practice there is often misorientation between neighbouring arms with respect to each other *within the same crystal*. This may be as much as 4° in some cases. It is due to the pressure of growing arms upon each other as they compete for space, rather like the branches of pine trees in a dense Forestry Commission plantation. Further misorientation of individual atoms occurs at crystal boundaries. Here is a fringe of metal about three atoms thick where the atoms are misfits with respect to both crystals. Years ago this was known as 'grain boundary cement' but fortunately this misleading term

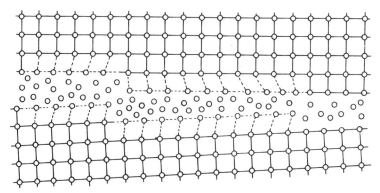

Fig. 3.25 The 'amorphous effect' due to misorientation at grain boundaries.

has fallen into disuse. The existence of this fringe of misfit atoms explains some of the mechanical properties of metals. Since it is of an amorphous nature (Fig. 3.25) it behaves rather like an extremely viscous liquid. Whilst at low temperatures metals generally fail by the propagation of *trans*crystalline cracks, at high temperature failure invariably follows *inter*crystalline paths, presumably because the amorphous film has become less viscous at the higher temperature.

Fine-grained metals, which will contain more grain boundary and, hence, more misfit atoms per unit volume, creep more readily than coarse-grained metals of similar composition. It is reasonable to suppose that creep (16.4) is a property closely associated with the amorphous state.

3.7.4 The overall size and, to some degree, the shape of crystals in a metal ingot varies with the rate of cooling. Rapid cooling gives rise to considerable undercooling and the presence of a mould surface helps to nucleate the liquid. As a result a dense shower of nuclei will form instantaneously and so a layer of very small or 'chill' crystals will result (Fig. 3.26(ii)). As the mould warms up progressively and the rate of

Fig. 3.26 Types of crystal structure in ingots. (i) Columnar crystals—excessive segregation of impurities at the core. (ii) Zones containing chill, columnar and equi-axed crystals. (iii) Large equi-axed crystals—less segregation because the cooling rate was low.

Fig. 3.27 Cross-section of a small aluminium ingot. The structure was developed by etching in 2% hydrofluoric acid. Actual size.

cooling is reduced a stage is reached when crystal growth inwards is balanced by heat flow outwards. Fresh nuclei are not formed and so the existing crystals grow in columnar shape. With large ingots the rate of cooling at the centre may be so low that few nuclei form since there is little undercooling. Consequently the resultant crystals tend to be large and 'equi-axed', that is, of equal axes.

An important effect which the process of solidification has upon an industrially cast metal is in the segregation of dissolved impurities at the boundaries of crystals. Pure metal tends to crystallise out first so that the metal in the region of the crystal boundaries may become so rich in impurities that the limit of solubility of the latter is exceeded causing it to separate out as films of impurity in the crystal boundary region. This may give rise to excessive brittleness in that region so that intercrystalline fracture results.

3.7.5 Rapid Solidification Processes (RSP) These have been investigated in recent decades partly with the aim of producing smaller crystals and so limiting the effects of segregation—the smaller the crystals the greater the amount of crystal boundary per unit volume of the metal and consequently the more 'dilute' the impurity. However, by using very high rates of cooling at the onset of solidification, for example in excess of 10^6 °C per second, it is found that with some metals and alloys a completely amorphous structure can be retained—the onset of crystallisation has been halted. Hence there will be no serious segregation since crystal boundaries have been eliminated. Further implications of RSP are mentioned in 13.6.8.

Crystal Structures of Non-metallic Materials

3.8   So far we have dealt with crystal structures which are dependent upon the formation of the ionic bond or the presence of the metallic bond between cores. The covalent bond and relatively weak van der Waals forces also play their parts in crystal formation. The covalent bond which is the basis of molecule formation may lead to the production of molecules so large that within these molecules exists a repetitive pattern common to crystalline structures. Similarly, the formation of crystals in which molecules are units relies on van der Waals forces operating between the molecules. Thus water crystallises to form ice. Here the forces of attraction are supplied by the hydrogen bond (2.4.4).

Giant Molecules

3.8.1 Three-dimensional Crystals   Although in a geometrical sense all crystals will be three-dimensional, the term is used here as a reference to axial directions. The best-known example of this type of structure is that of diamond. The crystal pattern is determined by the fact that in this allotrope carbon exhibits its normal valency of four, such that each carbon atom is joined to four of its immediate neighbours by covalent bonds. In order that this shall occur symmetrically each carbon atom is surrounded by four others whose centres lie at the corners of a regular tetrahedron. The central atom is in contact with the other four (in Fig. 3.28 only the centres of the atomic cores are indicated). If the structure is extended to include more carbon atoms (Fig. 3.28(ii)) it will be noticed that the whole figure is tetrahedral and will continue to extend 'downwards' such that each carbon atom is joined to four others by covalent bonds. The structure can also be regarded as being basically cubic (Fig. 3.28(iii)) and it can be shown that as such the structure is much less closely packed than a body-centred cubic

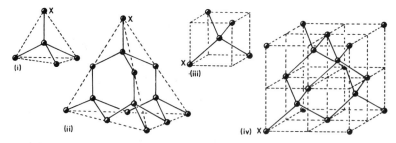

Fig. 3.28   The crystal structure of diamond. (i) The unit cell, indicating that here carbon has a coordination number of four; (iii) and (iv) show how diamond can be considered as a *cubic* structure. Note that in (ii) not all valencies are used up in the 'fringe' atoms of the crystal, which will of course be continuous in practice. (X represents the same atom in the system in each case.)

structure. Thus, in Fig. 3.28(iv), only four out of the eight cubic units has a carbon atom at its centre whilst many of the 'corner sites' are unused.

Diamond can therefore be regarded as a three-dimensional giant molecule (or macromolecule), C_n. In the crystal so produced the carbon atom has a coordination number of four. Diamond is mechanically strong and is the hardest substance known. Moreover it has an extremely high melting point—in fact at ordinary pressures it does not melt, but at about 3500°C thermal vibrations of the atoms are strong enough to overcome the powerful covalent bonds and vaporisation begins. Both of these factors illustrate the very great strength of the covalent bond. Since in diamond all carbon atoms are covalently bonded such that all available electrons are used up, diamond is not a conductor of electricity (18.3.2).

Other elements of similar valency (with carbon they constitute Group IV of the periodic classification), e.g. silicon, germanium and the 'grey' allotrope of tin, also crystallise in the diamond-type of structure.

3.8.2 Layer Structures   Each carbon atom in a diamond crystal achieves a stable octet of electrons by forming four covalent bonds with neighbouring carbon atoms. Another allotrope of carbon, graphite, also exists. In graphite the atoms are bonded in the form of a ring structure so that flat layers are produced in a regular hexagonal network (Fig. 3.29). Each

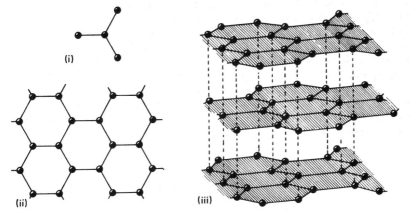

Fig. 3.29   The layer structure of graphite.

carbon atom is surrounded by three other carbon atoms (Fig. 3.29(i)) so that the coordination number of carbon in this instance is three. As Fig. 3.29(ii) indicates each carbon atom uses only three of its outer shell electrons to form covalent bonds. The fourth electron in each case can be regarded as being 'shared' to some extent in the nature of the electrons in the metallic bond. Thus, whilst diamond does not conduct electricity, graphite does. Whilst the carbon atoms within these layers or sheets are bonded covalently the layers themselves are held together only by weak van der Waals forces. For this reason whilst the sheets themselves are strong they slide over each other very easily so that graphite is a good lubricant. Materials like clay (un-fired) and talc are also of this type.

In the molecules of thermoplastic materials e.g. polythene, carbon atoms are covalently bonded to form long-chain molecules (2.3.3). Whilst these 'chains' are strong in themselves they are only bonded to neighbouring molecules by weak van der Waals forces so that under comparatively low stress these molecules will move relative to each other with ease, rendering polythene a soft flexible material.

3.8.3 Molecular Crystals Molecules are separate self-contained units as are atoms so that it is possible for them to arrange themselves in a crystalline form. Whilst the atoms within a molecule are held together by covalent bonding, the molecules themselves are bonded by van der Waals forces. The element tellurium forms long-chain molecules and these in turn arrange themselves in a pattern with like molecules to produce crystals.

Many polymers are regarded as being amorphous, that is, non-crystalline, yet many do possess some crystalline regions known as *fringed micelles*. These are often linked together by a chain molecule which extends from one micelle to another (Fig. 3.30). Since intermolecular

Fig. 3.30 Fringed *micelles* in polythene showing one molecule (X) extending between two micelles.

forces are only of the weak van der Waals type, crystallisation is not as perfect over a long range with these cumbersome chain molecules as it is when crystals grow from single atoms.

3.8.4 Fullerenes Lampblack—or soot—used to be classed as a third allotrope of carbon ('amorphous carbon') but it was later found to consist of very small graphite platelets. A short while ago, however, a third allotropic form of carbon was discovered—the *fullerene*. This allotropic form consists of a series of large molecules each containing an even number of carbon atoms arranged in the form of a closed hollow cage (Fig. 3.31(i)). When carbon atoms form a purely hexagonal network (Fig. 3.29) the structure consists of a flat sheet as in graphite but if *pentagons* are included at suitable sites then the structure can no longer remain flat but is 'drawn up' into globular form (Fig. 3.31(ii)).

This is similar to the type of construction used architecturally in geodesic domes such as that housing 'Science World' in Vancouver. In fact the name 'fullerene' is derived from that of the US engineer and architect Richard

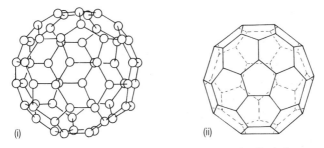

Fig. 3.31 The carbon allotrope fullerene-60 (C_{60}): (i) representing the hollow-cage structure of fullerene-60—as in graphite each carbon atom is bonded to three neighbours; (ii) showing the relationship between pentagon and hexagon positions in the structure of C_{60}. The modern soccer ball shows a similar distribution of pentagonal panels (black) and hexagonal panels (white).

Buckminster Fuller who developed geodesic design. Nearer to home, the panels of a modern soccer ball follow a similar pattern. Fullerenes contain even numbers of carbon atoms from 20 up to as many as 600 but the more symmetrical and stable molecules are those with between 28 and 70 carbon atoms.

The first fullerene to be produced and isolated in any quantity was C_{60}, known as buckminster fullerene. This can be manufactured by striking a carbon arc in a low-pressure argon atmosphere. The carbon vapour so formed condenses as a dust which contains a proportion of C_{60} molecules. This can be dissolved out of the residue with benzene in which it forms a magenta coloured solution. The fullerene will crystallise out from the benzene in the form of magenta face-centred cubic crystals.

In the fullerene molecule the even number of carbon atoms is arranged over the surface of a closed hollow cage. Each atom is bonded trigonally to its three neighbours (Fig. 3.31(i)) by bonds conforming to a polyhedral network consisting of 12 pentagons and n hexagons (where n is any number except unity). In C_{60} the 12 pentagons are isolated from each other and interspersed symmetrically among the 20 linked hexagons (Fig. 3.31(ii)). The C_{60} molecule is able to interact with some other atoms or groups, accepting up to three electrons in the process. Radicals and large atoms tend to be attached to the outside of the cage (forming *exo*hedral complexes) whilst smaller atoms such as those of the alkali metals are held inside the cage (forming *endo*hedral complexes). Interesting amongst these is $C_{60}K_3$ which is superconductive (18.4.2) up to 30 K suggesting that research might be worthwhile in this direction.

Fullerenes are such new materials that as yet their technology is in its infancy and large amounts of research remain to be done to assess their value in such wide fields as plastics materials, pharmaceutics, structural materials and electronics industries. One wonders what the sci-fi writers will make of them!

Fig. 3.32 A model of the molecule of fullerene, C_{60}, which I constructed with the aid of a child's soccer ball, chocolate drops and a felt pen. Each of the 12 pentagonal 'panels' (dark) is surrounded by five hexagonal 'panels' (light). Chocolate drops indicate the positions of nuclei of the carbon atoms. In the solid form these large molecules arrange themselves as face-centred cubic crystals.

Non-crystalline Substances

3.9 All true solids are crystalline in nature. However, a number of apparently solid materials are amorphous, that is, their atoms or molecules do not conform to any regular geometrical patterns. Solids such as glass and pitch are comparatively rigid and are broken by sudden impact, but if a steady load is applied over a long period of time the material will flow. A piece of pitch will flow very gradually, over a period of months, to take the shape of a vessel into which it has been placed. A glass rod, supported at each end, will sag under its own weight if left in this position for a long time. The deformation produced will be *permanent* and unlike the temporary elastic deformation produced when an elastic solid is lightly stressed within its elastic limit.

These amorphous 'solids' do not melt at a definite temperature as do pure crystalline materials. Instead they soften gradually and become more mobile thus resembling liquids of very high viscosity. The viscosity of a liquid increases as its temperature falls and if it is cooled below its freezing point without crystallisation taking place its viscosity may reach a value as much as 10^{10} times that of water. Viscosity is related to van der Waals forces. As temperature falls thermal vibrations of the molecules are reduced and so viscosity increases, since van der Waals forces will also increase.

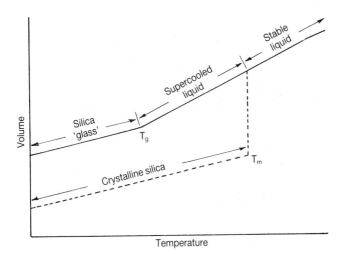

Fig. 3.33 The glass transition temperature (T_g) for silica. T_m is the freezing point at which slowly-cooled silica would crystallise.

The low mobility of large molecules in relation to their thermal activation was first investigated in glasses and so terms such as 'glass' and 'glassiness' are used to describe the properties and behaviour of amorphous 'solids' in general. At high temperatures glasses form normal liquids which flow under the application of shear stresses. As the glass cools normal contraction in volume takes place and, although crystallisation does not set in, a point is achieved where a change in the *rate* of contraction occurs (Fig. 3.33). Below this temperature, called the *glass transition temperature* (13.6), there is no further rearrangement of atom packing and any contraction as does take place is due only to a reduction in vibrations of the atoms as thermal energy is lost.

4

Mechanical Properties

4.1 Structural materials used in mechanical and civil engineering practice must generally have *strength*. This value is a measure of the externally applied forces which are necessary to overcome internal forces of attraction between fundamental particles within the material such as we have been discussing in the first three chapters of this book. Briefly, strength is due to the sum of forces of attraction between negatively charged electrons and positively charged protons within the material. When covalent bonds join large numbers of atoms to produce giant molecules as is the case of the carbon atoms in diamond or carbon fibre, the strength of the resultant material is great. Similarly the metallic bond, which in some respects can be regarded as a multiple covalent bond, ensures that for most metals strength is high.

Some materials such as polymers consist of long fibre-like molecules in which the strength within the fibres is high depending as it does on covalent bonds acting between the constituent atoms. However, only weak van der Waals forces hold one molecule to its neighbours and it is the sum of many such individually weak bonds which ensures that polymers are reasonably strong. Small molecules however have few van der Waals forces acting between them so substances such as H_2S, CO_2 and SO_2 remain as gases at ordinary temperatures and pressures where the thermal activity of such molecules overcomes the weak van der Waals forces acting between them.

4.1.1 Constructional materials generally must be able to withstand the action of considerable forces without undergoing other than very small amounts of distortion. In the field of what is loosely termed production engineering, however, very different properties may be desirable. Here a material must be capable of permanent deformation at the expense of as little energy as possible. That is, it must be *malleable* and *ductile*. In the case of metals a forming process causes the metal to lose its softness and become harder and stronger, that is, it *work hardens*. Thus the production engineer designs his forming process to utilise the malleability or ductility of the material and at the same time generate in it sufficient strength for subsequent service. Other mechanical properties include *elasticity*, *hardness*, *toughness* and also *creep* and *fatigue* properties (16.4; 16.5). In each case the property is associated with the behaviour of the material towards the application of force and the engineer is generally interested in the 'density of force' which is necessary to produce some definite amount of deformation, either temporary or permanent in the material.

4.1.2 Stress 'a measurement of density of forces'; defined as force per unit area of cross section. The SI unit of stress is the Pascal (Pa) which is equivalent to a force of one Newton acting on an area of one square metre, i.e. N/m^2 or Nm^{-2}. Since the Newton is a small unit of force relative to the square metre we shall generally be dealing with stresses measured in the region of millions of Pascals and the force required to break in tension a steel member one square metre in cross section is difficult to visualise. However, since

$$MPa = MN/m^2 = N/mm^2$$

we can think in terms of the latter unit (N/mm^2) since it will be equivalent to the number of units required to break in tension a wire or fibre slightly over 1 mm in diameter. *Numerically* of course it will be the same as that expressed in MPa.

Q | A steel rod 6 mm in diameter is under the action of a tensile force (P) of 400 Newtons. Calculate the tensile stress (σ) acting in the bar.

A

$$\sigma = \frac{P}{A}$$

$$= \frac{P}{\pi r^2}$$

$$= \frac{400}{\pi \times (0.003)^2} \, N/m^2$$

$$= 14\,147\,072 \, N/m^2$$

$$= 14.15 \, MPa$$

Units
$\sigma = N/(m \times m)$
$\quad = N/m^2$

4.1.3 Strain   refers to the proportional deformation produced in a material under the influence of stress. It is measured as the number of metres of deformation suffered per metre of original length and is a numerical ratio.

Q The original gauge length (L_o) marked on an aluminium test piece is 40 mm. The test piece is strained in tension so that the gauge length becomes 42.3 mm. Calculate the strain (ε).

A
$$\varepsilon = \frac{\text{Increase in length}}{L_o}$$

Units
$\varepsilon = (\text{mm} - \text{mm})/\text{mm}$
$= \text{mm/mm}$
$= \text{numerical ratio}$

$$= \frac{42.3 - 40}{40}$$

$$= \underline{0.0575}$$

Strain is usually quoted as a percentage, i.e.

$$\varepsilon = 0.0575 \times 100\%$$

$$= \underline{5.75\%}$$

Strain may be either *elastic* or *plastic*. Elastic strain is reversible and disappears when the stress is removed. Atoms are displaced from their initial positions by the application of stress but when this stress is removed they return to their initial positions relative to their neighbours provided that the strain has been of an elastic nature. Strain is roughly proportional to the applied stress (Fig. 4.1(i)), and, for practical purposes, the material obeys Hooke's Law. This states that, for an elastic body, strain produced is directly proportional to stress applied.

(i) Stress / Strain (ii) Stress / Strain — Permanent deformation

Fig. 4.1

4.1.4 Young's Modulus of Elasticity (E)   is the ratio between the stress applied and the elastic strain it produces. That is, it is the stress required to produce a unit quantity of elastic strain. It is related to the *rigidity* of the material and is a value of supreme importance to the constructional engineer. The modulus of elasticity is expressed in terms of either tensile or compressive stresses and its units are the same as those for stress, since:

$$E = \frac{\text{stress}}{\text{strain}} = \frac{\text{N/m}^2}{\text{m/m}} = \text{N/m}^2 = \text{Pa}$$

Because of the numerical magnitude of the value of E it is commonly expressed as GPa (or GN/m^2).

Q A steel wire 0.55 mm² in cross-sectional area (A) and 10 m long is extended elastically 1.68 mm by a force of 17.24 N. Calculate the modulus of elasticity (E) for the steel.

A

$$\sigma = \frac{P}{A}$$

$$= \frac{17.24}{0.55} \text{ N/mm}^2$$

$$= 31.34 \text{ MN/m}^2$$

$$\varepsilon = \frac{1.68 \times 10^{-3} \text{ m}}{10 \text{ m}}$$

$$= 0.000\,168$$

$$E = \frac{\sigma}{\varepsilon}$$

$$= \frac{31.34 \text{ MN/m}^2}{0.000\,168}$$

$$= 186\,547 \text{ MN/m}^2$$

$$= 186.5 \text{ GPa}$$

Alternatively E can be derived from the slope of the linear portion of the stress/strain diagram for the material, i.e.

$$E = \frac{\text{Slope} \times \text{gauge length}}{\text{Cross-sectional area}}$$

The sophisticated technology of these closing years of the twentieth century often involves consideration of the mass of material required to provide the necessary strength and rigidity in a structure. This is particularly so in the aero-space and other transport industries and in fact in any situation where work done against gravity must be paid for in terms of increasingly expensive fuel. Thus the modulus of elasticity is commonly expressed as a *specific modulus of elasticity* in which E is related to the relative density of the material:

$$\text{Specific modulus of elasticity} = \frac{E}{\text{relative density}}$$

Fig. 4.2 The principle of tensile testing. A threaded test piece is shown here but in most machines the test piece will be plain and held in friction grips.

4.1.5 Plastic Strain This results when a material is stressed to the extent where its elastic limit is exceeded. It coincides with a movement of the atoms within the structure of the material into permanent new positions with respect to neighbouring atoms. When the stress is removed only elastic strain disappears and any plastic strain produced is retained (Fig. 4.1(ii)). *Malleability* refers to the extent to which a material can undergo deformation in compression before failure occurs, whilst *ductility* refers to the degree of extension which takes place before failure of a material in tension. All ductile materials are malleable but malleable materials are not necessarily always ductile since a soft material may lack strength and thus tear apart very easily in tension.

Ductility is commonly expressed in practical terms as the percentage elongation in gauge length a standard test piece suffers during a tensile test to failure. Figure 4.3 illustrates the necessity for a standard relationship

$$\text{ELONGATION} \ \% \ = \frac{(L_u - L_0)}{L_0} \times 100$$

between gauge length and cross-sectional area of the test piece if results are to be comparable. Since the bulk of plastic deformation occurs at the 'neck' (between Z and Y) it is clear that a percentage elongation based on ZY as the gauge length would give different results from that based on XY as the gauge length. Consequently tensile test pieces should be geometrically similar and are known as *proportional* test pieces. They are usually circular in cross-section and BSI lays down (B.S. 18: Part 1 and 2*) that for proportional test pieces:

$$L_0 = 5.65\sqrt{S_0}$$

where L_0 is the gauge length and S_0 the original area of cross-section. This

* Superceded by British and European Standards (BSEN) 1002.

formula was adopted internationally and SI units are used. For test pieces of circular cross-section it gives a relationship:

$$L_0 = 5d$$

where d is the diameter at the gauge length. Thus a test piece of cross-sectional area 200 mm^2 will have a diameter of 15.96 mm (i.e. 16 mm) and a gauge length of 80 mm.

This is in fact in accordance with *Barba's Law* which states that the percentage elongation of a material at fracture is constant irrespective of specimen dimensions provided that: $L_0/S_0 = $ a constant. From this an empirical equation can be derived:

$$\% \text{ elongation} = 100\left(\frac{a\sqrt{S_0}}{L_0} + b\right)$$

where a and b are constants for the material.

Also related to ductility is *reduction in cross-sectional area* measured at the point of fracture. Very ductile materials are considerably reduced in cross-section before they break.

$$\% \text{ Reduction in area} = \frac{\begin{array}{c}\text{(Original area of} \\ \text{cross-section} - \text{final area of cross-section)}\end{array}}{\text{Original area of cross-section}} \times 100$$

4.1.6 Stress/strain Diagrams When corresponding values of stress and strain derived during a tensile test are plotted graphically it is found that each type of material is represented by a characteristic curve. Materials of negligible ductility, such as fully hardened steels, cast iron and concrete, undergo little or no plastic deformation before fracture (Fig. 4.4(i)). That is, there is no yield point and only elastic extension occurs. A ductile material, on the other hand, exhibits an *elastic limit* (or *limit of proportionality*) beyond which plastic deformation occurs. The maximum stress which a material can withstand before plastic flow sets in is known as its *yield strength*. In softer ferrous materials (wrought-iron and low-carbon steels) and some plastics materials the onset of plastic flow is marked by a very definite yield point (Fig. 4.4(iii)) and it is therefore a simple matter to calculate the yield stress. In other materials, comprising practically all ductile metals and alloys the elastic limit is not well defined (Fig. 4.4(iv)). In most respects the yield stress of a material is of greater importance to the design-engineer than is the maximum strength attained during plastic flow. Consequently a substitute value for yield strength is derived for those materials which show no obvious yield point. This is known as the *proof stress* and is that stress which will produce a permanent (plastic) extension of 0.1%[*] in the gauge length of the test piece. It is derived as indicated in Figs 4.4(ii) and (iv).

Materials which have received some treatment such as work-hardening

---

[*] Or up to 0.5% for some materials.

Fig. 4.4 Representative stress/strain diagrams for various types of material. (i) Non-ductile material. (ii) Semi-ductile material. (iii) and (iv) Ductile materials.
T = tensile strength; B = breaking strength; Y = yield stress and P = proof stress.

or, in the case of some alloys, suitable heat-treatment, are generally stronger but less ductile than those in the fully soft condition. This is indicated in the stress/strain curve Fig. 4.4(ii).

4.1.6.1 Secant modulus   Many soft materials particularly thermoplastics (12.5.4) do not obey Hooke's Law and the stress–strain diagram shows no region where a linear relationship applies so that it is not possible to make a simple measurement of Young's Modulus, E. As a representative alternative we can calculate the *secant modulus* of the material. This is the ratio of nominal stress to the corresponding strain at some specific point on the force/extension diagram.

Figure 4.5 shows that the secant modulus is in fact the slope of the line OS. If a force P is required to produce 0.2% strain in the gauge length, then:

$$\text{secant modulus} = \frac{\text{stress}}{\text{strain}}$$

$$= \frac{P}{A} \div 0.002$$

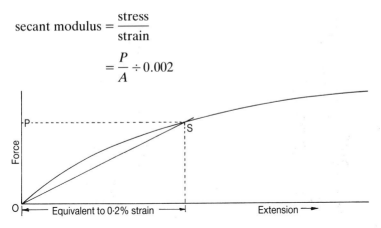

Fig. 4.5 Derivation of the secant modulus.

where A is the initial cross-sectional area of the test piece. For most materials the secant modulus associated with a strain of 0.2% is measured, though in some cases a strain of 0.1% is specified.

4.1.6.2 Tangent modulus Sometimes the *tangent modulus* is called for. This is determined by drawing a tangent to the force/extension diagram at a point corresponding to the specified strain (see exercise below). Then:

$$\text{tangent modulus} = \text{Slope of tangent} \times \frac{\text{Gauge length } (L_0)}{\text{Cross-sectional area } (A)}$$

Q A tensile test on a plastics material gave the results shown below during the initial stages of the test:

Extension (mm)	0	0.03	0.055	0.09	0.14	0.20	0.37	0.61
Force (N)	0	100	150	200	250	300	400	500

The test piece had a cross-sectional area of 50 mm^2 and a gauge length of 50 mm.

Plot the force/extension diagram for the material over the range of readings given and determine:

(i) the tangent modulus at 0.2% strain;
(ii) the secant modulus at 0.5% strain.

A (i) *Tangent modulus at 0.2% strain*
0.2% strain on a gauge length of 50 mm = 0.1 mm. Hence tangent, *AT*, is drawn at an extension equivalent to 0.1 mm. The slope of tangent *AT* is given by:

$$\frac{TB}{AB} = \frac{(440 - 95)\ N}{0.3\ mm}$$

$$= 1150\ N/mm$$

Hence the tangent modulus at 0.2% strain is equal to:

$$\frac{1150 \times 50 \text{ mm}}{50 \text{ mm}^2} = 1150 \text{ N/mm}^2$$

$$= 1.15 \text{ GPa}$$

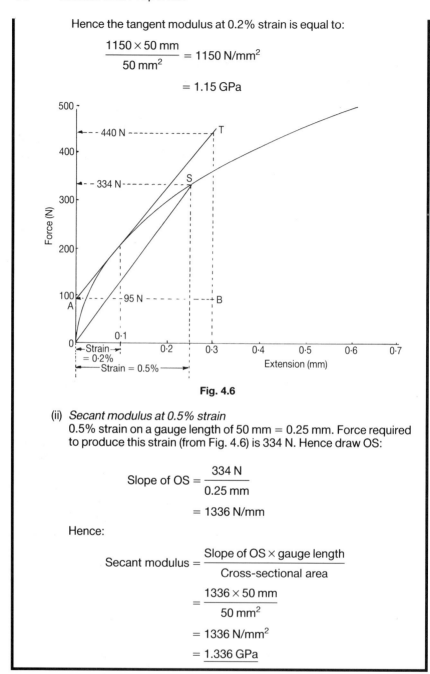

Fig. 4.6

(ii) *Secant modulus at 0.5% strain*
0.5% strain on a gauge length of 50 mm = 0.25 mm. Force required to produce this strain (from Fig. 4.6) is 334 N. Hence draw OS:

$$\text{Slope of OS} = \frac{334 \text{ N}}{0.25 \text{ mm}}$$

$$= 1336 \text{ N/mm}$$

Hence:

$$\text{Secant modulus} = \frac{\text{Slope of OS} \times \text{gauge length}}{\text{Cross-sectional area}}$$

$$= \frac{1336 \times 50 \text{ mm}}{50 \text{ mm}^2}$$

$$= 1336 \text{ N/mm}^2$$

$$= 1.336 \text{ GPa}$$

4.1.7 The *tensile strength* of a material is derived by dividing the maximum force sustained during the test by the *original* cross-sectional area of the test piece. The units involved are those of stress and generally MN/m^2 or N/mm^2 are the most convenient to use though MPa will be the appropriate derived SI unit. They will of course be numerically the same. It is important to note that at all stages during a tensile test *stress is calculated on the basis of the original area of cross-section*. That is, it does not take into account the diminishing cross-sectional area at the narrowest point in the 'neck' during the final stages of plastic deformation. For this reason these so-called 'stress/strain' diagrams are really modified force/extension diagrams and to plot a *true* stress/strain diagram it would be necessary to take into account the diminishing cross-section by measuring the minimum diameter at the neck with each reading of the force used (Fig. 4.7).

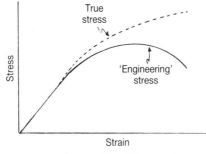

Fig. 4.7

Generally measurement of true stress in this manner is impracticable and the value referred to as the *engineering stress* is of more use in practice.

$$\text{Engineering stress} = \frac{\text{Force}}{\text{Original area of cross-section}}$$

However, it should be appreciated that the ordinate usually labelled 'stress' in the majority of published diagrams nearly always refers to this 'engineering stress' rather than to true stress. The reduction in cross-section of ductile materials during plastic flow leads to the apparent anomaly that the breaking strength is less than the tensile strength. In fact of course the *true* breaking stress is the greater as indicated in Fig. 4.7.

Q | **The following results were obtained during the tensile testing of a heat-treated nickel-chromium-molybdenum steel:**

Force (kN)			50	100	150	175	200
Extension (mm)			0.009	0.19	0.28	0.33	0.39

225	250	275	300	310	310	290
0.47	0.59	1.10	3.40	5.00	6.40	7.40 (Break)

The original diameter of the test piece was 16 mm and the gauge length 80 mm. The diameter at fracture was 12.4 mm. Determine

(i)   **tensile strength;**
(ii) **Young's Modulus;**
(iii) **0.2% proof stress;**
(iv) **% elongation;**
(v)   **% reduction in area at fracture.**

A

Fig. 4.8

(i) $T = \dfrac{P_{max}\,(kN)}{A_0\,(mm^2)}$

$\quad = \dfrac{311}{201}\ kN/mm^2$

$\quad = 1.547\ kN/mm^2$

$\quad = \underline{1.547\ GPa}$

From Fig. 4.8(i)
$P_{max} = 311\ kN$
$A_0 = \pi d^2/4$
$\quad = [(3.142) \times (16)^2]/4$
$\quad = 201\ mm^2$

(ii) $E = \text{Slope of OE} \times \dfrac{L}{A}$

Slope of 'linear' section OE in
Fig. 4.8(ii) is 150 kN/0.28 mm

$\quad = \dfrac{150\,(kN) \times 80\,(mm)}{0.28\,(mm) \times 201\,(mm^2)}$

$\quad = 213.22\ kN/mm^2$

$\quad = \underline{213.22\ GPa}$

(iii) *0.2% proof stress*
0.2% of 80 mm = 0.16 mm. Hence draw a line parallel to the 'linear'
portion of the force/extension diagram from a point representing a
permanent extension of 0.16 mm. This meets the diagram at P
indicating that a force of 257 kN would produce this amount of
permanent extension.

$0.2\%\ \sigma_p = \dfrac{P\,(kN)}{A\,(mm^2)}$

$$= \frac{257}{201} \text{ kN/mm}^2$$

$$= 1.279 \text{ kN/mm}^2$$

$$= 1.279 \text{ GPa}$$

(iv) % Elongation $= \dfrac{\text{Extension at fracture} \times 100\%}{L_0}$

$$= \frac{7.40 \text{ mm} \times 100\%}{80 \text{ mm}}$$

$$= \underline{9.25\%}$$

(v) $\% A_R = \dfrac{A_R}{A_0} \times 100\%$

$$= \frac{\left(201 - \dfrac{\pi \times (12.4)^2}{4}\right)}{201 \text{ mm}^2} \text{ mm}^2 \times 100\%$$

$$= \underline{39.91\%}$$

4.1.8 The phenomenon described above may be explained as follows. During a tensile test the elongation produced in the test piece is at first *elastic* and the force used must increase as it pulls atoms further from their positions of equilibrium. When the yield stress is reached elongation takes place more rapidly as *plastic flow* takes over and atoms move into new positions and a return to the original shape of the test piece is impossible. Nevertheless as plastic flow proceeds the force necessary to produce it increases as *work-hardening*, the nature of which is discussed later (5.7), sets in. The stress–strain curve follows an approximately parabolic form according to the empirical relationship:

$$\sigma_t = C\varepsilon_t^n$$

where σ_t is the true stress and ε_t the true strain and C and n are constants for the metal. n is termed the 'strain-hardening exponent' (usually between 0.1 and 0.4 for most metals).

As the test piece extends so its strength increases due to work-hardening according to the above expression but a stage is reached where the *uniform* reduction in cross-sectional area, which has accompanied the *uniform* increase in gauge length, ceases and a degree of plastic instability is introduced by the reduction in cross-sectional area. At this point, coinciding with maximum force on the nominal stress/nominal strain diagram, $\sigma_t = n$. The plastic deformation then becomes concentrated over a small part of the gauge length and 'necking' occurs. Since the cross-sectional area decreases rapidly the stress at the neck increases and with old-type mechanically loaded testing machines this led to the rapid stretch and break of the test piece at the neck. With modern hydraulically- or spring-loaded machines force is automatically relaxed and the test can proceed to a measured breaking force but of course during this final stage of the test, though the applied force decreases the *true stress* at the neck is continually increasing as the diameter at the neck decreases.

4.1.9 When a nominal stress/nominal strain diagram is plotted the nominal stress $\sigma_n = P/A_0$ where P is the force and A_0 the original area of cross section, whilst the nominal strain, $\varepsilon_n = (L - L_0)/L_0$ where L is the length of the original gauge length under force P, and L_0 is the original gauge length. This can be written:

$$\varepsilon_n = \frac{L}{L_0} - 1 \quad \text{or} \quad \varepsilon_n + 1 = \frac{L}{L_0}$$

The true stress $\sigma_t = P/A_t$ where A_t is the true cross-sectional area at the instant when force P is operating. The true strain is given by:

$$\varepsilon_t = \int_{L_0}^{L} \frac{dL}{L}$$

$$= \ln \frac{L}{L_0}$$

Since

$$\frac{L}{L_0} = (1 + \varepsilon_n)$$

$$\varepsilon_t = \ln(1 + \varepsilon_n)$$

If we assume that the volume change during plastic deformation can be neglected and

$$LA_t = L_0 A_0$$

or

$$A_t = \frac{L_0 A_0}{L}$$

then

$$\sigma_t = \frac{PL}{L_0 A_0}$$

but

$$\frac{L}{L_0} = (1 + \varepsilon_n)$$

therefore

$$\sigma_t = \frac{P}{A_0}(1 + \varepsilon_n)$$

Hardness

4.2   In general terms hardness is defined as the ability of a material to resist surface abrasion. The relative hardnesses of minerals have in fact long been assessed by reference to Moh's Scale (Table 4.1). This consists of a list of materials arranged in order such that any mineral in the list will scratch any one below it. Thus diamond, the hardest substance known, heads the list with a hardness index of 10 whilst talc is at the foot with a hardness index of 1. The surface hardness of any substance can be related to Moh's Scale by determining which of these standard substances will just scratch it.

Table 4.1   *Moh's Scale*

Mineral	Hardness index
Diamond	10
Corundum	9
Topaz	8
Quartz	7
Orthoclase feldspar	6
Apatite	5
Fluorite	4
Calcite	3
Gypsum	2
Talc	1

Obviously Moh's Scale would be inadequate in the *accurate* determination of the hardness of such materials as metallic alloys, and rather different types of hardness test have been developed for such substances. Such instruments as the Turner Sclerometer (which attempted to measure surface 'scratchability') were soon abandoned in favour of machines which measure the resistance of the surface layers of a material to penetration by some form of indenter rather than surface hardness defined in terms of abrasion resistance. Whilst these current methods are less representative of true surface hardness they offer means by which surface properties can be compared much more accurately.

4.2.1   The *Brinell test* was devised by a Swede, Dr Johann Brinell in 1900. In this test a hardened steel ball is forced into the surface of the test piece using an appropriate specified force. The diameter of the impression produced is measured and the Brinell Hardness Index, H_B, derived from:

$$H_B = \frac{P}{\text{Surface area of impression}}$$

The surface area of the impression can be shown to be:

$$\frac{\pi D}{2}(D - \sqrt{D^2 - d^2})$$

where D is the diameter of the indenter used and d the diameter of the impression produced, so:

$$H_B = \frac{2P}{\pi D \left(D - \sqrt{D^2 - d^2}\right)}$$

The units are kgf/mm^2 but hardness values are always quoted as a simple index. (In the real world of course this formula is little used, H_B being derived from tables relating H_B with D and d; or in modern testing machines, direct reading by electronic means.)

In this test it is important that P shall bear a correct relationship to the diameter of the indenter, D, for reasons indicated in Fig. 4.9. Obviously as

Fig. 4.9 The influence of depth of impression on the accuracy of measurement of Brinell hardness: (i) force, P_1, too great; (ii) force, P_2, too small; (iii) force, P_3, correct.

d approaches D in value (Fig. 4.9(i)), d becomes insensitive to variations in P and the result is meaningless. In Fig. 4.9(ii) variations in d will be erratic. Thus for different materials the ratio P/d^2 is standardised in order to obtain comparable results:

Material	P/d^2	Approximate values of H_B
Steels and cast irons	30	Over 100
Copper alloys and aluminium alloys	10	30–200
Pure aluminium	5	15–100
Tin, lead and their alloys	1	3–20

Thus in testing steel a 10 mm ball would be used in conjunction with a 3000 kgf force; or a 1 mm ball with a 30 kgf force. The diameter of the ball used will be governed by the width and thickness of the test piece (Fig. 4.10) always bearing in mind that the larger the diameter of ball which can be used the more accurate the result is likely to be.

Fig. 4.10 The relationship between ball diameter, D, depth of impression and test piece dimensions in Brinell test.

4.2.2 Meyer Index   This is related to the susceptibility of a material to work harden, in this case during the application of the hardness test. It can be determined by plotting a series of suitable forces, P, against the diameter, d, of the indentation produced from a single ball indenter and determining the Meyer Index, n, from:

$$P = \lambda d^n$$

or $\log P = \log \lambda + n \log d$

where P is the force (kgf), d is the diameter of the impression (mm) and λ is a constant. For most metals n varies between 2 and 2.5. Since n represents the capacity of a metal for absorbing cold work before becoming work-hardened, it gives insight into the mechanical properties of the material.

Q   Test pieces of 'fire-refined' copper in both the soft (annealed) and cold-worked conditions were Brinell tested using a 1 mm diameter ball and a series of different loads as shown:

Material	Force, P (kgf)	Diameter of impression (mm)
Annealed copper	5	0.369
	10	0.514
	15	0.614
Cold-worked copper	10	0.365
	20	0.514
	30	0.596

(i)   Calculate the average Brinell Hardness Index for each specimen.
(ii) Carry out a Meyer Analysis to determine the Meyer constants for each specimen.

A

(i) Use the formula: $H_B = \dfrac{2P}{\pi D(D - \sqrt{D^2 - d^2})}$.

For annealed copper:

with 5 kgf force, $H_B = 45.1$
10 kgf force, $H_B = 44.8$
15 kgf force, $H_B = 45.3$
Average value of $H_B = \underline{45.1}$

For cold-worked copper:

with 10 kgf force, $H_B = 92.3$
20 kgf force, $H_B = 89.5$
30 kgf force, $H_B = 96.9$
Average value of $H_B = \underline{92.9}$

(ii) Meyer's Index, n, is derived from $P = \lambda d^n$ where n and λ are the constants to be determined. The Meyer equation can be written:

$$\text{Log } P = \log \lambda + n \log d$$

The most convenient way of deriving the constants n and λ is to plot graphs of $\log P$ against $\log d$ for each specimen.

Material	P (kgf)	Log P	d (mm)	Log d
Annealed copper	5	0.699	0.369	−0.433
	10	1.000	0.514	−0.289
	15	1.176	0.614	−0.212
Work-hardened copper	10	1.000	0.365	−0.437
	20	1.301	0.516	−0.287
	30	1.477	0.596	−0.225

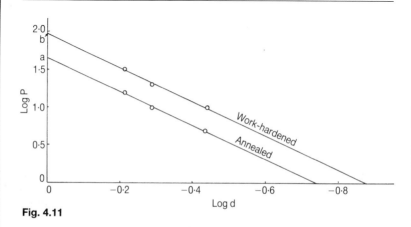

Fig. 4.11

Slope of graph is given by n which is equal to $\log P / \log d$.

For annealed copper:

$$n = \frac{1.65}{0.7375} = \underline{2.2}$$

For work-hardened copper:

$$n = \frac{1.98}{0.875} = \underline{2.3}$$

For annealed copper:

$$\log \lambda = 1.65 \quad \text{(intercept with ordinate at a)}$$

Hence:

$$\lambda = \underline{44.7}$$

For work-hardened copper:

$$\log \lambda = 1.98 \quad \text{(intercept with ordinate at b)}$$

Hence:

$$\lambda = \underline{95.5}$$

4.2.3 The *Vicker's Diamond Pyramid Test* (or Diamond Pyramid Hardness Test) uses a diamond pyramid indenter of geometrical shape indicated in Fig. 4.12. This has the following advantages over methods which use a spherical steel indenter:

(i)   the pyramid is of constant angle making the choice of a suitable P/D^2 ratio unnecessary;
(ii) the diamond pyramid is less likely to distort under high forces than is a steel ball.

The Vicker's Hardness, H_V, is related to the applied force, P, and the surface area of impression in the same manner as is Brinell Hardness, viz:

$$H_V = \frac{\text{Force } (P)}{\text{Surface area of impression}}$$

$$= \frac{2P \sin \theta/2}{d^2}$$

$$= \frac{1.8544\,P}{d^2}\ \text{kgf/mm}^2$$

where θ is the angle (136°) between opposite faces in the diamond pyramid and d is the mean length of diagonals (d_1 and d_2) of the impression (mm).
Hardness indices, H_V, vary between approximately 8000 for diamond and 5 for lead.
A number of other hardness tests such as Rockwell (which uses either a diamond cone or a steel ball) are also in use.
For most metallic alloys tensile strength is approximately proportional to hardness though there is no fundamental connection between them other than in the general stiffness of the material.

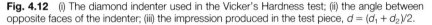

Fig. 4.12   (i) The diamond indenter used in the Vicker's Hardness test; (ii) the angle between opposite faces of the indenter; (iii) the impression produced in the test piece, $d = (d_1 + d_2)/2$.

Fig. 4.13   The principle of most hardness-measuring machines. The indenter may be a steel ball as shown; a diamond pyramid or a diamond cone.

Toughness

4.3   Toughness is measured in terms of the amount of *energy* required to fracture a standard test piece. For this reason it must not be confused with strength which is measured in terms of *stress* required to break a standard test piece. Since energy is the product of average force and the distance through which it acts, the area under the stress/strain diagram is related directly to the energy necessary to break a material.

In fact some materials, when hardened and strengthened by a suitable process, lack toughness whereas in their softest and most ductile state they are extremely tough. This relationship is indicated by the area under the stress/strain curve (Fig. 4.14) in each case.

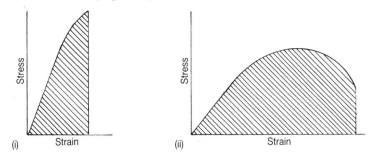

Fig. 4.14   Stress/strain diagrams for (i) an alloy which has been treated to increase its strength, (ii) the same alloy in the soft, ductile condition. The energy, indicated by the area under the curve, required to break the test piece, is greater in the case of the less strong but more ductile material.

4.3.1   The practical methods used for measuring toughness differ from those associated with the stress/strain diagram in that they employ shock loading. A portion of the kinetic energy of a swinging pendulum is absorbed in fracturing a standard, suitably notched, test piece. In both the

Fig. 4.15 The principle of impact testing machines. The energy required to fracture the specimen is measured on the scale in joules.

Izod and Charpy methods of measuring impact toughness the unit of energy is the Joule. These tests offer a useful practical indication of how a material will react to conditions of shock loading, and in many circumstances toughness will be more important than tensile strength as a criterion of suitability of a material.

Other Mechanical Tests

4.4 Many other tests have been devised in order to assess specific mechanical properties of interest in practical engineering. For example the Erichsen test gives a rather better estimation of a material's suitability for deep drawing than does a formal % elongation. Various bend tests and compression tests are also useful in some circumstances. Simple descriptions of some of these tests will be found in *Materials for the Engineering Technician* by this author.

It may have been noticed that no mention has been made here of creep- or fatigue-testing. These topics are associated with metal failure and will be dealt with later in this book (16.4 and 16.5).

5

The Deformation of Materials

5.1 A well-worn cliché of yesteryear, much too old to have originated from any of today's pseudo-scientific journalists, speculated upon the possible effect of 'an irresistible force meeting an immovable object'. In fact any force may be regarded as irresistible in so far that no object is completely immovable, since all substances when subjected to mechanical stress suffer some change in shape. Many return to their original form when the stress is relaxed, providing the stress has not been too great; such substances possess *elasticity*. When force is applied the ions or molecules of which the material is composed are moved small distances from their positions of equilibrium and when the displacing force is removed they return to their original positions (Fig. 5.1(i)).

If the displacing force is increased steadily the substance ultimately reaches its elastic limit. Increase of stress beyond this limit leads to one of two alternative results:

  (i)   the material may shatter into fragments (Fig. 5.1(iii));
 (ii)   it may remain as a coherent whole but fail to return to its original form when the displacing forces have been removed. It has suffered *plastic deformation* (Fig. 5.1(ii)).

5.1.1 In general non-metallic elements and most ionically bonded compounds belong to the first group, that is, they possess elasticity but no plasticity. Metals and many organic polymers, however, possess high degrees of both elasticity and plasticity. They can be shaped by compression and tension and after such treatment their structures are substantially unaffected since they have suffered deformation but have retained their continuity.

A simplified explanation of the differences in behaviour between metals and ionic crystals can be made in terms of their respective structures when stressed beyond their elastic limits. A metallic crystal consists of positively

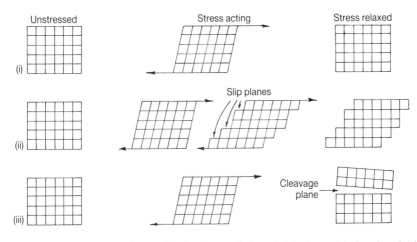

Fig. 5.1 The behaviour of materials in stress. (i) A material stressed below its elastic limit—removal of stress leads to disappearance of strain. (ii) A ductile material, such as a metal, undergoes elastic and plastic strain. The latter does not disappear when stress is removed. (iii) Completely brittle materials do not strain plastically.

charged ions surrounded by an 'electron cloud' (3.5). Whilst the positive ions will repel each other they are held in equilibrium positions by forces of mutual attraction between each ion and the negatively charged 'cloud'. When a displacing force is applied to a plane of ions (Fig. 5.2(i)) this tends to move positive ions in one plane nearer to those in the next adjacent plane. The forces of repulsion between ions will increase to a maximum at the elastic limit but are then overcome so that one layer of ions moves with respect to its neighbour. There is no disruption, however, at this stage because of the mutual attraction between positive ions and the surrounding electron cloud. The plane along which movement occurs is called a *slip* plane.

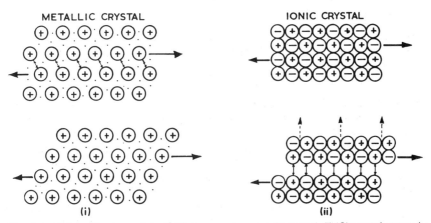

Fig. 5.2 Deformation in crystals. (i) 'Slip' occurs in a metallic crystal. (ii) 'Cleavage' occurs in ionic crystals.

A crystal of a substance such as sodium chloride consists of positively charged sodium ions and negatively charged chlorine ions arranged so that each ion is surrounded by those of opposite charge. Any attempt to produce slip in the [100] direction (Fig. 5.2(ii)) will fail since this would bring like ions into closer contact. Like ions would immediately repel each other so that the two halves of the crystal separate. Thus the crystal shatters along a *cleavage* plane.

Because of this severe restriction in the availability of slip systems in ionically bonded materials, such materials tend to fracture at fairly low stresses—lower than those at which slip would otherwise occur. In ceramics both ionic and strong covalent bonds are present. Consequently one might expect them to have high strengths and high elastic moduli. Whilst moduli are reasonably high and compressive strengths demonstrate the bond strengths, tensile strengths tend to be low. This is due principally to the propagation of micro-cracks which act as stress raisers.

Plastic Deformation in Metals

5.2 The fact that a metal can undergo both elastic and plastic deformation is demonstrated during a tensile test in which the test piece is tested to destruction. That plastic deformation takes place by some form of slip is observable using an ordinary metallurgical microscope (Figs 5.3 and 5.4).

Fig. 5.3 The formation of slip bands in a metal stressed beyond its yield point. (i) Before stressing. (ii) After stressing—blocks of slipped atoms some 40 atoms wide and up to 400 atoms high cast shadows on the surface of the metal giving an appearance as in (iii).

If a ductile metal or alloy is polished and etched to reveal its crystal structure and is then squeezed laterally in a vice, taking care not to damage the etched surface, the shadows cast by the ridges formed as slip occurs can be seen as hair-like lines (Fig. 5.3(iii)) at a magnification of 100× or so.

Slip takes place on specific crystallographic planes in a given metal. Generally these slip planes are the atom planes of greatest inter-planar spacing since then forces between the sliding planes will be at a minimum. Within the planes the greatest density of atom packing is present. Thus slip occurs in the direction of the greatest density of atom packing.

5.2.1 In FCC metals like copper slip usually occurs along the {111} planes (Fig. 3.15(iv)) since the perpendicular distance between these planes is greater than that for any other set of planes in the crystal and the atomic population in these planes is denser than in any other family of planes. Thus the planes of easiest slip are also those of densest packing.

Fig. 5.4 'Slip-bands' in copper. The specimen was polished, etched and then squeezed laterally in a vice. Slip-bands are shown clearly in the crystal in the centre of the field of view. ×200.

The slip *directions* in the FCC structure are the ⟨110⟩ family (face diagonals). There are four effective sets of planes but each plane has three possible slip directions, giving a total of twelve directions in which slip could take place with equal ease in a FCC crystal. In Fig. 5.5(i) only two of the four planes are shown for the sake of clarity.

5.2.2 In the CPH structure the (0001) or basal planes are of similar arrangement to the (111) planes in the FCC structure. However, in CPH there is but one such set of basal planes instead of four (111) planes in FCC. Hence there are only three easy slip systems in CPH (Fig. 5.5(ii)). This is reflected in the differences in mechanical properties between the malleable and ductile FCC metals like aluminium and copper, and the relatively brittle CPH metals like zinc where slip is restricted.

Fig. 5.5 There are twelve directions in which slip can take place in the FCC structure (i), but only three in which slip can take place in CPH structures with equal ease (ii).

As indicated in Fig. 3.16 the BCC structure does not contain any planes of closest packing as are present in the FCC or CPH structures. However, slip occurs along those planes where the density of packing is greatest. There are six such planes (110) where slip can take place with two directions [111] in each, giving a total of 12 slip systems.

Table 5.1   *The Relationship between Crystal Structure and Slip Systems*

Crystal structure	Slip plane	Slip direction	Number of slip planes	Number of slip directions	Total number of slip systems
FCC	(111)	[110]	4	3	12
CPH	(0001)	[1000]	1	3	3
BCC	(110)	[111]	6	3	12

5.2.3   Normally we are dealing with polycrystalline metals but much fundamental information on the nature of slip has been obtained by studying the behaviour of *single* crystals of metals in stress. These single crystals can be grown under carefully controlled laboratory conditions and then machined as test pieces. During a tensile test slip occurs along parallel slip planes (Fig. 5.6) the precise direction of which can be observed. By

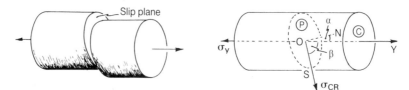

Fig. 5.6   Slip in a single metallic crystal.

determining the yield stress of such a crystal experimentally we can then calculate the critical resolved shear stress acting along a slip plane which will just produce slip.

If σ_y is the yield stress and σ_{CR} the critical resolved stress acting along OS on the slip plane ⓟ then,

Component of σ_y along OS $= \sigma_y . \cos\beta$
The area of projection
on ⓒ of unit area of ⓟ $= \cos\alpha$
Hence, $\sigma_{CR} = \sigma_y . \cos\alpha . \cos\beta$

In different single crystals of the same material values of α and β will vary, that is crystallographic planes will be differently orientated with respect to the axis of the test piece from one specimen to the next. Thus single crystals of the same material will slip at different angles to the axis and consequently at different values of σ_y, but when the stress is resolved along the slip plane all crystals of the same material slip at the same critical resolved shear stress.

It is possible to calculate theoretically this critical resolved shear stress from a knowledge of the forces which act between fundamental particles within the metallic bond. If we do this we find that the 'theoretical strength' of a metal is generally *one thousand times or more* the value obtained experimentally as outlined above. Obviously, therefore, slip is not a simple process in which all of the atoms in a complete plane move *simultaneously* with reference to the atoms in an adjoining plane, rather like one penny in a pile sliding over its neighbour. Moreover, the assumption that slip takes place in this simple manner does not explain why metals work harden progressively during a cold-working operation. Only in recent decades has the low strength of a metal relative to its 'theoretical strength' been successfully explained.

Imperfections in Crystals

5.3   During the early days of the twentieth century a great deal of investigation into the crystal structures of metals took place and there was a tendency for metallurgists to regard metallic crystals as being relatively perfect in form. Such a concept, however, made it difficult to explain some of the properties of metals and alloys, particularly such phenomena as diffusion in solid solutions (7.4) and the fact mentioned above, namely, that the practical value obtained for the critical shear stress of a metal is several orders of magnitude less than that deduced from theoretical considerations based on instantaneous 'block' slip.

The suggestion that atoms in a metallic crystal are arranged in a rigid pattern from which there is no deviation has long since been rejected as evidence to the contrary has accumulated over the years. Currently the metallurgist thinks of a metallic crystal as having atoms arranged according to some general overall pattern but in which all manner of local faults and deficiencies can exist causing distortions and irregularities in the surrounding crystal lattice (Figs 5.7 and 5.8). Whilst X-ray investigations (3.4) will

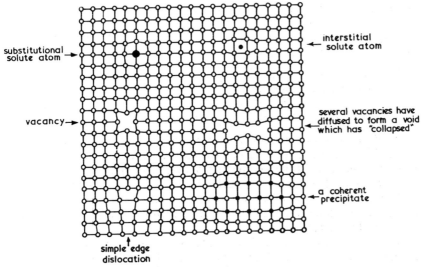

Fig. 5.7   The imperfect nature of a metallic crystal.

Fig. 5.8 Further crystal defects. (i) A *Frenkel defect* is a vacancy produced when an atom moves (as a result of an increase in thermal vibration) into a *self interstitial* position *within* the crystal. (ii) A *Schottky defect* is a vacancy formed in a similar manner but the displaced atom takes up a lattice position on the *surface* of the crystal.

reveal the general crystal pattern of a metal they cannot detect single lattice faults and the existence of many of these is still largely a matter of hypothesis.

5.3.1   One of the most important of these lattice faults was postulated more or less independently in 1934 by Taylor, Polanyi and Orowan. In its simplest form it consists of a region containing an extra half-plane of atoms (Fig. 5.9). This is termed an *edge dislocation* and it plays a very important

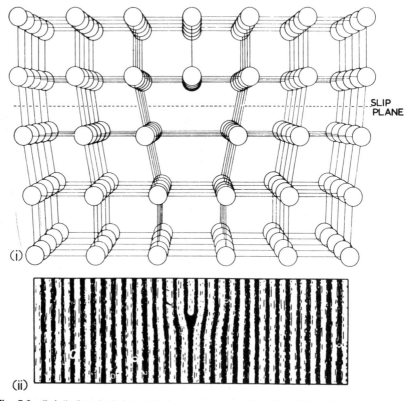

Fig. 5.9   (i) A 'ball-and-wire' model of a simple edge dislocation. (ii) Crystallographic planes containing an edge dislocation as they appear in aluminium at a magnification of several millions.

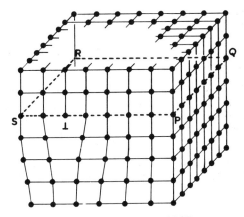

Fig. 5.10   An edge dislocation and associated slip plane SPQR.

role in determining the mechanical properties of metals. In the early days of the Dislocation Theory there was no direct proof that such faults did in fact exist in crystals but over the years more and more supporting evidence accumulated so that during the post-war years the theory blossomed and became widely accepted. More recently the existence of dislocations was verified metallographically using very high-power modifications of the electron microscope.

The arrangement of atoms in the region of an edge dislocation is shown in Figs 5.9 and 5.10. The presence of dislocations in such diagrams is generally indicated by the sign ⊥. Under the action of adequate stress the dislocation will move progressively through the crystal to the right (Fig. 5.11) until it ultimately forms a *slip step* as indicated. In practice the movement may be halted by some other fault or discontinuity within the crystal or, alternatively, by the crystal boundary.

5.3.2   Possibly Prof. N. Mott's classical analogy of the methods available for smoothing wrinkles from a heavy carpet explains most clearly the relationship between slip and the force necessary to produce it. Imagine that on laying a wall-to-wall carpet a small wrinkle remains in the carpet near to an edge which is touching a wall (Fig. 5.12(i)). Attempts to remove the wrinkle by pulling on the opposite edge of the carpet will achieve little—save the possibility of broken finger nails—because of the necessity of overcoming friction between the whole of the carpet and the floor. In a similar way instantaneous block slip in a metallic crystal over a complete plane would necessitate overcoming inter-atomic forces over the whole of the plane at the same instant.

Returning to our carpet, one can with very little effort coax the wrinkle along with one's toe so that it moves, step by step, across the floor (Fig. 5.12(ii) to (iv)). Only local friction in the immediate vicinity of the wrinkle then needs to be overcome. This is analogous to the progressive movement of an edge dislocation where the only inter-atomic forces being overcome at any instant are those acting in the locality of the dislocation. In this way

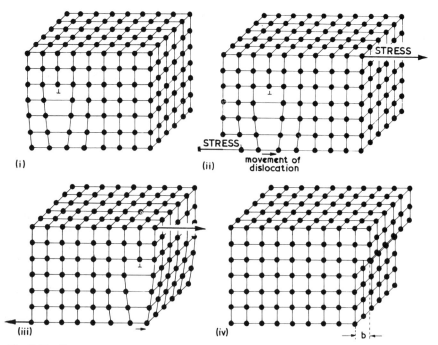

Fig. 5.11   The movement of an edge dislocation under the influence of stress. The unit slip, b, is known as the Burgers vector of slip.

we can explain why the yield stress of a metal is so small in comparison with that calculated on the assumption of block slip.

5.3.3   Slip may also occur by the movement of a *screw dislocation*, the principle of which is shown in Fig. 5.13. Under the influence of shear stress this type of dislocation also moves in the slip plane. However, dislocations are generally of a more complex nature which can nevertheless be resolved into a combination of the edge and screw types (Fig. 5.14).

Fig. 5.12   Mott's analogy of the carpet.

Fig. 5.13   The movement of a screw dislocation.

The Generation of Dislocations

5.4   We have seen that when a metal is plastically deformed slip takes place on a number of planes within any crystal (Fig. 5.3). The fact that these slip planes are visible under the microscope at fairly low magnification indicates that not only are the planes quite far apart but that the 'steps' are of considerable thickness.* However, a dislocation moving along a slip plane and running out at the observed surface of the metal can only produce a surface step of one atomic spacing in depth (Fig. 5.11) and it would require a large number of dislocations, all on the same plane, to produce the resultant large step of 400 atoms observed experimentally. It seems, therefore, that an initial dislocation must in some way be able to reproduce itself and so lead to a continuation of the process of slip. The Frank–Read source offers a plausible explanation of this possibility. The original source is visualised as a dislocation line anchored at its ends, possible by other faults. The application of shear stress causes the line to bow outwards, turning in upon itself (Fig. 5.15) and eventually forming a complete dislocation loop. Since the original dislocation line still remains the process can repeat producing a series of dislocation loops or ripples radiating from the original source.

5.4.1   So far we have been dealing only with the movement of existing dislocations and their permitting of extensive slip by means of Frank–Read sources. It has been assumed that such dislocations were 'built into' the structure during the solidification process. It is reasonable to suppose that this will occur as dendrite arms interfere with each other's growth during the competition for diminishing inter-dendritic space. The formation of dislocations along grain boundaries where there is a small angle of misfit (5.5) arises in a similar manner.

The movement of dislocations will eventually be halted by such obstacles as grain boundaries, other groups of dislocations and imperfections along the slip plane. If the acting stress is increased new dislocations will be initiated in the structure and slip will continue until all available slip planes have been used up so that the movement of dislocations is no longer possible. Further increase of stress will ultimately lead to fracture.

---

* Approximately 400 atoms high as mentioned earlier.

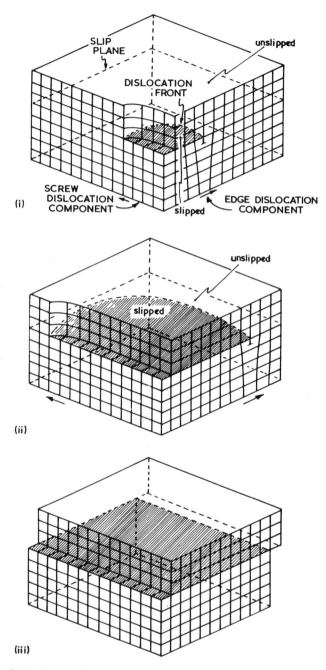

Fig. 5.14 Slip by the movement of a dislocation loop over a slip plane.

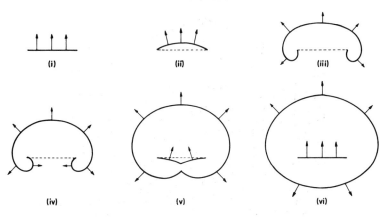

Fig. 5.15   The operation of a Frank–Read Source.

The stress necessary to produce new dislocations is much greater than that required to move those already present. This relationship is illustrated by the tensile properties of metallic 'whiskers'. These are very small hair-like single crystals grown under carefully controlled conditions and generally containing only a single dislocation which runs along the central axis. A tensile stress along this axis will cause no movement of the dislocation. As no other dislocations are available, slip cannot take place until a new dislocation is initiated at E (Fig. 5.16). Immediately, the stress falls to that necessary to move this new dislocation (Y) and further dislocations are propagated by the Frank–Read mechanism so permitting plastic flow to proceed.

5.4.2   In 1956 D. Marsh developed an extremely sensitive micro testing machine capable of determining the tensile properties of specimens of such small dimensions as 10^{-7} mm^2 cross-sectional area and 0.25 mm long. This machine could detect extensions of the order of 0.5 nm. In addition to

Fig. 5.16   The relationship between the tensile properties of a single copper 'whisker' and ordinary polycrystalline copper. The whisker was stressed parallel to the {111} direction. The whisker yielded suddenly at E and the load was relaxed and then re-applied at Y.

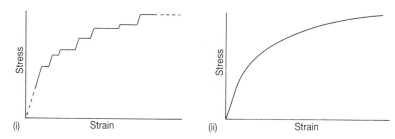

Fig. 5.17 Using a sensitive 'micro' testing machine on a very small specimen containing a finite number of dislocation centres reveals the incremental nature of slip (i). In a normal tensile test on a large specimen (ii) slip is taking place continually at a very large number of centres so that a regular curve is obtained.

being useful for determinations on single whiskers other small diameter specimens containing a limited number of dislocation centres could also be tested. Figure 5.17(i) shows the stress/strain diagram for a small thin specimen tested in a Marsh machine. Here the stress/strain curve is in the form of a series of steps since the machine is sufficiently sensitive to detect the generation of dislocations from an individual source. Whilst *single* dislocations could not be detected even by the sensitive Marsh machine the production of a dislocation source releases a number of dislocations simultaneously so that the total amount of slip associated with one source is measurable.

With further increases of stress there are potential dislocation sources ready to generate dislocations and these are presumably triggered by the inflow of thermal energy. Sources are triggered quite erratically and the slip produced is virtually instantaneous giving a series of irregular steps illustrated. In the case of a large test piece (Fig. 5.17(ii)) so many centres of dislocation are being triggered continuously throughout the specimen that the typical smooth curve of a stress/strain diagram is produced.

5.4.3 Interaction between dislocations moving on the same or on nearby planes must be considered particularly in so far as they affect the distribution of strain energy. The presence of the extra half plane of atoms in a simple edge dislocation gives rise to an increase in strain energy in that region. Imagine an axe blade (representing the extra half plane of atoms) driven into a log of wood (Fig. 5.18). This will produce locked-up strain

Fig. 5.18

Fig. 5.19   Dislocations of the same sign will repel each other.

energy in the surrounding wood in a manner similar to the extra half-plane of atoms in an edge dislocation. Consequently two edge dislocations of the *same* sign (Fig. 5.19), moving on the same or nearby planes, will tend to repel each other in order to reduce the concentration of elastic strain energy in the region. Increased stress will be required to move them closer together so that, under such circumstances, not only does the total strain energy increase but also the tensile strength which is a measure of the stress required to produce slip. Normally dislocations of the same sign will pile up against obstacles such as grain boundaries when the applied stress is great enough.

Conversely, if edge dislocations of *opposite* sign (Fig. 5.20) move into

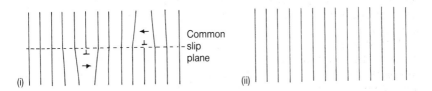

Fig. 5.20   Dislocations of opposite sign will tend to attract (i) and so annihilate each other (ii).

close proximity they will tend to attract each other. Assuming that they are on the same slip plane they will annihilate each other with a consequent reduction in strain energy. If their slip planes are separated by a few atomic spacings a row of vacancies will remain (Fig. 5.21) but the total strain energy will be reduced due to the annihilation of the dislocations.

Fig. 5.21   Dislocations of opposite sign not moving on the same plane. Here a row of vacancies (ii) remains.

Fig. 5.22 A grain boundary where the angle of misfit is small. (i) Indicating the position of the crystal boundary. (ii) Crystals joined with edge dislocations present. (After Read, 'Dislocations in Crystals'.)

Dislocations within Crystal Boundaries

5.5 That grain boundaries are a region of some disorder has already been suggested (3.7.3). However, when the angle of 'misfit' between adjacent crystals is small an atomic arrangement similar to that in Fig. 5.22 may be produced. Here the two crystals are able to join more or less continuously along most of their common boundary with, of course, a certain amount of elastic strain. Since adjacent planes in the two crystals are not parallel some of these planes must terminate at the boundary giving rise to line imperfections passing through the crystal normal to the plane of the paper. These faults manifest themselves as simple edge dislocations.

If D is the spacing between these dislocations, b the interatomic spacing and θ the angle of misfit, then:

$$\sin\frac{\theta}{2} = \frac{b/2}{D}$$

$$= \frac{b}{2D}$$

Thus

$$D = \frac{b}{2 \sin \theta/2}$$

Since θ is small

$$D = \frac{b}{\theta}$$

As θ increases the value of D decreases, i.e. the dislocations move closer together until their individual identity is lost and the boundary becomes a general disordered region (Fig. 3.25).

5.5.1 The passage of a dislocation across a grain boundary must be extremely rare since it would require a high degree of 'coherency' between the adjacent crystal lattices. Normally, grain boundaries of what ever sort will act as barriers to the movement of dislocations and, during plastic deformation, will become regions of very high strain energy as dislocations pile up from there to form a complex intercrystalline 'traffic jam'. This high concentration of strain energy at the grain boundaries of cold-worked metal gives rise to the preferred intercrystalline corrosion prevalent in heavily cold-worked metals (19.5).

5.5.2 Since a grain boundary impedes the movement of dislocations by causing them to 'pile up' there, it follows that fine-grained metals are *stronger* than coarse-grained metals of similar composition since they contain more grain boundary per unit volume of metal. The yield strength, σ_y, is related to grain size by the *Hall–Petch equation*:

$$\sigma_y = \sigma_i + k_y d^{-1/2}$$

where d is the average grain diameter, σ_i and k_y being constants for the metal at a given temperature. (σ_i is related to the intrinsic resistance of the metal to the movement of dislocations whilst k_y is associated with the generation of dislocations.)

Fig. 5.23 Experimental verification of the Hall–Petch equation which relates grain size with yield stress.

Deformation by Twinning

5.6   Although slip is the more significant process by which plastic deformation occurs in metals brief mention must be made of the phenomenon known as *twinning*. Whilst slip is a process associated with a *line* defect (the dislocation), twinning is related to a *plane* defect (the twin boundary).

In slip all atoms in a block have moved the same distance when slipping is complete (Fig. 5.24(i)) but in deformation by twinning (Fig. 5.24(ii))

Fig. 5.24   Deformation by (i) slip and (ii) twinning.

atoms within each successive plane in a block will have moved different distances. When twinning is complete the lattice direction will have altered such that one half of the twin is a mirror image of the other half, the twinning line corresponding to the position of a mirror. Twinning, like slip, also proceeds by the movement of dislocations. The net result of a twinning process on the lattice of a crystal is depicted in Fig. 5.25.

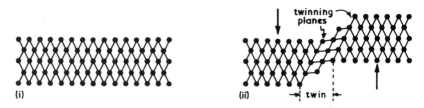

Fig. 5.25   An example of deformation by twinning. (i) Before stressing. (ii) After stressing.

The stress required to produce deformation by twinning tends to be higher than that necessary to produce deformation by slip. Twinning is more likely to occur in metals which are shock loaded at low temperatures. Thus in BCC iron sudden loading at low temperatures produces thin lamellar twins which are generally known as Neumann bands. Twinning is commonly encountered in CPH metals like zinc, whilst when a bar of tin is bent suddenly the formation of twins can be heard taking place. The Ancients knew this as 'the cry of tin'.

5.6.1   Twin-formation takes place during the annealing of some cold-worked metals. In this case recrystallisation is initiated at a plane instead of at a point source and twin bands are observed within the new crystals as a result. These 'annealing twins' are common in copper, brasses, bronzes and austenitic steels—alloys with low stacking fault energies—and are different in this respect from the mechanical twins under consideration here. They are formed not by mechanical shear but as part of the process of grain growth.

Work Hardening

5.7   As deformation proceeds—whether by slip or by twinning—the metal becomes harder and stronger and a stage is reached when further deformation is impossible. Any increase in stress will lead only to fracture. At this stage, when tensile strength and hardness are at a maximum and ductility is at a minimum, the material is said to be work-hardened.

We can summarise the effects of plastic deformation briefly as follows— if sufficient stress is applied to a metal, slip (or twinning) will take place in individual crystals. As deformation proceeds the capacity for further cold work diminishes as dislocations find themselves in positions from which further movement is impossible. An increase in stress is then necessary to initiate new dislocations and move these until they too become blocked. Ultimately a point is reached when no more mobile dislocations are available. This coincides with the point of maximum resistance to slip (the maximum strength and hardness) where no more deformation is possible and fracture will take place if an increase in stress is attempted. Further plastic deformation can only be carried out if the material is annealed.

5.7.1   During a cold-worked process some 90% of the mechanical energy employed is converted to heat whilst the remaining 10% is retained in the material as potential energy in the form of elastically balanced strains in the lattice structure. The introduction of heat energy to the system during an annealing process permits sufficient mobility of the atoms so that dissipation of strain energy can take place. The initial relief of stress is followed by recrystallisation of the metal.

5.7.2 Strengthening Mechanisms   There are three principal strengthening mechanisms in metals. They are: *strain hardening* (5.7); *solution hardening* (7.5) and *dispersion hardening* (14.1.5 and 14.3). In solid solutions the movement of dislocations is impeded by irregularities in the lattice structure caused by the presence of solute atoms (Fig. 5.7), whilst the presence of small but strong particles dispersed as separate phases through the metal will also halt the movement of a dislocation 'front'.

Stress-relief and Recrystallisation

5.8   As we have seen a cold-worked metal is in a state of considerable mechanical stress resulting from elastic strains internally balanced. Much of the 'locked up' strain energy is associated with the presence of

Fig. 5.26

dislocations. In the region of P (Fig. 5.26) the lattice is in tension and since atoms there are displaced further apart the region will possess potential energy. The high energy associated with the congregation of dislocations at grain boundaries has been mentioned as a reason why corrosion occurs more readily there than elsewhere in the structure. These high-energy regions also initiate recrystallisation during an annealing process so that new 'seed' crystals appear first at the old grain boundaries of the original distorted structure (Fig. 5.30).

In the early stages of an annealing process some degree of stress relief occurs as atoms move over limited distances into positions nearer to equilibrium. At this stage, however, there is no alteration in the distorted appearance of the structure and in fact hardness and tensile strength remain at the high value produced by cold work.

At low temperatures the movement of dislocations is restricted to glide along slip planes, but at higher temperatures an edge dislocation is able to move out of its slip plane by a process known as *climb*. If the terminal row of atoms (normal to the plane of the paper) in the extra half-plane is removed this constitutes *positive climb* and the strain energy of the immediate vicinity will be reduced. If an extra row is added to the half-plane this results in *negative climb* and the strain energy of the region will be increased.

In this instance we are interested mainly in positive climb and the reduction of strain associated with it. It can take place most easily by the diffusion (7.4) of vacancies towards the dislocation as suggested in Fig. 5.27. Here atoms migrate from the end of the half-plane to fill vacancies which approach them by means of the diffusion mechanism.

5.8.1 The diffusion of interstitial atoms towards an edge dislocation (Fig. 5.28(i)) can give rise to *negative climb* whilst the diffusion of atoms away

Fig. 5.27   'Positive climb' by two examples of vacancy diffusion.

Fig. 5.28 Dislocation climb associated with interstitial atoms.

from the half-plane to become interstitials will cause *positive climb* (Fig. 5.28(ii)). All of these processes necessitate the mass migration of atoms by diffusion and climb is therefore possible only by thermal activation. The reduction of strain energy during annealing will most likely involve positive climb by the diffusion of vacancies.

The reader may possibly be worried by the fact that in the above description of the mechanism of climb the assumption has been made that a whole row of atoms is removed (or added) simultaneously, whereas in practice, individual vacancies or small groups of vacancies diffuse to (or from) the dislocation. Figure 5.29 illustrates climb involving a short section

Fig. 5.29 Dislocation 'jogs'.

of a dislocation line, resulting in the formation of two steps generally described as *jogs*.

During an annealing process elastic strains are first dissipated by limited movements of atoms in the manner of those mentioned above. Then, at a higher temperature, wholesale recrystallisation of the distorted structure occurs and is accompanied by a fall in tensile strength and hardness to approximately their original values, whilst the capacity to accept cold-work returns. This type of annealing process, which is used principally when a metal is to undergo further cold deformation, takes place in three stages:

5.8.2 The Relief of Stress   This often takes place at relatively low temperatures at which atoms nevertheless are able to move small distances into positions where they will incur less strain. Local movement of interstitial atoms or vacancies may be involved as outlined above. There is, however, no observable alteration in the distorted structure which was produced by mechanical work and both tensile strength and hardness will remain at a high level.

Low-temperature annealing which does not cause recrystallisation and which consequently does not lead to a fall in strength and stiffness engendered by cold-work, is useful in dissipating local strain which might otherwise give rise to stress corrosion (19.5).

5.8.3 Recrystallisation   Although low-temperature annealing is often used to relieve internal stress as outlined above, most annealing processes involve complete recrystallisation of the distorted cold-worked structure. As the annealing temperature is increased a point is reached where new crystals begin to grow from nuclei initiated within the most heavily deformed regions. These nuclei form at points of high energy such as crystal boundaries and other localities rich in piled-up and entangled dislocations. The new crystals so formed are very small at first but grow steadily until they have absorbed the whole of the distorted structure produced originally by mechanical work (Fig. 5.30). The new crystals are

(i)   (ii)   (iii)   (iv)

Fig. 5.30   The recrystallisation and grain growth of a cold-worked metal.

equi-axed in shape, that is they do not show any directionality as did the distorted crystals which they replace.

The minimum temperature at which recrystallisation will take place is called the *recrystallisation temperature*. This temperature is lowest for pure metals and is raised significantly by even small amounts of impurities. Thus, whilst commercial-grade aluminium recrystallises at about 150°C following mechanical work, that of 'six nines' purity (99.9999% pure) apparently recrystallises below room temperature and consequently does not cold-work when deformed at ambient temperatures.

Tin and lead normally recrystallise below room temperature so that it is impossible to work-harden them since they recrystallise whilst the working process is taking place. Hot-working is, of course, a process which is carried out above the recrystallisation temperature. Normally the working temperature must be well above the recrystallisation temperature so that recovery is rapid enough to keep pace with the rate of deformation.

It is not possible to assign a precise recrystallisation temperature to any metal since the temperature at which recrystallisation begins is governed largely by the amount of cold-work which the metal has undergone. The greater the amount of cold-work, the greater will be the amount of strain energy locked up in the structure and so the lower the temperature at which recrystallisation may begin when heat energy is supplied. For most metals the recrystallisation temperature is between one third and one half of the melting point (Kelvin scale). That is the mobilities of all metallic atoms are roughly equal at the same fraction of their melting points (K).

5.8.4 Grain Growth At temperatures above that of recrystallisation of the metal, the newly formed crystals will tend to grow by absorbing each other in cannibal fashion. This will result in the formation of a relatively coarse-grained structure, the ultimate grain size depending jointly upon the time of treatment and the temperature employed (Fig. 5.31).

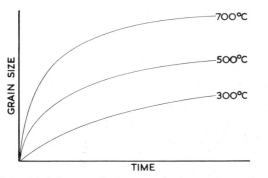

Fig. 5.31   The relationship between grain size, annealing temperature and annealing time.

Since boundaries have higher energies than any other region of the crystal, a polycrystalline mass will be able to reduce its energy if some of the grain boundaries disappear. Thus at temperatures in excess of that of recrystallisation large crystals grow by absorbing small ones. A crystal boundary tends to move towards its centre of curvature (Fig. 5.32) in order

Fig. 5.32   The growth of a crystal at the expense of its neighbours.

to shorten its length. To accommodate this atoms move across the boundary to positions of greater stability where they will be surrounded by more neighbours in the concave face of the growing crystal.

5.8.5 The Measurement of Grain Size   Crystal size—or 'grain' size—has a direct influence on the ductility of many metals and alloys and is of great importance in selecting a material for deep drawing (6.4.3) and similar processes. A method of assessing grain size is therefore necessary. In all methods some form of microexamination is involved. The most direct method is to count the number of crystals observed in a known area of field of view so that grain size can be expressed as the number of grains/mm^2. The ASTM grain size index number is derived from the number of grains/in^2 when counted at a magnification of $\times 100$. The ASTM* index, N, is given by:

$$n = 2^{(N-1)}$$

where n is the number of grains/in^2 at a magnification of $\times 100$. This can be written:

$$\operatorname{Log} n = (N - 1) \log 2$$

or
$$N = \frac{\operatorname{Log} n}{0.3010} + 1$$

For ductile non-ferrous metals used in deep drawing operations grain size is often expressed in terms of 'Average Grain Diameter' (AGD) (in mm). This is more easily and quickly assessed using the ground glass screen of a microscope equipped for photomicrography. A line 100 mm long is drawn on the screen and an image of the test specimen projected on the screen at a magnification of $\times 100$. (The 100 mm line therefore represents a length of 1 mm on the surface of the test specimen.) The number of crystals cut by the line is then counted so that:

$$\text{AGD} = \frac{1 \text{ mm}}{\text{Number of crystals cut}}$$

(This assumes that on average crystals will be regular polygons in shape, or, very roughly, circular.)

For most metals destined for deep drawing optimum AGD is generally between 0.04 and 0.06 mm.

5.8.6 Arrhenius Law   This is an equation of a semi-empirical type which relates to many rate-controlled processes such as are met in physical and chemical reactions. In materials science thermally-activated phenomena like recrystallisation, creep and diffusion are governed by this Law which states that the rate, v, at which a process occurs varies exponentially with the temperature, T (K).

$$v = A e^{-Q/RT}$$

where Q is the activation energy of the reaction (J k mol^{-1});
R is the universal gas constant (8.3143 J mol^{-1} K^{-1});
A is a constant for the reaction.

* American Society for Testing Materials.

The equation is usually expressed as:

$$v = A \exp(-Q/RT)$$

which can be written in logarithmic form as:

$$\ln v = \ln A - (Q/RT)$$

Q Assuming that cold-worked aluminium will recrystallise completely in 100 hours at 260°C but that at 330°C the process is complete in only six minutes, calculate the activation energy of the recrystallisation process, given that the universal gas constant (R_0) is 8314.3 J k mol^{-1} K^{-1}.
In what time would recrystallisation be complete at 290°C?

A In the Arrhenius equation:

$$v = A \exp(-Q/RT)$$

v represents the rate of recrystallisation whilst Q is the activation energy of the process (J k mol^{-1}). A is a constant for the system.
Let v_1 = relative recrystallisation rate at 260°C (533 K)

$$= \frac{k}{6000} \text{ min}^{-1}$$

and v_2 = relative recrystallisation rate at 330°C (603 K)

$$= \frac{k}{6} \text{ min}^{-1}$$

where k is constant for the system.
Thus,

$$v_1 = A \exp(-Q/R_0 T_1) \tag{1}$$

and

$$v_2 = A \exp(-Q/R_0 T_2) \tag{2}$$

Dividing (2) by (1):

$$\frac{v_2}{v_1} = \exp(-Q/R_0 T_2 + Q/R_0 T_1)$$

$$= \exp\frac{Q}{R_0}\left(\frac{T_2 - T_1}{T_1 T_2}\right)$$

Therefore:

$$\frac{k}{6} \div \frac{k}{6000} = \exp\frac{Q}{8314.3}\left(\frac{603 - 533}{321\,399}\right)$$

$$\frac{6000}{6} = \exp Q\left(\frac{70}{2.672 \times 10^9}\right)$$

and: $$10^3 = \exp Q \times 2.62 \times 10^{-8}$$

Hence:

$$\ln 10^3 = 2.62 \times 10^{-8} Q$$

$$6.9078 = 2.62 \times 10^{-8} Q$$

$\boxed{\log_e 10^3 = 6.9078}$

Therefore:

$$Q = 2.637 \times 10^8 \text{ J k mol}^{-1} \quad \textbf{or } 263.7 \text{ MJ k mol}^{-1}$$

Let v_3 be the recrystallisation rate at T_3 i.e. 260°C (533 K)

$$v_3 = A\exp(-Q/R_0 T_3) \tag{3}$$

Dividing (3) by (1):

$$\frac{v_3}{v_1} = \exp\frac{Q}{R_0}\left(\frac{T_3 - T_1}{T_1 T_3}\right)$$

or: $\quad \dfrac{v_3}{v_1} = \dfrac{k/t}{k/6000} = \dfrac{6000}{t} = \exp\dfrac{2.637 \times 10^8}{8314.3}\left(\dfrac{563 - 533}{533 \times 563}\right)$

where t is the time necessary for recrystallisation to be complete at 290°C (563 K). Hence:

$$\frac{6000}{t} = \exp 3.1708$$

$$6000 = t(\exp 3.1708)$$

$$\ln 6000 = \ln t + 3.1708$$

$$8.6995 = \ln t + 3.1708$$

$$\ln t = 5.5287$$

$$\underline{t = 251.8 \text{ minutes}}$$

The Deformation of Polymeric Materials

5.9   The arrangement of atoms in the crystal structure of a metal follows a regular three-dimensional periodicity but the atoms in a polymer are generally arranged in a far more complex and less-regular manner, particularly since the molecules themselves are often not geometrically aligned. For this reason it has been less easy to explain the deformation of polymers in terms of their structures.

Single bonds in a polymer chain are capable of *rotation*. This is illustrated in Fig. 5.33 which represents the last twelve atoms of a linear polymer molecule. Suppose a rotation of 180° took place about *a* only. The rest of the molecule, containing possibly thousands of carbon atoms, would be displaced relative to the bond *a* and it is unlikely that sufficient space would be available for this even if enough molecular energy were available. If, however, the rotation at *a* were accompanied by a simultaneous 180° rotation at *b*, only a short section of the chain would be displaced. Not only

Fig. 5.33 The effect of two 180° bond rotations on the relative spacing of atoms in a carbon chain.

would less space be needed but less molecular energy would be required. Nevertheless the position of a short section of the molecule would be changed with respect to its neighbours. If sufficient rotational energy is supplied thermally this type of rotational transition will be taking place continuously in all molecules throughout the material and the molecular chains will be in constant motion relative to each other. The system could be visualised as something like a tub full of intertwined wriggling eels.

The activity of the molecular chain will depend upon temperature but also upon the size and nature of the intermolecular forces opposing rotation. It will also depend upon the complexity of the chain units. So, whilst molecules of simple 'section' like polythene can rotate easily, molecules with large complex units would require extra space in which to rotate, thus making rotation correspondingly difficult.

5.9.1 In materials of low molecular weight atoms are held in fixed positions either by chemical bonds or by inter-molecular forces acting within the crystal structure. When stress is applied the atoms are displaced from their equilibrium positions and distances between them alter until the new forces which are set up balance the applied stress. Assuming that distortion is within the elastic range, atoms will return to their original positions when stress is relaxed. Greater stresses cause permanent distortion. In metals this is achieved by the movement of dislocations whilst in other materials, such as ceramics, sudden rupture occurs with little or no plastic deformation.

When stress is applied to a polymer material elastic deformation will occur wherever two atoms are held in fixed positions relative to one another by chemical bonds or inter-molecular forces. Thus in cross-linked polymers (12.3) and in thermoplastics at temperatures below their glass-transition points (12.4.3) all atoms can be regarded as being fixed in this way so that only elastic deformation is possible. Some elastic deformation is also possible in thermoplastics at temperatures above their glass-transition points as the atoms within the molecules are held by covalent bonds, whilst if crystallites are present atoms will be held by inter-molecular forces also. Elastic deformation of this type occurs instantaneously.

In amorphous regions of the polymer material the molecules are held only by the relatively weak van der Waals forces operating between them. Hence they can move relative to each other without difficulty. This molecular slip—or plastic deformation—does not occur instantaneously but is time dependent since it is related to the viscosity. Consequently the combined elastic and plastic distortion is said to be *viscoelastic* (12.5.1).

5.9.2 In a linear polymer a large number of different chain arrangements are possible varying from the tightly coiled to the fully extended form. Since rotational transitions are taking place continuously the chain shapes will be constantly changing. For each molecule which tends to curl up another will straighten so that the *average* extension of the chains remains constant. However, when stress is applied the extensions are no longer random but occur in the direction in which the stress is acting. Thus tensile stress produces extension in the material (Fig. 5.34).

MOLECULAR STRAIGHTENING   MOLECULAR SLIPPING

Fig. 5.34   The effect of stress on molecular shape and arrangement in a linear polymer.

When stress is removed the changes in shape of the molecules continue but those changes which involve curling up are now in the preponderance and the original average extension is regained so that the material returns to its original size. Thus chain straightening is an elastic process but it is viscoelastic because the changes in shape are time dependent in so far as the time dependence is determined by the magnitude of those inter-molecular forces opposing the rotational transitions.

The application of stress also causes molecular slip (Fig. 5.34) as the relatively weak van der Waals forces operating between the linear molecules are overcome. This is a non-reversible process so that the extension produced is non-elastic. It is also a time-dependent process being influenced by the rotational transitions. The extent of rotational transitions to accommodate slip must be greater than that to allow for the type of segment shift which facilitates chain straightening. Straightening is therefore quicker than slipping. The phenomenon is discussed further in Chapter Twelve (12.5).

6

Shaping Processes

6.1 The arts and crafts of early Man were involved mainly with the shaping of wood and stone, and metals such as gold, copper and later—iron. The use of primitive hand tools gave way over the centuries to manufacture by relatively sophisticated processes which take into account the accumulated knowledge of physical and chemical properties of the substances being shaped. Forming processes are now based on the use of moulds and dies—often of complex shape—for shaping a wide range of materials. The development of such processes made possible the mass production of large numbers of identical articles with a consequent reduction in the *real* costs of manufactured goods generally.

Many solid substances possess sufficient plasticity to allow them to be shaped by either tensile or compressive forces. Generally plasticity increases with temperature and, so, many such processes are hot-working operations. Some solid materials, however, possess negligible plasticity and are relatively brittle at all temperatures. These materials, notably concrete, cast iron and others, can most easily be shaped by allowing them to solidify from the liquid state in a suitably shaped mould; or by using some machining process in which particles of the material are chipped away by a lathe or even a mason's chisel. Nevertheless, recent work on processes such as hydrostatic extrusion has shown that the reduction of surface friction between work piece and die, leading to the maximum utilisation of what plasticity a material does possess, allows relatively brittle materials like cast iron to be deformed mechanically without fracture.

Methods of shaping materials are therefore varied and often very ingenious and a single chapter in a book of this type can do little more than catalogue the more important methods used.

Shaping by Solidification from the Liquid State

6.2   What are commonly called 'casting processes' are used to shape a number of materials but more especially metals. In fact well over 99% of the total mass of metals used are melted and cast at some stage in the manufacturing process. A considerable quantity of both ferrous and non-ferrous alloys are of course cast as ingots which are then forged, rolled or extruded, whilst large amounts of iron, steel and light alloys are cast into some form of shaped mould to produce a component which will be used in substantially this form. Metallic alloys are strongly crystalline and pass from a mobile liquid to a rigid solid over a temperature range which is generally small. The resultant solid is strong because of its crystalline nature and the operation of the metallic bond within crystals.

Organic polymers are generally unsuitable for casting because many of them decompose at temperatures well below those necessary to produce adequate liquid mobility. Van der Waals forces operating between long-chain molecules oppose mobility. Such plastics materials as are cast therefore depend upon a *chemical* rather than a *physical* change taking place to effect solidification. Cold-setting plastics of the type used with glass-fibre or carbon-fibre reinforcement, consist of *two* mobile liquids which are mixed and which will then proceed to polymerise (12.2.1 and 12.3.2) and so, harden. The same principles apply in many respects to cement, concrete, plaster and other ceramic products which are cast into moulds. Crystallisation, accompanying chemical change, produces rigidity in these cases.

6.2.1 Casting Metals   Since the bulk of metallic materials is cast as either ingots or finished shapes, a melting process is an integral stage in most metal-forming operations. Furnaces used vary considerably in design but those burning fuel can be classified broadly into three groups:

(i)   Furnaces, like the foundry cupola, in which the charge is in intimate contact with the burning fuel. Although thermal efficiency is high because of the continuous operation of the furnace and the intimate contact between burning fuel and charge, direct transfer of impurities from fuel to charge can take place.

(ii)   Furnaces in which the charge is out of contact with the fuel but is in contact with its products of combustion. Transfer of impurities will be limited to those of a gaseous nature and thermal efficiency will still be reasonably high. Reverberatory furnaces (Fig. 6.1(ii)) of many individual designs—including the open-hearth furnace formerly used so widely in steel making—fall into this category.

(iii) Furnaces in which the charge is isolated both from the fuel and its products of combustion. Crucible furnaces constitute this group. Maximum purity of charge is maintained because contact between charge and fuel is at a minimum. Nevertheless thermal efficiency will be very low since most of the heat by-passes the charge and leaves with the flue gases.

Fig. 6.1 Typical fuel-burning furnaces for melting metals. Group (i) The foundry cupola. Group (ii) A reverberatory furnace using pulverised coal, oil or gas. Group (iii) An oil- or gas-fired crucible furnace.

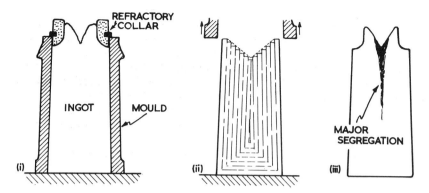

Fig. 6.2 The structure of steel ingots. (i) A typical 'big-end-down' ingot mould showing how the 'pipe' can be restricted. (ii) The development of the 'pipe' by successive solidification of elemental 'shells'. (iii) Major segregation of impurities in the central pipe.

Obviously electricity is a chemically clean fuel whether used for resistance heating, induction heating or in an arc furnace, but it is relatively expensive. It is therefore apparent that in any melting process where purity of charge is important melting costs will be high.

6.2.1.1 Ingot Casting Steel ingots of several tonnes in mass are commonly cast into cast-iron moulds. These are usually cast 'big-end-down' (Fig. 6.2(i)) to facilitate the stripping of the mould. This can be lifted off the solidified ingot with a minimum disturbance of the latter (Fig. 6.2(ii)), an important factor since the ingot is very easily damaged by cracking when in the hot 'tender' state.

As a metal solidifies it shrinks and so the level of the remaining liquid falls progressively leading to the formation of a *pipe* in the top of the ingot (Fig. 6.2(ii)). This is minimised by the use of a *hot top*—a refractory collar at the top of the ingot mould which retains a reservoir of molten metal which feeds into the ingot as it solidifies and shrinks. The pipe, as it forms, is filled from the molten reservoir. The absence of any feeding arrangement (Fig. 6.2(ii)) would lead to excessive piping.

Segregation of impurities is also troublesome in steel ingots. During the initial stages of solidification relatively pure metal crystallises first so that any dissolved impurities tend to accumulate in that metal which solidifies last. Hence the metal in the central pipe will be more impure than that which crystallises initially at the surface of the ingot (Fig. 6.2(iii)). This is generally termed *major* segregation. *Minor* segregation on the other hand refers to the accumulation—by the coring mechanism (7.3.3)—of impurities at crystal boundaries throughout the ingot. Of the two, minor segregation is the more deleterious in its effects since it causes overall intercrystalline weakness.

Non-ferrous alloys may be cast as flat slabs for rolling to sheet, or as cylinders for extrusion of rod, tube and other sections of varying complexity (6.3.3).

6.2.1.2 Continuous Casting   It will be obvious that the casting of steel by the ingot process is inefficient. Major segregation in the 'piped' region means that some of the ingot 'top' (Fig. 6.2(iii)) must be cropped during the early stages of rolling (and subsequently remelted). With the elimination of this waste in mind the development of *continuous casting processes* has been going on for many years and—aided by the expansion of modern electronic remote control—has reached the stage where the bulk of ordinary steel produced in the developed world comes from continuous-casting plants.

The elements of continuous casting are illustrated in Fig. 6.3. Molten

Fig. 6.3   The principles of the continuous casting of steel slab. Once started, casting can proceed virtually non-stop. The product is cut into coiled lengths, as required at the end of the rolling process by a flying saw.

metal is fed into a short assistance-cooled mould at a controlled rate and as soon as solidification is sufficiently advanced the base of the mould is withdrawn taking with it the embryo ingot. This is engaged by a series of support and withdrawal rolls which convey the continuous ingot to the reduction mill proper. Such segregation of impurities as takes place will be uniform and will be further dispersed by the rolling process. Metal loss is extremely small.

The production of castings to some finished and often intricate shape may be achieved by one of a number of different processes. The particular process used will be governed by such criteria as:

(i)   the size of the casting—large castings will be most economically made in sand moulds;

(ii)   the surface finish desired—a better surface finish is obtained using metal, 'synthetic sand' or plaster moulds, but these processes are more expensive than sand casting;

(iii)   the dimensional accuracy required—again greater accuracy is poss-ible using methods mentioned in (ii) as against sand casting;

(iv)   complexity of design, e.g. the presence of re-entrant sections and internal cavities. In many instances sand casting will be the only process possible;

(v)   the mechanical properties specified—generally the use of metal moulds gives finer grain due to rapid cooling so that properties are rather better;

(vi)   the total number of castings required—generally it will be uneconomical to produce a metal mould unless several thousands of castings are required.

6.2.1.3 Sand Casting   Most readers will be familiar with the principles of this process. A wooden pattern is made such that when sand is rammed round it in a moulding box (Fig. 6.4) the mould can be split and the pattern

Fig. 6.4   A simple two-part sand mould with the wooden pattern removed and ready to receive the charge of molten metal.

withdrawn leaving a cavity into which metal is poured. Castings of great complexity can be produced by this process using multi-part boxes and, where necessary, sand *cores* to form holes and other internal cavities in the component. Moreover the process is cheap to operate particularly when only small numbers of castings are required, since the wooden pattern is relatively inexpensive to produce. As against these advantages dimensional accuracy is inferior to that attainable by other processes whilst surface finish is sometimes poor. The slow rate of solidification which prevails in a sand mould gives rise to coarse grain and this in turn results in a relatively weak brittle structure.

6.2.1.4 Shell Moulding   In ordinary sand casting the moulding sand contains sufficient natural clay to act as a binding material for the silica particles. In shell-moulding processes this natural clay bond is replaced by a synthetic bonding material of the thermosetting-resin (phenol formaldehyde) type. The moulding mixture therefore consists of clean, clay-free silica sand mixed with 5% of the plastic bonding agent.

Each half of the shell mould is made on a metal pattern plate which is heated to about 250°C before being placed on top of a 'dump box'. The box is then inverted (Fig. 6.5(i)) so that the plate is covered with the sand/resin

Fig. 6.5 The principles of shell moulding.

mixture. The resin melts and in about thirty seconds the hot pattern plate becomes coated with a shell of resin-bonded sand. This shell hardens as the resin sets and the box is turned to its original position (Fig. 6.5(ii)) so that surplus sand/resin mix falls back into the box. The shell is then stripped from the pattern plate with the help of ejector pins. The two halves of the mould thus produced are then clamped or cemented together to receive their molten charge.

The main advantage of this process lies in the high degree of dimensional accuracy obtained. Surface finish is also far superior to that associated with an ordinary sand casting. Shell moulds store well and are quite rigid so that they can be transported as necessary. The principal disadvantage of the process arises from the high cost of metal patterns though this is somewhat offset by the fact that relatively unskilled labour can be utilised so that labour costs are much less than those associated with green-sand moulding.

6.2.1.5 Investment Casting   Whilst shell moulding was developed in Germany during the Second World War, the origins of investment casting lie in pre-history when ancient craftsmen produced moulds by pressing clay around a pattern carved from beeswax. Subsequent baking of the clay mould melted out the wax leaving a cavity to receive the metal charge.

For modern investment casting wax patterns are produced in precision metal moulds and then affixed to a 'tree' (Fig. 6.6). The wax assembly is then 'invested' with a mixture of finely powdered sillimanite and ethyl silicate. The action of water and heat on the latter cause it to form a strong silica bond between the sillimanite particles. The heating process also melts out the wax pattern leaving a mould cavity which will receive the charge of metal. To improve the accuracy of the mould impression metal is cast into the mould under pneumatic pressure or by centrifuging.

One advantage of the process is that extremely complex shapes can be cast since the pattern is not withdrawn. Much of the intricate work in gold

Fig. 6.6 The production of an 'investment' mould.

ornamentation carried out in the sixteenth century by Benvenuto Cellini was by this process. From the aesthetic point of view there is no disfiguring 'parting line' on the finished casting as is apparent with other casting processes. Extreme precision in dimensions is possible and the process is of engineering importance for making small components from very hard, strong materials which cannot be shaped by forging or machining. Blades for gas-turbines and jet engines can be cast by this process. The principal disadvantage of investment casting is its high operational cost.

6.2.1.6 Die-casting   This process involves the use of a metal mould. Since the cost of machining the mould cavity is high, die-casting is economical only when large numbers of castings—generally several thousands—are required. In *gravity* die-casting, now termed *permanent-mould* casting, molten metal is poured into the metal mould in a similar manner to ordinary sand casting. In *pressure* die-casting, often called just 'die-casting', the molten metal is injected into the mould cavity under pressure (Fig. 6.7) so that a much more accurate impression of the mould cavity is obtained.

Pressure die-casting is the more common process and 'cycling' is rapid. As soon as the casting is solid the die is parted. Generally it is so designed that the casting moves away with the moving part of the die from which it is detached by a system of ejector pins.

Fig. 6.7 A pressure die-casting machine.

The main advantages of die-casting are accuracy of dimensions, good surface finish and improved mechanical properties arising from the fine grain produced by rapid solidification. Although the cost of the mould is high, labour costs in operation are low relative to sand casting, so that the process is economical for the manufacture of large numbers of zinc-base and aluminium-base components in the automobile and other industries.

6.2.1.7 The Lost-foam Casting Process   This was preceded by what was called the *full-mould process*. In this a pattern was carved from a block of 'expanded' polystyrene (EPS) (12.8). This pattern was then placed in a flask and covered with dry resin-bonded silica sand. When molten metal was poured directly on to the EPS pattern the latter evaporated and the space so provided was occupied by the metal. Collapse of the sand mould was prevented by the thermosetting-resin bonding material.

It was found that if *unbonded* dry silica sand was used to surround the EPS pattern in a steel flask and the flask then vibrated or tapped to consolidate the sand, bonding agents could be dispensed with. This was named the *lost-foam process*. Whilst the original full-mould process was essentially a 'one-off' operation used for making single castings, e.g. for ornamental purposes, the lost-foam process has been adapted for mass production. The patterns are made by injecting EPS 'beads' into a heated split aluminium die which is further heated so that the 'beads' expand and fuse together. The process is used in the manufacture of aluminium alloy and iron castings.

6.2.2 Casting Non-metallic Materials   Not many non-metallic materials are suitable for casting because most of them lack a well-defined point where a mobile liquid transforms to a rigid solid as is the case with metals.

6.2.2.1 Organic Polymers   The disorder→order change which occurs in metals on crystallisation results in a complete and sudden change in properties from a mobile liquid to a very rigid solid. No such well-defined change occurs in plastics materials since only very limited crystallisation may occur. Because of the presence of large chain molecules in a polymer considerable van der Waals forces operate between them so that mobility in plastics is never very high and most of them in fact decompose on heating to temperatures well below those at which any significant mobility is attained.

Such polymer materials as are cast are in fact those of the cold-setting type (12.3.2). Here two substances of relatively small molecular size are mixed and allowed to polymerise so that long-chain molecules subsequently form and the viscosity of the material increases progressively until a rigid structure is obtained (Fig. 12.4). This change is chemical rather than physical since valency links form between small molecules to produce large ones. Casting processes involving such changes are used for 'potting' small specimens in a transparent plastic case and also for encasing electronics equipment in a protective insulating capsule. The process is also used to produce structures in which cold-setting resins are strengthened with glass- or carbon-fibre.

6.2.2.2 Ceramics   Plaster of Paris and sillimanite-ethyl silicate mixtures (6.2.1.5) are often cast for various purposes. Some cement mixtures are also cast, generally mixed with some form of aggregate. In all of these instances hardening is accompanied by a chemical reaction though in a few cases crystallisation may also occur. Again, the change producing rigidity is chemical rather than physical.

Concrete (14.6.7) is cast into simple shapes in building construction. The mould is generally constructed of simple wooden 'shuttering' which is removed when the concrete has set. Such casting is carried out *in situ*. Various concrete mixtures may also be cast in such forms as paving flags, pipes or garden 'furniture' (gnomes, toads or Aphrodites). This is termed *pre-cast* concrete in the sense that it is not cast *in situ*, but is transported after it has set. The more complex shapes and ornamental work are often cast into Plaster of Paris or gelatine moulds. Much of the cast concrete used in building construction is either of the reinforced or the pre-stressed variety (15.10.1/2).

Hot-forming of Solids

6.3   As might be expected, these processes are applied to those materials the plasticity of which increases appreciably as the temperature rises. Many metals and all thermoplastic polymers fall into this class. However, whilst increase in temperature gives rise to an increase in malleability, this is usually accompanied—particularly in metals—by a *decrease* in ductility, because tensile strength is reduced and the material tears apart more readily. Consequently most hot-forming processes involve the use of compressive forces so that the material is rolled, forged in a die or forced through a shaped orifice.

A metal is hot-worked above its recrystallisation temperature (5.8.3) so that recrystallisation occurs simultaneously with deformation and the material does not work harden. In metals the as-cast structure is relatively weak and brittle because of the coarse crystal structure present and also the segregation of impurities at the crystal boundaries. Mechanical working redistributes these impurities more uniformly throughout the metal so that the embrittling effect is less marked. At the same time recrystallisation produces much finer grain which further leads to better mechanical properties. Impurities tend to be elongated in a fibrous manner along the main direction of working.

6.3.1 Forging   Blacksmiths have been shaping heated pieces of metal during the last 6000 years or so using simple tools such as a hammer and an anvil. Although some hand work is still carried out by smiths particularly in repairs to agricultural machinery and the like, most engineering forgings are produced in some form of 'closed die'.

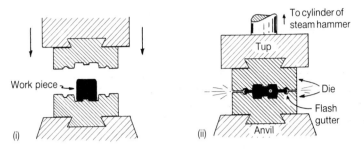

Fig. 6.8 'Closed-die' forging using a double-acting steam hammer.

6.3.1.1 Drop Forging   This involves the use of a closed die, one half of which is fixed to a massive anvil, whilst the other half is attached to the 'tup' of a guided hammer (Fig. 6.8). A heated work piece is interposed between the die faces as they come together. In order that the die cavity shall be filled a small excess of metal must be available and this is squeezed outwards to form a *flash* which is easily trimmed away from the resultant forging.

6.3.1.2 Hot Pressing   is a development of drop forging which is generally used for producing simple shapes. Here the hammer is replaced by a hydraulically driven ram so that the work piece is shaped by steady pressure rather than by a succession of hammer blows. Deformation is then more uniform throughout the work piece. Large forgings can be made direct from ingots of suitable size. During forging refinement of grain and redistribution of impurities lead to a big improvement in mechanical properties. Thus a forging is always stronger and tougher than a casting of similar shape and composition. Naturally a forging is far more expensive to produce than an equivalent casting. Moreover relatively simple shapes only can be produced by forging since solid metal, even under high pressures, flows to the shape of a die much less easily than does a liquid metal.

Smaller forgings are generally produced from stock in the form of hot-rolled bar. In such stock impurities will already have been distributed directionally by the rolling process producing *fibre* as indicated in Fig. 6.9(i). During forging this fibre will follow the contours of the die and so strength will be maintained. If for some reason of bad die design, fibre 'outcrops' to the surface producing a situation similar to that indicated for the *machined* component in Fig. 6.9(iii), this may give rise to planes of weakness along which fracture can occur.

Fibre distribution in a forging can be examined quite easily. A representative section is first cut—this is the most tiresome part of the job if no mechanical aid is available—and then ground flat to a reasonably smooth finish (about '280' grade emery paper). The section is then etched for about 15 minutes in hot 50% hydrochloric acid when the fibre will be clearly revealed (Fig. 6.10).

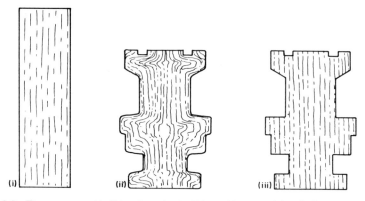

Fig. 6.9 The component in (ii) has been 'up-set' forged from stock bar (i). Flow lines follow the contours of the component. In (iii) the same component has been machined from bar stock and in this case the 'fibre' is cut. High stress concentration is likely to cause failure along these exposed fibres and so weaken the component.

Fig. 6.10 'Flow lines' in forged components. The fibrous structures were developed by deep-etching the cross-sections in boiling 50% hydrochloric acid for 30 minutes ×0.6.

Fig. 6.11 Hot-rolling steel strip. The ingot is first 'broken down' in a two-high reversing mill (the piped top is cropped after several passes through the mill). The resultant work piece then passes, still hot, to the train of finishing rolls which roll it down to strip.

6.3.2 Hot Rolling Metal plate, sheet, strip and rod are all produced by hot-rolling processes. In most cases a simple 'two-high' reversing mill (Fig. 6.11) is used in the initial stages to 'break down' the ingot. The resultant slab or *bloom* is then successively reduced in cross-section in a suitable train of finishing rolls. Much hot rolling is now integral with a continuous-casting unit (6.2.1.2).

Plate, sheet and other flat material may be produced in plain rolls but rod and other sections such as rails and RSJ must be passed through rolls which are appropriately grooved. Rolled steel products are manufactured by hot-rolling since the FCC structure (present above about 900°C in mild steel) is much more malleable than is the BCC structure. However, most metals and alloys are significantly more malleable at high temperatures and are therefore hot rolled in the initial stages from the ingot. Cold working is only applied as a finishing process in the interests of surface finish, accuracy of dimensions and the control of mechanical properties.

6.3.3 Extrusion The extrusion of both simple and complex sections is achieved by forcing the *solid* material through a suitably shaped die rather in the manner in which toothpaste is forced from a tube. Both ferrous and non-ferrous metals are extruded from heated billets though in many cases cold extrusion is possible. A few metals are extruded from hot or cold powders whilst thermoplastic polymers are extruded as heated powders, sometimes around a steel reinforcement strip as in the case of plastic curtain rail.

The cast billet is heated to the required temperature (350°–500°C for aluminium alloys; 1100°–1250°C for steels) and transferred to the container of the extrusion press. The ram is driven hydraulically with sufficient force to extrude the metal through an alloy steel die (Fig. 6.12).

Fig. 6.12 The principles of extrusion of a solid billet.

The principal advantage of the process is that, in a single operation from a cast ingot, quite complex sections with reasonably accurate dimensions can be obtained. Typical products include round rod; hexagonal brass rod (for parting off as nuts); stress-bearing sections in aluminium alloys (for aircraft construction); carbon and alloy-steel tubes as well as small diameter rod for drawing down to wire.

6.3.4 Blow-moulding of Plastics This process is used to manufacture hollow articles such as polythene bottles. Extruded tube (parison) is heated and then blown by air pressure against the walls of a steel mould. The two halves of the mould then part to release the moulding.

6.3.4.1 Film-blowing of Plastics Film can be blown continuously from extruded thermoplastics as indicated in Fig. 6.13. Such material has a nominal thickness of not more than 0.25 mm and the process is widely used in the manufacture of polythene film and tube for parting off as bags and other forms of packaging.

6.3.5 Calendering This process is used to produce thermoplastic sheet, particularly of PVC and polythene. The calendering machine consists of two or more heated rolls into which the plastics material is fed as a heated dough. The formed sheet is cooled on a chill roll (Fig. 6.14). Calendering machines are also used to coat paper, fabric and foil with a film of plastics material.

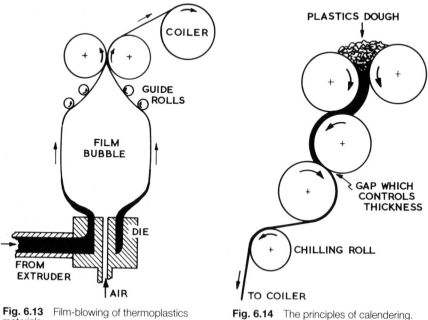

Fig. 6.13 Film-blowing of thermoplastics materials.

Fig. 6.14 The principles of calendering.

Fig. 6.15   The principles of the vacuum-forming of plastics materials.

6.3.6 Vacuum-forming of Plastics   This process is used to produce simple shapes from thermoplastic sheet stock. The heated sheet is held at its edges and then stretched by the advancing mould section (Fig. 6.15). The final shape is produced by applying a vacuum so that the work piece is moulded to the shape of the die by external atmospheric pressure.

Cold-forming of Solids

6.4   Most solid materials can be shaped more easily at high temperatures. The cost of the heating process is more than offset by a big saving both in time and mechanical energy required for the subsequent shaping of the work piece. Nevertheless a large number of cold-working processes are used in the forming of ductile metals and alloys. These are generally finishing processes except in the case of drawing and deep-drawing operations where the applied forces are tensile in nature and hot-working cannot therefore be used because of the low tensile strength of most metals at high temperatures.

Cold-working is a terminal operation in many metal-shaping processes for one or more of the following reasons:

(i)   a clean, smooth or polished finish can be obtained;
(ii)   closer dimensional tolerances are possible;
(iii) the mechanical properties of many alloys can be adjusted only by the amount of cold-work applied to them. Strength, hardness and rigidity increase with the amount of cold-work.

6.4.1 Cold-rolling   Most sheet and strip metal is cold-rolled in the final stage for one or all of the reasons just noted. Simple two-high mills with highly polished rolls are often used to produce a smooth dense finish. When rolling wide sheets of thin foil considerable roll pressures are required. Since small diameter rolls are necessary in order to prevent slip between rolls and work piece, the rolls tend to bend under such pressures thus frustrating the attempt to produce very thin hard foil. The situation is resolved by the use of backing rolls (Fig. 6.16).

6.4.2 Drawing   Wire, rod and tubes are cold-drawn from stock which was initially either extruded or rolled from the ingot. In wire manufacture this often involves a series of drawing stages interposed with annealing operations to soften the material as it becomes work-hardened.

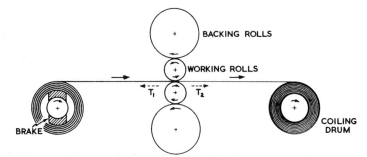

Fig. 6.16 Cold-rolling with a four-high mill (by rolling the strip whilst in tension, T_1 and T_2, roll pressures can be reduced).

The dies used (Fig. 6.17) are of hardened tool steel or tungsten carbide. The drawing force in wire manufacture is provided by winding the wire on to a rotating drum or 'block' (Fig. 6.17(i)) whilst for the production of rods or tubes some form of draw bench is required (Fig. 6.17(ii)).

In tube drawing it is generally necessary to support the internal bore of the tube. This is often done by drawing the tube on to a steel rod—or *mandrel*—which passes through the die along with the tube from which it is subsequently extracted.

6.4.3 Deep-drawing This process differs from simple metal presswork in that some degree of wall-thinning always takes place. Generally some form of cup is first produced from a flat blank and this cup is then drawn so that its walls become progressively thinner. The tools used comprise some form of punch and die system.

Deep-drawn components range in size from small brass cartridge cases to aluminium milk churns. Generally the more ductile materials such as 70–30 brass, cupro-nickel, copper, aluminium and mild steel are used.

The forces operating in deep-drawing are largely tensile so that it is

Fig. 6.17 Drawing processes: (i) wire; (ii) tube—a similar draw bench is used for rod.

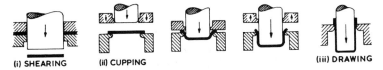

Fig. 6.18 Stages in a deep-drawing process.

inevitably a cold-working process. Inter-stage annealing is therefore used when the work piece becomes work-hardened. As a result of annealing, regions of weakness are often encountered in a deep-drawn component. These are due to variations in grain size arising during the recrystallisation process which accompanies annealing. Thus, in the drawn cup (Fig. 6.19),

Fig. 6.19 Variations in grain-size coincident with different amounts of cold-work in a deep-drawn cup which was subsequently annealed.

the grain size at A is satisfactory since the material has received no cold-work and the grain size is more or less that of the original stock, no recrystallisation having taken place. At B the grain size is satisfactory because heavy cold-work followed by annealing has induced complete recrystallisation from many nuclei. The metal at C, however, received only the *critical* amount of cold-work necessary just to initiate recrystallisation. Hence very few nuclei formed in this region of very little deformation so that the grain size was large and ductility correspondingly low.

6.4.4 Cold-forging This includes a number of processes where a cold work piece is forged in some form of closed die.

6.4.4.1 Coining (Fig. 6.20)   is a process of this type in which variation in

Fig. 6.20 The difference between coining and embossing. In embossing no variation in thickness of the work-piece occurs.

thickness occurs in the component during forging and hence some degree of metal flow relative to the die. This consumes energy as friction and wear on dies is considerable. Nevertheless a hard, wear-resistant component results.

6.4.5 Impact Extrusion   This is a cold-working process used in a number of variations. Possibly the best known of these is the method used to produce collapsible tubes for toothpaste, shaving cream and the like (Fig. 6.21). These may be in lead, tin or aluminium, though blow-moulded thermoplastics materials are now the most widely used for toiletries.

Fig. 6.21   Impact extrusion of a disposable tube.

A small unheated blank is fed into the die cavity and as the punch descends very rapidly metal is squeezed upwards into the gap between punch and die to form a cylindrical shell. The threaded nozzle may be formed during impact or it may be produced in a separate process. Other articles made by impact extrusion include aluminium canisters and capsules for food, medical products and photographic films; and shielding cans for radio components.

6.4.6 Cold-forming from Sheet Metal   In these processes little or no variation in the thickness of the work piece occurs during forming though diverse methods are used to achieve the required shape.

6.4.6.1 Embossing (Fig. 6.20)   differs from coining in that the thickness of the original stock is more or less maintained. Since little flow of metal occurs much less energy is lost as friction as compared with the coining process. Embossing is therefore used in the manufacture of hollow buttons, badges and the like.

6.4.6.2 Rubber Pressing (Fig. 6.22)   used for the forming of car-body panels, hub caps and other re-entrant shapes, stainless-steel tableware, aircraft panels and pots and pans. A rubber press usually consists of a stationary upper container which holds the rubber pad and a lower platen which carries the die and work piece. This lower platen is forced upwards into the rubber by means of hydraulic cylinders, the flexible rubber forcing the work piece into the die cavity and thus acting as a universal punch.

Fig. 6.22 Rubber pressing.

For pressing soft aluminium alloys dies can be made relatively cheaply from paper-reinforced synthetic resin, though zinc alloys and improved plywoods are also useful die materials. Some materials such as titanium sheet and magnesium alloy sheet are hot pressed. Then mild-steel or cast-iron dies are used together with heat-resisting rubber pads (generally protected by a 'slave' mat).

6.4.6.3 Stretch Forming (Fig. 6.23)   of panels both in the aircraft industry and in coach building is carried out principally on ductile heat-treatable aluminium alloys, though some stainless steel and titanium alloys are also stretch formed. The earliest type of machine was of the rising-table model as shown. Here the work piece is gripped between jaws and the forming block is mounted on a rising table actuated by a hydraulic ram.

As with the other cold-forming processes stresses used must exceed the elastic limit so that the work piece is deformed within its plastic range. Stretch blocks are generally of wood or compressed resin-bonded plywoods.

6.4.6.4 Spinning   a process used to produce a hollow shape by applying lateral pressure to a revolving blank so that it is forced on to a suitable former which rotates with it (Fig. 6.24). In its simplest form the equipment consists of a lathe in which the blank is held between a chuck and a tail plate, and simple forming tools which are generally bars with rounded ends. For external spinning a former is usually attached to the chuck.

The equipment required is simple and hardwood formers are cheap to

Fig. 6.23 Stretch forming.

Fig. 6.24 Hand-operated spinning.

produce. Small numbers of components can therefore be produced much more cheaply than by an equivalent deep-drawing operation. Moreover complex components of re-entrant shape can be made easily by spinning, the former (as in Fig. 6.24) then being segmented to facilitate withdrawal. Conical shapes, difficult to make by deep drawing, can also be spun.

Typical products are ornaments in copper and brass, musical instruments (including 'antique' posthorns currently being produced in large numbers in B*rm*ng*am!), cooking utensils and stainless-steel dairy utensils. Mechanised spinning has been developed in recent years under the name of 'flow turning'.

6.4.6.5 Explosion Forming   During recent years a number of techniques have been developed for the forming of metals in which the energy used for forming is released over a very short time interval. These are loosely called high-energy-rate forming—or HERF—processes. In these HERF processes energy is released at such a rate that local deformation occurs before energy can be dissipated by 'plastic waves' through the material. Mechanical efficiency is thus improved.

Energy can be released quickly either by a sudden change in pneumatic pressure; by the use of high-voltage electrical discharge; or by the use of high explosives. Thus by detonating no more than 0.1 kg of a suitable high explosive, energy can be dissipated at a rate equivalent to 6×10^6 kW.

A method of explosion forming in which water is used to transmit the shock waves is illustrated in Fig. 6.25. This is generally termed the 'stand-off' method, the charge being suspended above the blank-die assembly. Dies can be of clay or plaster for one-off jobs or of concrete, wood, resin or metal for longer 'runs'. The process is used extensively at present for prototype production or for very short runs. Obviously labour costs of the process are relatively high since skilled and responsible

Fig. 6.25 Explosion forming.

operatives must be employed to handle and prevent the loss of such dangerous explosives to terrorist organisations.

Forming from Powders

6.5 All finished shapes in thermosetting polymers and many of those in thermoplastic polymers are produced by compressing heated powders or granules. In thermosetting polymers cross-linking bonds formed during the setting process render the change irreversible and the component once formed, cannot subsequently be softened. With thermoplastic polymers van der Waals forces operating between molecules give rise to plasticity which varies with temperature.

Whilst most metals are cast either as ingots or as shaped castings, increasing use is made of powder-metallurgy processes. Initially these processes were used to deal with those metals with very high melting points and which could not be melted satisfactorily on an industrial scale for lack of a suitable high-temperature refractory. Powder metallurgy was further developed to introduce such useful features as controlled porosity and the formation of alloys from metals which are not miscible in the liquid state. The bulk of engineering ceramic products are shaped by similar processes.

6.5.1. Powder Metallurgy The principles of powder metallurgy are mentioned elsewhere (7.8) and here the technical details will be briefly described. A typical process may be divided into four stages:

(i)   production of a suitable powder and mixing it with necessary additives;

(ii) compacting the powder in a shaped die;

(iii)   sintering the resulting compact at a temperature high enough to effect continuity between the particles;

(iv)   sizing and finishing.

Powders may be produced by a number of processes. For example, tungsten powder is obtained by the chemical reduction of powdered tungsten oxide, whilst copper powder can be electro-deposited from solution. Some hard, brittle metals are powdered by grinding or pulverisation.

The powder is then compacted in a die of suitable shape (Fig. 6.26).

Fig. 6.26   Stages in the compacting of a bronze-bearing bush, from a mixture of copper and tin powders.

Pressures used may vary between about 5 N/mm^2 for soft metals like tin and 1500 N/mm^2 for very hard metals like tungsten. Dies of tool steel or tungsten carbide are generally used but even so wear on dies is severe. To reduce this wear some lubricant such as graphite, oil, zinc stearate or paraffin wax (in a solvent) is usually mixed with the powder prior to compacting it.

The compacts are then sintered by heating them to a temperature which is generally above that of recrystallisation. Diffusion and recrystallisation across adjacent particle boundaries knit the particles firmly together so that a strongly continuous structure is produced. In some cases, as in the manufacture of oil-less bearings from mixed copper and tin powders, one of the metals melts and is drawn by capillary action into the spaces between the particles of the other metal virtually 'soldering' them together. Sintering is carried out in some form of protective atmosphere which will limit surface oxidation to a minimum. Some refractory metals like platinum and tungsten are given treatments in the region of $1250°$–$2500°C$ in an atmosphere of hydrogen.

Sintering inevitably produces some distortion and alteration in size. A sizing operation is therefore used to achieve dimensional precision and improve surface finish of the component. Sizing is carried out on mechanically or hydraulically operated presses, using tools which are basically similar to the original compacting tools.

6.5.2 Forming Polymers from Powders   In most of these processes the powdered or granulated polymer is compressed in a heated mould or die, though in injection moulding the polymer powder is heated in an external compression chamber and transferred to a *cold* mould. This has the advantage that the moulding hardens quickly thus shortening the production cycle.

6.5.2.1 Extrusion   Plastics are generally extruded by means of a 'screw pump' (Fig. 6.27). The powder is carried forward by the screw mechanism

Fig. 6.27   The extrusion of plastics materials—using a 'screw pump'.

into the heated zone where it becomes soft enough to be forced through the die, the aperture of which is shaped in accordance with the required cross-section.

Man-made fibres are extruded using a multi-hole die, or *spinneret*. Wire can be coated with plastics by extrusion, the wire being passed into the die aperture in the manner of a mandrel. Plastics curtain rail is similarly extruded on to a steel strip which provides added strength.

6.5.2.2 Injection Moulding (Fig. 6.28)   This is a rapid process widely used for moulding in polythene and polystyrene. The powder is heated in the injection nozzle so that it softens and is transferred to the mould under pressure. Since the mould itself is *cold* the plastics material soon hardens and can be quickly ejected. A 'spray' of components is moulded at one 'shot'. Individual components are then broken away from the system of runners.

Fig. 6.28   Injection moulding of plastics materials.

Fig. 6.29   Compression moulding.

6.5.2.3 Compression Moulding (Fig. 6.29)   is very widely used for both thermoplastic and thermosetting polymers, but is particularly suitable for the latter. The mould must be heated in either case but must be cooled again for thermoplastic mouldings before they can be ejected. A carefully measured quantity of powder is used and provisions made to squeeze out the slight excess necessary to ensure filling of the mould cavity.

6.5.2.4 Transfer Moulding (Fig. 6.30)   is used for thermosetting polymers. Both powder chamber and die are heated and the resultant moulding allowed to remain in the mould long enough to set. This process is capable of producing rather more intricate shapes than is compression moulding.

Fig. 6.30   Transfer moulding.

7

Alloys

7.1   The first metals to be used by Man were those found 'native', namely gold and copper, and it was not until the advent of the Bronze Age that the widescale use of a metallic alloy began. It is thought that bronze was discovered accidentally by some primitive hunter who inadvertently smelted a mixed copper–tin ore in his camp fire. Nevertheless it was soon realised that this new material could be much harder and stronger than pure copper and therefore useful for many more purposes. In the modern world we use pure metals only when properties such as high electrical or thermal conductivity or high ductility are required. By alloying we are able to extend the range of useful metallic properties and often to introduce entirely new ones.

7.1.1   An alloy can be defined as a metallic solid or liquid consisting of an intimate association of two or more elements.The elements concerned mingle together on the atomic scale and whilst the principal constituents are generally metals some non-metallic elements are also important constituents of engineering alloys. Thus carbon as the alloying addition to iron forms our most important alloy, steel.

Normally a useful alloy can only be produced if the elements concerned are soluble in each other in the molten state, that is they form a single homogeneous *solution* in the crucible. Some molten metals do not dissolve in each other but instead form two separate layers—as do oil and water when an attempt is made to mix them. Thus molten lead and molten zinc will not dissolve in each other completely (unless the temperature is in excess of 798°C (8.6.1)). The lead atoms are too large relative to those of zinc to allow much solubility. Instead a layer of molten zinc (containing some dissolved lead) will float on top of a layer of molten lead (which will contain some dissolved zinc). Such a situation cannot be expected to give rise to the formation of a useful alloy. Clearly, when cast the lighter metal will tend to float to the surface before solidification is complete though there may be limited 'entanglement' of the two component metals. Often a useful material would result from a *uniform conglomerate* of particles of two insoluble metals, for example iron and copper, as a cheap bearing metal—then powder metallurgy (7.8.2 and 6.5.1) comes to our aid.

7.1.2 Generally, then, a prerequisite to the formation of a useful alloy is that all of the components of the alloy shall be mingled intimately together (Fig. 7.1(ii)) in the liquid state. Whilst an alloy frequently contains more

Fig. 7.1 The extent of liquid solubility. (i) Limited solubility in which two separate liquid layers will form but each layer containing a little of the other metal. (ii) Complete mutual solubility in which atoms of each substance are intimately mingled so that a single 'phase' is present in the container.

than two elements we shall be dealing in the main with a *binary* mixture, that is one containing *two* components. The term *component* refers in this case to a metallic element though it could equally imply a chemical compound as would be the case if we were dealing with the solution of some substance in water.

The metal which is present in the larger proportion is often referred to as the *parent metal* or *solvent*, whilst the metal (or non-metal) present in the smaller proportion is known as the *solute*. This nomenclature is used in a study of solutions generally, for example, solutions of substances in water or in organic solvents.

When the molten solution begins to solidify the affinities which different types of atoms have for each other govern the type of arrangement which results in the subsequent solid structure. Thus if two different metals are similar in chemical properties so that they have little chemical affinity for each other, they will, provided their atoms are of similar order of size, 'coexist peacefully' forming mixed crystals, or what is termed a *solid solution*. If, on the other hand, they are rather different in respect of chemical characteristics they may attract each other to the extent that they form a type of chemical compound—often referred to as an *intermetallic compound*. In a situation where unlike atoms attract each other less than like atoms, the two types tend to separate into different crystals which meet at a mutual grain boundary. A heterogeneous mixture of this type is termed a *phase mixture*. The *eutectic* structure found in many alloys is such a mixture.

7.1.3 On the face of it the way two metals behave when we attempt to alloy them seems to result in a very hit-or-miss situation. In fact nothing could be further from the truth and the type of structure produced is a function of:

(i)   the relative positions of the two metals in the Periodic Classification (Table 1.3)—that is the number of valency electrons in their respective outer shells;

(ii)   the sizes of the two atoms relative to each other. These influences will become apparent in this chapter and the one which follows.

Solid Solutions

7.2   A fundamental characteristic of all solid solutions is that the complete intermingling of the atoms of both metals which prevail in the liquid solution is retained in the solid state. The *range* of compositions can vary in both liquid and solid solutions. Some materials are soluble in each other in all proportions, others only partially soluble one in the other.

7.2.1   Alcohol and water furnish an example of two liquids which show complete mutual solubility in all proportions. Thus beer is, alas, a relatively weak solution of alcohol in water whilst whisky and vodka are relatively strong solutions of alcohol and water. In all alcoholic beverages a

Fig. 7.2   Cored dendrites of solid solution in a 60Ni–40Cu alloy which has been chill cast. The dendrite cores (light) are nickel-rich whilst the outer fringes (dark) are copper-rich. (Etched in Acid ferric chloride solution.) ×80.

single phase is present in the bottle. The parallel case in metallic solid solutions is illustrated by the metals copper and nickel. They are said to be *completely miscible* as solids since they form single-phase alloys in all proportions. That is, whatever the composition of a copper–nickel alloy, its microstructure will consist of crystals of one type only, containing copper and nickel atoms intimately mingled. Only a limited number of pairs of metals are completely mutually soluble in this way. They include, in addition to copper and nickel, gold–silver; gold–platinum and antimony–bismuth (7.3.1).

7.2.2 Some liquids are partially soluble in each other. Thus water will dissolve a small amount of acetone giving a single liquid solution rich in water, whilst acetone will similarly dissolve a small amount of water giving a single liquid solution rich in acetone; but if approximately equal volumes of water and acetone are shaken together, two separate layers will remain. The upper layer will be acetone *saturated* with water and the lower layer will be water *saturated* with acetone. The compositions of these two layers will be different, the upper being acetone-rich and the lower water-rich. A great many pairs of metals show limited solid solubility in parallel manner. They are said to be *partially miscible*.

7.2.3 A very few pairs of metals mutually soluble as liquids, become completely insoluble as solids. Even then it is possible that some very slight solid solubility exists but is so small as to be difficult to measure.* Thus cadmium and bismuth which form a single homogeneous liquid solution in all proportions, separate out, to all intents and purposes, completely on solidification so that the microstructure contains two phases—crystals of pure cadmium and crystals of pure bismuth.

7.2.4 Solid solutions are either *substitutional* or *interstitial* (Fig. 7.3). Whilst atomic size is not the only criterion governing the type of solid

Fig. 7.3 Types of solid solution: (i) Disordered substitutional. (ii) Ordered substitutional. (iii) Interstitial.

solution to be formed it is clear that interstitial solid solutions are generally produced only when the solute atom is small compared with the solvent atom. The solute ion is then able to fit into the *interstices*, or gaps, between the sites occupied by the solvent ions. Whilst an interstitial solid solution

---

*In a similar manner we often loosely refer to substances as being insoluble in water, whereas in fact they are very sparingly soluble. Most substances have some degree of solubility in water.

may be formed during the solidification of an alloy it will be apparent that such a solution can also form in a parent metal which is already solid, by 'infiltration' of the spaces between its ions. Non-metallic elements of small atomic size are able to dissolve in this way in a number of metals. Thus steel can be carburised by permitting carbon atoms to diffuse into the surface of FCC iron (9.9.1), whilst it can be nitrided by allowing nitrogen atoms to be absorbed by BCC iron (9.9.2). On the debit side, hydrogen embrittlement is suffered by a number of metals since the very small hydrogen atom can be absorbed interstitially by the lattices of many solid metals.

In substitutional solid solutions atoms of the solute metal take up sites in the lattice structure of the solvent metal. This substitution can be either *ordered* or *disordered*. In the former, atoms of the solute metal take up certain fixed positions (Fig. 7.3(ii)). Such solid solutions tend to be hard and brittle, whilst those of the disordered type are tough and ductile and therefore more useful as constituent phases of engineering alloys. Fortunately the greater proportion of substitutional solid solutions are of the disordered type.

7.2.5 The formation of solid solutions in metallic alloys is to be expected since atoms in metals are held together principally by the mutual attraction between their positive ions and the common 'electron cloud' (3.5) permeating them. Such a bond is relatively indifferent both to the exact proportions of the different types of component ions and also their precise distribution in the crystalline pattern. Variable solid solutions of wide ranges of comparison are therefore common.

7.3 We are indebted to Prof. W. Hume-Rothery of Oxford more than to any other metallurgist for our fundamental knowledge of the structure of alloys. He concluded that the following factors have the greatest influence on the formation of substitutional solid solutions:

(i)   *atomic size*   the greater the difference in size between the atoms of the two metals involved the smaller will be the range over which they are soluble. If atom diameters differ by more than 15% of that of the solvent metal then solid solubility is generally extremely small but if diameters differ by less than 7%, then, other things being equal, they will be likely to form solid solutions in all proportions. Any difference in atomic size will of course produce some strain in the resultant crystal structure;

(ii) *electrochemical properties*   the ions of all metals are electropositive but some are more electropositive than others. Generally, the greater the difference in electropositivity the greater will be the chemical affinity of one ion for another so that they will tend to form a *compound* rather than a solid solution. If on the other hand their electropositivities are similar then, other things being equal, they will very probably form a solid solution;

(iii) *valency*   a metal of lower valency is more likely to dissolve one of higher valency than *vice versa* always assuming that other conditions are favourable. This holds true particularly for alloys of the monovalent metals copper, silver and gold with many metals of higher valency.

7.3.1 From the above it can be concluded that two metals are most likely to form substitutional solid solutions over a wide range of compositions if their atomic sizes are about equal and if their electrochemical properties are similar. These conditions are fulfilled when two metals are very close together in the periodic classification of the elements—usually side by side in the same period (nickel and copper) or one above the other in the same group (silver and gold). Also if singly they crystallise in the same pattern this will assist simple disordered substitution. Pairs of metals which fulfil these conditions and dissolve in each other in all proportions in the solid state include: Cu–Ni; Ni–Pd; Mo–W; Ir–Pt; Fe–Cr; Fe–Ni; Au–Ag; Au–Pt; Ag–Pd; Mo–Ti; Se–Te; Re–Ru; Nb–Ti; K–Rb; Cr–W; Co–Pd; Bi–Sb; As–Sb and Ti–Zr. Many of these—but not all—are FCC pairs of elements.

7.3.2 The valency effect is related to the ability of the metallic bond to accept changes in the electron concentration as compared with the rigid valency laws associated with covalent compounds. If, say, aluminium is dissolved in copper the electron concentration is increased since, whilst each copper atom contributes one electron to the common 'cloud' each aluminium atom will contribute three electrons. Solubility of aluminium in copper reaches a limit with about 20% atoms of aluminium—that is, about 1.4 valency electrons per atom.* This ratio holds good for the solubility of many other metals (of different valencies) in copper.

7.3.3 The atomic size factor plays a significant part in the formation of many alloys of industrial importance. Thus both copper and iron are roughly in the middle of the metals as far as size factor is concerned. Consequently they will dissolve many metals in fairly large quantities, so that copper is the basis of brasses, bronzes and cupro-nickels; whilst iron will dissolve large amounts of nickel, chromium, manganese, tungsten, vanadium and cobalt giving a wide range of alloy steels.

Primary substitutional solid solutions based on FCC metals are very suitable for use as wrought alloys since they can be shaped by plastic deformation because of their high ductilities. The process also increases strength and hardness. When solubility decreases with fall in temperature to a very low value this may form the basis for *precipitation hardening* treatment (7.5.2).

7.3.4 Whilst a pure metal solidifies at a single fixed temperature a solid solution solidifies over a range of temperatures (8.6.3.1). As crystallisation

---

*In every 100 atoms, 80 valency electrons will be provided by 80 atoms of copper and 60 valency electrons by 20 atoms of aluminium.

commences there is a tendency for the metal with the higher melting point to solidify more quickly than that of lower melting point. Consequently the core of the resultant crystal will contain rather more atoms of the metal of higher melting point whilst the outer fringes and boundary regions of the crystal will contain a larger proportion of atoms of the metal of lower melting point. This change in composition from core to outer fringes reveals the dendritic structure of such an alloy when a prepared section is examined under the microscope. The difference in composition between the core of the crystal and its boundary regions is referred to as *coring* (Fig. 7.4), and is most pronounced in solid solutions which have solidified rapidly and cooled quickly to room temperature. Slow cooling on the other hand permits diffusion to occur so that coring is totally or partially lost.

7.3.5 The presence of solute atoms in a lattice structure imposes strain. This strain will be at a minimum if the solute atoms are evenly dispersed throughout the crystal. Hence there is a tendency for diffusion to occur until the concentration gradient of solute atoms within the crystal is zero. If rapid cooling of the alloy has resulted in extensive coring then prolonged annealing at a temperature just below the freezing range will generally supply sufficient thermal activation to permit diffusion and the dispersion of coring.

Diffusion in Solid Solutions

7.4 It is a feasible assumption that diffusion of interstitial solute atoms will take place with ease since the relatively small atoms involved can move easily through the structure of the parent metal. Carburising and nitriding have already been mentioned in this context. We can only attempt to explain diffusion in substitutional solid solutions, however, if we postulate the presence of *vacant sites* (or *vacancies*) (5.3) in the structure. We have already seen that the existence of dislocations explains some of the mechanical properties of metals and this is a further instance where an otherwise unexplained physical property relies on the speculation that metallic crystals contain many imperfections (5.3).

7.4.1 Since the presence of a solute atom imposes strain it is reasonable to suppose that a solute atom and a vacancy would migrate to each other and so reduce strain to a minimum. The associated solute atom and vacant site could then migrate through the structure, each movement of one site spacing being achieved by a series of 'moves' as indicated in Fig. 7.5. The rate at which diffusion takes place depends presumably upon the availability of vacant sites. It also depends upon the concentration gradient of the solute atoms within the structure of the parent metal and upon the temperature at which diffusion is taking place.

In 1855 Adolph Fick enunciated laws governing the diffusion of substances generally on a quantitative basis. Here we are concerned with the application of these laws to metallic solid solutions.

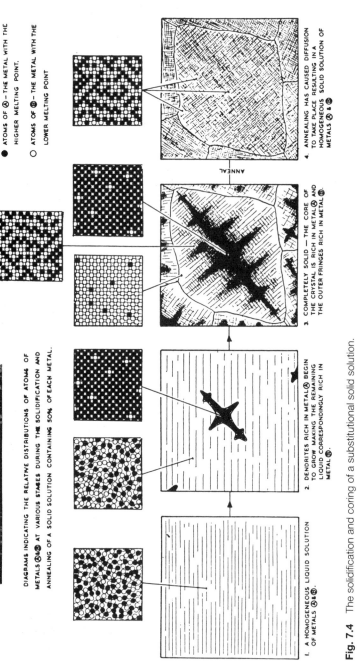

THE CORING OF SOLID SOLUTIONS.

DIAGRAMS INDICATING THE RELATIVE DISTRIBUTIONS OF ATOMS OF METALS Ⓐ&Ⓑ AT VARIOUS STAGES DURING THE SOLIDIFICATION AND ANNEALING OF A SOLID SOLUTION CONTAINING 50% OF EACH METAL.

● ATOMS OF Ⓐ – THE METAL WITH THE HIGHER MELTING POINT.

○ ATOMS OF Ⓑ – THE METAL WITH THE LOWER MELTING POINT

1. A HOMOGENEOUS LIQUID SOLUTION OF METALS Ⓐ&Ⓑ.

2. DENDRITES RICH IN METAL Ⓐ BEGIN TO GROW MAKING THE REMAINING LIQUID CORRESPONDINGLY RICH IN METAL Ⓑ

3. COMPLETELY SOLID — THE CORE OF THE CRYSTAL IS RICH IN METAL Ⓐ AND THE OUTER FRINGES RICH IN METAL Ⓑ.

4. ANNEALING HAS CAUSED DIFFUSION TO TAKE PLACE RESULTING IN A HOMOGENEOUS SOLID SOLUTION OF METALS Ⓐ & Ⓑ

ANNEAL

Fig. 7.4 The solidification and coring of a substitutional solid solution.

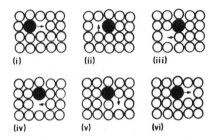

Fig. 7.5   The diffusion of a substitutional solute atom associated with an adjacent vacant site.

7.4.2   *Fick's First Law* states that the amount (J) of material moving across a unit area of a plane in unit time is proportional to the concentration gradient (dc/dx) at the same instant but of opposite sign, i.e.

$$J = -D \cdot \left(\frac{dc}{dx}\right)$$

(This assumes that the x-axis is parallel to the direction in which the concentration gradient is operating.)

Fig. 7.6   AB is the initial concentration gradient (at A) and CD the final gradient ($dc/dx = 0$) after a long time during which diffusion was in progress.

The constant D is the *diffusion coefficient* and varies with different solid solution systems and other prevailing conditions such as temperature. It is expressed in units m²/s, i.e.

$$\frac{atoms}{m^2/s} = -D \cdot \left(\frac{atoms/m^3}{m}\right) \qquad \left[J = -D \cdot \left(\frac{dc}{dx}\right)\right]$$

or $-D \equiv m^2/s$

The negative sign indicates that movement is in a down-gradient direction. When $dc/dx = 0$, $J = 0$ which means that when the 'driving force' provided by the concentration gradient reaches zero (CD in Fig. 7.6 is horizontal) movement of atoms will cease.

Fick's Law is of the same general type which governs for example, the flow of heat along a metal bar. In fact Fick's Law was really an extension of observations made by Fourier in 1822. Ohm's Law is also of this type and deals with the flow of electrons from a region of high concentration to one of lower concentration where the driving force is provided by the mutual repulsion between like-charged electrons which can travel easily through the lattice structure of a metal. The flow of *ions* as governed by Fick's Law generally requires thermal activation since lattice distortion caused by the presence of solute atoms does not give rise to sufficient energy to permit diffusion unaided. Moreover, as we have seen (7.4.1), ions (atoms) are not able to move easily through the lattice of the solvent metal.

The situation dealt with by Fick's Law is one where the concentration gradient remains *constant* thus providing a steady driving force and consequently *steady-state* diffusion. In practice this situation rarely obtains and generally the concentration gradient at a plane alters as soon as diffusion begins so that both concentration gradient and atom flow change with time so that *transient-state* diffusion prevails.

7.4.3 *Fick's Second Law* (also known as the *Diffusion Transport Equation*) deals with transient-state diffusion. It derives from Fick's First Law and from the fact that matter is conserved within the system. It relates the change in concentration with time ($\partial c / \partial t$). It can be expressed:

$$\frac{\partial c}{\partial t} = \frac{\partial}{\partial x}\left(D\frac{\partial c}{\partial x}\right)$$

which will delight the hearts of *aficionados* of the differential calculus. The above expression deals with the general case but if D refers to a specific case and is constant then:

$$\frac{\partial c}{\partial t} = D\frac{\partial^2 c}{\partial x^2}$$

From this it can be seen why the final stages of diffusion during a homogenisation annealing process are very slow indeed. The rate of diffusion decreases as the concentration gradient diminishes and even after prolonged annealing a trace of coring may be visible in the microstructure of the alloy.

We can write Fick's Second Law in the form:

$$\frac{\partial C_x}{\partial t} = D\cdot\frac{\partial^2 C_x}{\partial x^2}$$

where C_x is the concentration of a diffusing species of atom at some distance x from a surface whilst t is the time (seconds).

The usual mathematical solutions to differential equations of this form may be used* but assuming that the concentration of C_s of the diffusing

---

*Dab hands with the calculus may wish to pursue this further by referring to *The Mathematics of Diffusion* by J. Crank (Oxford University Press, 1975). This book deals with mathematical solutions to differential equations.

species at the surface remains constant—and this is common when dealing with the penetration by gases or small atoms like those of carbon into a solid metal surface as in carburising (9.9.1)—then the following solution can be used:

$$\frac{C_s - C_x}{C_s - C_0} = erf\left(\frac{x}{2\sqrt{(Dt)}}\right)$$

This is the *Grube* solution to the Diffusion Transport Equation in which C_0 is the initial concentration of the diffusion species (when $t = 0$). 'Erf' refers to the 'error function' involved in such a solution and is tabulated in some mathematical tables (a limited range of values of the error function is given in Appendix I for corresponding values of the independent variable, y).

When values are such that:

$$erf\ y \simeq y$$

then:

$$\frac{C_s - C_x}{C_s - C_0} \simeq \frac{x}{2\sqrt{(Dt)}}$$

and this expression often provides values which are sufficiently accurate for many purposes.

Q A steel plate of uniform carbon content 0.6% was inadvertently exposed to a strongly oxidising atmosphere for 10 minutes at 1000°C.

(i) What will now be the carbon content at a depth of 0.25 mm below the surface?

(ii) At what depth will the carbon content have fallen to 0.35%?

(The diffusion coefficient (D) is 3.10×10^{-11} m²/s at 1000°C.)

A (i) The presence of the strongly oxidising atmosphere at the surface means that the carbon content there is 0. Using the Grube solution of the equation:

$$\frac{C_s - C_x}{C_s - C_0} = erf\frac{x}{2\sqrt{(Dt)}}$$

$x = 0.25$ mm
$= 0.25 \times 10^{-3}$ m
$t = 10$ minutes
$= 600$ s

or:

$$\frac{0 - C_x}{0 - 0.60} = erf\frac{0.00025}{2\sqrt{3.1 \times 10^{-11} \times 600}}$$

$$= erf\ (0.917)$$

When $y = 0.917$, erf $y = 0.8055$ (extrapolating values from Appendix I). Hence:

$$\frac{0 - C_x}{0 - 0.60} = 0.8055$$

Rearranging:

$$-C_x = -0.60 \times 0.8055$$

and:

$$C_x = 0.48\% \text{ carbon}$$

(ii) Since in this case C_x is 0.35% carbon at a distance x below the surface:

$$\frac{0 - 0.35}{0 - 0.60} = \text{erf} \frac{x}{2\sqrt{(3.1 \times 10^{-11} \times 600)}}$$

$$\frac{-0.35}{-0.60} = \text{erf}\left(\frac{x}{2(1.36 \times 10^{-4})}\right)$$

$$0.5833 = \text{erf}\left(\frac{x}{2.72 \times 10^{-4}}\right)$$

When erf $y = 0.5833$, $y = 0.574$. So:

$$0.574 = \frac{x}{2.72 \times 10^{-4}}$$

and

$$x = 1.56 \times 10^{-4} \text{ m}$$

$$= 0.156 \text{ mm}$$

A further exercise involving the Diffusion Transport Equation will be found in Chapter 9 (9.9.1).

7.4.4 Diffusion in a mixture of gases takes place without any significant energy input to the system since all atoms (or molecules) are in a state of motion. With solid metals, however, an input of thermal energy is required to cause an atom to 'jump' out of its lattice position and move to a new position. If E_d is the energy required to produce that 'jump' then it can be shown that:

$$D = D_0 e^{-E_d/RT}$$

or $D = D_0 \exp(-E_d/RT)$

where D_0 is a constant. This Arrhenius type equation indicates that in solids diffusion rates are influenced considerably by temperature whilst in gases, where there is virtually no energy barrier ($E_d \approx 0$), diffusion coefficients are generally independent of temperature, i.e. $D \approx D_0$.

Strengthening Mechanisms Dependent on Solid Solubility

7.5 The most common reason for alloying is to increase the strength of a metal and generally the required properties result from the formation of a solid solution. The presence of solute atoms will impede the movement of dislocations and so increase the yield stress of the material.

7.5.1 'Solute Atmospheres' In any metal the presence of a dislocation causes distortion of the lattice structure near it. In a solid solution this distortion—and the strain energy associated with it—can be reduced by the presence of solute atoms. Thus, if the solute atoms are larger than those of the parent metal they will reduce strain in the lattice if they take up sites where the lattice is in tension (Fig. 7.7(i)). If on the other hand the solute

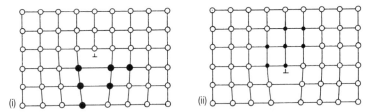

Fig. 7.7   The 'pinning down' of dislocations by 'solute atmospheres'.

atoms are smaller than those of the parent metal they will reduce strain if they occupy sites where the lattice is in compression (Fig. 7.7(ii)). The necessary migration of solute atoms to suitable sites to produce this situation will occur during solidification or during an annealing process which will permit diffusion. Such an array of solute atoms in the region of a dislocation is often referred to as a *solute atmosphere* or *Cottrell atmosphere* after Sir Alan Cottrell the eminent scientist who in his chosen discipline of metallurgy did so much to establish dislocation theories.

7.5.1.1 Having thus migrated to these positions the solute atoms, whether large or small, will effectively 'pin down' the dislocation and oppose its movement since any attempt to move the dislocation away from the solute atmosphere will increase considerably the distortion and strain energy of that region of the lattice. This effect is indicated in Fig. 7.8 which

Fig. 7.8   Distortion of the lattice caused by the presence of a residual solute atmosphere.

shows the distortion remaining when the dislocation has moved away from the solute atmosphere in each case. In this way solute atmospheres increase the yield stress of an alloy. Annealing, following a cold-deformation process, will permit solute atoms to align themselves once more near to a dislocation and so produce a new solute atmosphere.

7.5.1.2 In interstitial solid solutions the small solute atoms will occupy positions where the lattice is in tension (Fig. 7.9) since interstitial gaps will

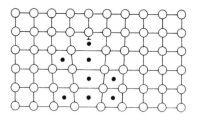

Fig. 7.9   Interstitial solute atmosphere, e.g. carbon in BCC iron.

be larger in these regions. This hypothesis affords a convincing explanation of the yield-point effect in mild steel and did much to help establish the dislocation theory at a time when very limited evidence was available.

In Fig. 7.10 the yield point A corresponds to the stress required to move dislocations away from their associated solute atmospheres of interstitial

Fig. 7.10   The 'yield point effect' in mild steel.

carbon atoms by which they were pegged down. The stress B is that necessary to keep the dislocations moving once they have been separated from their solute atmospheres. (The stress required for slip to continue subsequently increases as the steel work-hardens, that is, dislocations become jammed against each other or against other obstacles.)

7.5.2 Coherent Precipitates   If salt is dissolved in water in increasing amounts a *saturated* solution is ultimately formed and no more salt will dissolve at that temperature. As the temperature is raised the solution becomes unsaturated and so more salt will dissolve until saturation is again reached. When the temperature is allowed to fall the solution tends to become super-saturated so that solid salt is precipitated from the solution as small crystals.

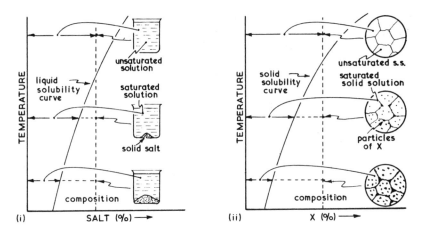

Fig. 7.11   The analogy between limited liquid solution (i) and limited solid solution (ii).

7.5.2.1   A similar relationship usually prevails in metallic *solid* solutions in cases where solubility is limited. That is, the solid solubility of one metal in the other increases as the temperature rises (Fig. 7.11). If, however, the temperature is allowed to fall quickly, precipitation cannot occur as easily as in the case of liquid solutions since diffusion in solids is very slow as compared with liquids. Consequently a supersaturated solid solution may be retained at room temperature. This supersaturated solid solution will not be in a state of equilibrium and the tendency for the solute to precipitate will remain. Whether or not precipitation does occur will depend upon the degree of thermodynamical imbalance in the system. In many cases the supersaturated solid solution will be retained completely and indefinitely whilst in others the solute will separate out as isolated particles visible at high magnifications under the microscope. There is, however, an intermediate stage, where, by means of diffusion, solute atoms begin to congregate into groups so that their concentration in some regions is greater than it was initially (Fig. 7.12). Nevertheless they still occupy positions within the lattice of the parent metal and the lattice is still *continuous* or *coherent*, albeit distorted by the presence of the high concentration of solute atoms. Such an arrangement of solute atoms is termed a *coherent precipitate* and is known to have a considerable strengthening effect on the structure presumably because it will oppose the movement of dislocations.

7.5.2.2   If precipitation continues the solute atoms ultimately leave the lattice of the parent metal to form their own particular lattice pattern. Since this lattice does not 'match' that of the parent metal the structures are no longer coherent so that a grain boundary forms between the precipitated particle and the 'mother' solid solution (Fig. 7.12(iv)). Thus a *non-coherent precipitate** is produced and will be observed as separate

*Some writers describe these as '*incoherent precipitates*'.

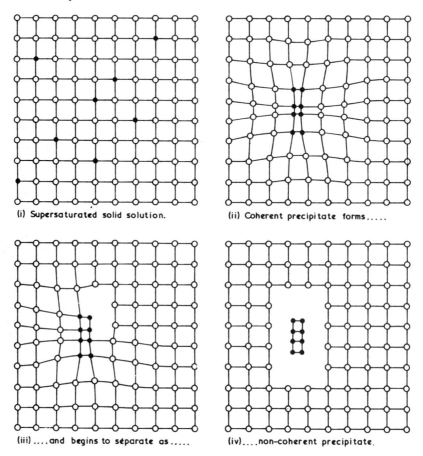

(i) Supersaturated solid solution.

(ii) Coherent precipitate forms.....

(iii) ....and begins to separate as .....

(iv)....non-coherent precipitate.

Fig. 7.12 This attempts to illustrate in simple terms precipitation from a *supersaturated* solid solution. First, solute atoms (i) migrate to form a coherent precipitate which is continuous with the lattice of the parent metal (ii). The solute atoms ultimately begin to leave the lattice of the parent metal (iii) and form crystals of non-coherent precipitate (iv) now separated from the parent metal at a grain boundary. (In practice this precipitate will rarely be of pure metal as shown in this simplified diagram. Generally the solute atoms will take with them some atoms of the parent metal to form crystals of an intermetallic compound or another solid solution.)

small crystals, even though they are so small as to require an electron microscope to render them visible. When non-coherent precipitates have been formed the strength of the alloy falls considerably since large numbers of solute atoms have been lost from the initial solid solution. Dislocations will therefore be able to move more easily.

In Fig. 7.12(iv) the non-coherent precipitate has been depicted, for the sake of simplicity, as being of a pure element. In practice it is much more likely to be either an intermetallic compound or another solid solution. That is, it will take with it some atoms of the parent metal to produce a crystal structure which is not coherent with that of the original solid solution.

Solid Solutions in Non-metallic Materials

7.6   Solid solutions are formed in many ionic compounds. Thus mixed crystals can be grown involving those inorganic salts which have geometrically similar crystal forms. Such salts are said to be *isomorphous* and are typified by the group known as 'alums'. These are double sulphates of sodium, potassium, aluminium and ammonium.

However, not all ionic solid solutions are derived from isomorphous salts. Thus nickel oxide, NiO, and magnesium oxide, MgO, dissolve in each other in all proportions in the solid state whilst silver bromide and silver iodide, though of very different crystal structures will nevertheless form solid solutions. Partial solid solubility is common with ionic compounds as it is with metallic solid solutions.

7.6.1   Although organic polymers are not necessarily crystalline in form an analogy can be drawn between co-polymers (12.2.1) and metallic solid solutions. In these co-polymers different 'mers' (12.1.3) are linked together by covalent bonds which correspond to the free electron bonds in metallic solid solutions. By using different mers to build up a polymer chain materials of modified mechanical properties can be produced. The extremely tough plastics material ABS—a co-polymer of acrylonitrile, butadiene and styrene mers—is such a substance. Whilst these co-polymers are *not* solid solutions the use of different mers in the polymer chain modifies the properties as does the introduction of solute atoms into a ductile metal.

Intermediate Phases

7.7   These are phases which in general have different lattice structures from those of the parent metals and are formed in the intermediate ranges of composition. Several different forms of intermediate phase occur.

7.7.1 Secondary Solid Solutions   Sometimes solid solutions are formed in intermediate ranges of composition which do not include the pure metals as components. Generally they have crystal patterns which are different from those of the component metals and are termed *secondary solid solutions*. Except that the range of compositions does not include one of the pure metals, the secondary solid solution is not otherwise very different in properties from ordinary primary solid solutions. In some secondary solid solutions the distribution of atoms amongst the available sites follows a disordered arrangement with a fairly wide range of solubility. In others, the atoms are in an ordered pattern and in some respects resemble intermetallic compounds more than disordered secondary solid solutions.

7.7.2 Size-factor Compounds   are those in which composition and crystal structure adjust themselves in order to permit component atoms to pack themselves together in a coherent pattern. This state of affairs is typified by the *Laves phases* in which the general composition is based on the formula AB_2, e.g. $MgCu_2$ (cubic); $MgZn_2$ and $MgNi_2$ (both hexagonal). These compounds form mainly because of a substantial difference in size between the component atoms. This size difference is of the order of 30% and allows the component atoms to pack together in a crystal pattern of higher coordination number (3.6.5) than the maximum (twelve) which is possible in lattices of close-packed equal spheres. In the formula AB_2 each 'A' atom has sixteen neighbours (4A + 12B) whilst each 'B' atom has twelve neighbours (6A + 6B), giving an average coordination number of 13.33. These high coordination numbers are made possible by the presence of the free-electron bond in metallic structures.

7.7.3 Intermetallic Compounds　These fall into two separate groups.

7.7.3.1 Valency Compounds　When one metal is very strongly electropositive and the other only weakly electropositive they may combine to form a compound in which the ordinary rules of chemical valency apply. Thus, the strongly electropositive metals, towards the left-hand side of the Periodic Table, form compounds such as these with the very weakly electropositive metals adjacent to the non-metals on the right-hand side of the Table, e.g. Mg_3Sb_2, Mg_2Sn and Mg_2Pb. In these compounds the range of solubility of any excess metal is generally very small and often they have simple lattice structures like ionic crystals (3.2.2). Sometimes such a compound is so stable that it will have a higher melting point than either of its parent metals. Thus Mg_2Sn melts at 780°C whereas the pure parent metals magnesium and tin melt at 650°C and 232°C respectively.

These appear to be true chemical compounds whereas the other intermediate phases mentioned here are virtually transitional between these compounds and primary solid solutions since in all other cases atoms are held in position by the free electrons of the metallic bond.

7.7.3.2 'Electron Compounds'　or Hume–Rothery compounds—in which the normal valency laws are not observed but in which there is often a fixed ratio between the total number of valency electrons of all the atoms involved and the total number of atoms in the 'molecular formula' of the intermetallic compound in question. There are three such ratios:

Ratio	Type of structure	Representative phases
(i) 3/2 (21/14)	BCC (β-brass type)	$CuZn$; Cu_5Sn; Cu_3Al; $CuBe$; Ag_3Al.
(ii) 21/13	Complex cubic (γ-brass type)	Cu_5Zn_8; $Cu_{31}Sn_8$; Cu_9Al_4; Ag_5Zn_8.
(iii) 7/4 (21/12)	CPH (ε-brass type)	$CuZn_3$; Cu_3Sn; $CuBe_3$; Ag_5Al_3.

Thus in the compound $CuZn$ the copper atom provides one valency electron whilst the zinc atom provides two, giving a total of three valency electrons to two atoms, i.e. a ratio of 3/2. Similarly in the compound $Cu_{31}Sn_8$, each copper atom contributes one electron and each tin atom four electrons. Hence 31 electrons are contributed by copper and 32 (4×8) by tin making a total of 63 electrons. Altogether 39 atoms are involved in the molecular formula $Cu_{31}Sn_8$. Hence the ratio is 63/39 or 21/13.

These Hume–Rothery ratios have been useful in relating structures which apparently had little else in common.

7.7.4 In general intermetallic compounds are less useful as a basis for engineering alloys than are solid solutions. In particular intermetallic compounds are usually very hard and brittle, having negligible strength or ductility. Often the compound has none of the physical characteristics of its parent metals. Thus $Cu_{31}Sn_8$ is a pale blue phase as seen under the microscope, and a small ingot of it will crumble to powder under quite gentle pressure in the jaws of a vice. The unexpected colour is due to the different lattice characteristics as compared with those of the parent metals.

Since intermetallic compounds are of fixed compositions, and will generally dissolve only small amounts of their parent metals, they do not exhibit coring when viewed under the microscope. This fact—and, often, unusual colour—helps to identify them microscopically. Moreover the lack of variability in composition imposed by the formation of compounds gives a possible explanation of the brittleness of these materials. Because crystals of *fixed composition* are formed there are no amorphous regions at the grain boundaries and hence a lack of grain-boundary cohesion (3.7.3). Thus the compound Ni_3Al has an ordered FCC structure so that whilst a *single crystal* of Ni_3Al is extremely ductile a polycrystalline mass of Ni_3Al is very brittle and fails *along the grain boundaries*. This can be overcome by adding 0.05% boron which raises the % elongation from 0 to over 50%. This is possibly due to the observed segregation of boron at the grain boundaries causing disorder there and allowing the pile-up of dislocations to be relieved by slip *across* boundaries rather than by cracking.

7.7.5 Very often, small amounts of these intermetallic compounds are present as essential phases in engineering alloys. For example, a bearing metal must be hard and wear-resistant whilst at the same time be tough and ductile. These are diverse properties not to be encountered in a single-phase alloy since, whilst primary solid solutions are tough and ductile they are relatively soft and do not resist wear and, on the other hand, whilst intermetallic compounds are hard and wear-resistant with a low coefficient of friction, they are weak and brittle. The metallurgist, however, formulates an alloy which will contain small isolated particles of intermetallic compound held firmly in a tough ductile matrix of solid solution. The latter wears leaving the particles of intermetallic compound standing proud so that they carry the load at a very low coefficient of friction. Thus bearing bronzes (10–15% Sn) contain particles of $Cu_{31}Sn_8$ embedded in a matrix of primary solid solution α (11.7.2). Similarly 'white' bearing metals (Babbitts) contain crystals of the intermetallic compound SbSn embedded in a matrix of antimony–tin solid solution (11.7.1).

It was mentioned earlier that some of these compounds have higher melting points than their parent metals. Thus NiAl melts at 1640°C and is being used as a coating for turbine blades since, in addition to this high melting point, the film of Al_2O_3 which forms on the surface acts as a barrier (19.2.1) to further oxidation.

7.7.6 Interstitial Compounds   formed when the solubility of an interstitially-dissolved element (7.2.4) is exceeded. Then, an intermediate phase is precipitated in which the former solute atoms are again held in interstitial positions, but now in a compound often of more complex crystal structure. Thus, under normal conditions, BCC iron will hold no more than 0.02% carbon in interstitial solid solution and when this amount is exceeded the interstitial compound Fe_3C (*Cementite*) is precipitated.

These intersitial compounds are generally the carbides, nitrides, hydrides and borides of the transition metals, e.g. TaC, TiN, TiH_2, W_2C and WC. Most of them have metallic properties. They include the hardest and most refractory of substances with melting points as high as 3500°C (NbC) and 3800°C (TaC). Some of the carbides are extremely important and form the basis of high-speed cutting tools (WC, W_2C, TiC, MoC and TaC) (10.4.1). Since many of these interstitial compounds are strong at high temperatures they are used in a finely-dispersed form to strengthen many heat-resisting steels (10.5.2) and alloys (11.5.2).

Powder Metallurgy

7.8   It is sometimes desirable to make an 'alloy' of two metals which are either completely or partially *in*soluble in each other in both the solid *and liquid* states. This is the oil–water concept mentioned earlier. Obviously any attempt to make an alloy by the usual procedure of melting and casting will fail since the two metals will form separate layers during the fusion process. A wide dispersion of the two metals can be produced however by mixing them thoroughly as extremely fine powders. The powder mixture is then 'compacted' in a closed die using a pressure which varies between 4.0 and 800 N/mm^2 depending upon the strength and ductility of the materials being used. After such treatment the compacted shape is strong enough to permit handling since some adhesion will have taken place between individual particles providing that their surfaces are clean and unoxidised. Metal-to-metal contact under high pressure apparently leads to some degree of 'cold-welding' between sliding adjacent surfaces.

7.8.1   The compact is then 'sintered', that is, heated to some temperature generally below the melting points of both metals (but sometimes above that of one of them as in the case of sintered-bronze bearings where the tin powder melts and infiltrates the copper particles). This treatment produces a mechanically strong material as the particles bond together across the interface where only limited adhesion was effected by compacting. In some cases this bonding is probably due to grain-growth across the cold-worked interfaces but it is more generally agreed that diffusion is the principal mechanism by which surfaces knit together.

7.8.2 Initially, powder metallurgy was used to produce ingots of refractory metals like platinum and tungsten which have such high melting points that they cannot be melted conveniently. Later the process was used as a means of 'alloying' metals, which, as mentioned above, do not dissolve in each other in the liquid state. The controlled degree of porosity possible with compacted powders is also utilised in such components as stainless-steel filters for corrosive liquids, and, of course, oil-less bronze bearings. Useful low-cost oil-less bearings can also be made from mixtures of copper and iron powders since these two metals are only partially soluble in each other in either the liquid or solid states.

As the relative costs of metallic powders continue to fall, powder metallurgy will be used more and more as a standard production method. The negligible amount of process scrap as compared with casting operations will make the method competitive for uses other than the rather specialised ones outlined above.

8

Phase Equilibrium

8.1 In a study of metallic alloys, or indeed other materials which contain more than one component, a most important concept is that concerning the term 'phase'. A *phase* can be defined as a *homogeneous body of matter existing in some prescribed physical form*. Thus the chemical substance water, H_2O, can exist in three different phase forms—solid, liquid and vapour; whilst pure iron can exist as four different phases (Table 8.1). (Note that there are only two different solid phases in iron since α and δ are identical. The terms α and δ are only used to differentiate between the temperature ranges over which BCC iron can exist.)

Table 8.1   *Phases in Pure Iron*

Temperature range (°C) over which the phase is stable at 760 mm pressure	Phase
Above 3070	gas
1536 to 3070	liquid
1400 to 1536	BCC solid (δ)
910 to 1400	FCC solid (γ)
Below 910	BCC solid (α)

8.1.1 Many solid substances are *multiphase* in character. For example, ordinary concrete contains cement, sand and aggregate as identifiable phases in the structure whilst, on the microscopical scale, the structure of steel consists of two phases—a solid solution we call ferrite, and iron carbide, Fe_3C. The various types of phase likely to be encountered in metallic alloys were discussed in Chapter Seven. These phases can be either pure metals, intermetallic compounds or solid solutions for, despite the effects of coring, a solid solution is a single continuous phase which will become homogeneous if it is allowed to reach thermodynamic equilibrium by a process of diffusion. In any multiphase structure two phases have a common boundary separating them. Thus, a drop of oil placed in water shares a common phase boundary with the water which surrounds it.

The Thermodynamic Basis of Phase Equilibrium

8.2   The entire study of phase equilibrium is dependent upon the fundamental laws of thermodynamics. Whilst a comprehensive treatment of the discipline cannot be undertaken here an understanding of some of the terms involved will be useful.

Knowledge of the energy content of a phase is necessary to a study of phase relationships. Although this energy content may be expressed in a number of ways it is perhaps most easily defined in terms of the internal energy, E. This comprises the thermal energy of the body together with vibrational energy, electronic energy and other forms which may be internally stored in the structure of the body.

8.2.1   The kinetic theory of gases proclaims that the constituent particles[*] are in a state of constant motion and that the pressure exerted by a gas is due to the resultant changes in momentum arising from the sum of impacts which these particles make with the walls of the container. If heat energy is supplied to the system this is stored, since energy is conserved, as an increase in the kinetic energy of the particles. Nevertheless the kinetic energy of the particles is not the only energy possessed by the gas. Consider a quantity of gas enclosed in a container by means of a mass-less, friction-less piston (Fig. 8.1). If external heat energy is supplied to the gas

Fig. 8.1

then it will expand and the piston will be raised so that the pressure both inside and outside the contaniter remains constant. Since the volume of the gas increases and moves the piston against atmospheric pressure, work will be done. This work will be equal to the product of the total force acting on the end of the piston and the distance through which it moves, i.e.

$$\text{Work done by expanding gas} = P.A.L$$
$$= P.V \text{ (where the volume, } V = A.L\text{)}$$

Thus the heat supplied to the gas not only raises its temperature but also

---

[*] Either atoms or molecules depending upon the nature of the gas.

does work equivalent to PV. Hence the total energy supplied is $E + PV$ where E is the increase in internal energy.

If we now consider the process operating in reverse so that energy is stored in the gas this energy will be available subsequently. Further if we imagine that this reverse process ends when both volume and internal energy approach zero then the total stored energy is given by:

$$H = E + PV$$

This total stored energy is called the *enthalpy* (H).

8.2.2 The First Law of Thermodynamics deals essentially with the conservation of either internal energy or of enthalpy depending upon how the Law is applied. When we consider an isolated quantity of a substance in a *closed* system only internal energy is involved but when the mass of the substance we are dealing with is allowed to vary, that is, enter or leave the system then enthalpy is involved.

8.2.3 The Second Law of Thermodynamics deals with the conservation of another thermodynamical quantity known as *entropy*, S. Entropy is associated with the degree of 'randomness' or disorder of the particles constituting a mass of matter. The greater the degree of disorder the greater the entropy, the more precise the order the lower the entropy. Thus when a metal melts its atoms pass from a state of order to one of relative disorder and the entropy increases. Whilst most materials expand slightly as they melt a few, such as antimony, bismuth and ice, shrink slightly. Thus entropy is not connected with volume change and the term PV is not involved.

We can conclude therefore that latent heat supplied in melting a solid is stored in the degree of disorder which is produced. This stored energy however is not available for doing work and is referred to as 'transformation energy' or 'entropy energy'. It can be expressed as the product of temperature (T) and the entropy (S). Because the liquid phase which is produced contains non-available energy this must be subtracted from the total energy—or enthalpy (H)—in order to derive the amount of energy available for doing work. Thus the free energy (G) can be derived from:

$$G = H - T.S$$
$$= E + PV - TS$$

The quantity TS represents the energy which is 'bound up' in the solid. The free energy (G) is at a minimum when the system is in equilibrium. It is this thermodynamic quantity which becomes important in studying phase relationships. Since free energy incorporates both the thermal energy content of a material through H and the degree of order (or disorder) of the material through S it defines the nature of the material absolutely.

Phase Equilibrium of Single Substances

8.3 The phenomenon of vapour condensation has already been mentioned in terms of the relationship between the kinetic energy of particles and van der Waals forces operating between them. As a vapour is cooled the average kinetic energy of its particles decreases. Ultimately a temperature is reached where the kinetic energy of most of the particles is insufficient to resist the van der Waals forces acting between them. Those particles with less than average kinetic energy stick together, that is, condense to form droplets of the liquid phase. The temperature at which this condensation begins is often referred to as the *dew point* of the vapour.

8.3.1 As further kinetic energy is dissipated particles condense in increasing numbers. This occurs at a constant temperature, the kinetic energy being given up in the form of heat energy (the latent heat of vaporisation). Some of the particles with kinetic energy greater than the average do not condense but are able to remain in the vapour phase, this state of affairs obtaining when these remaining particles exert a pressure equal to that of the prevailing ambient pressure. Once the majority of the particles have condensed the temperature begins to fall again as further heat is withdrawn from the system and further condensation of some of the remaining vapour particles occurs. The process is illustrated by the cooling curve for water (Fig. 8.2) in which the condensing particles will be molecules of H_2O.

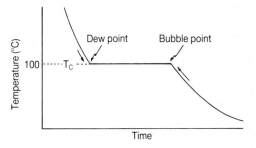

Fig. 8.2 The cooling curve for steam/water at 760 mmHg pressure.

This process is of course reversible. If the liquid is heated the first bubbles of vapour would appear at the temperature T_c. Further addition of heat—the latent heat of vaporisation—would cause the liquid to continue to vaporise isothermally* until all of the liquid had been vaporised. If heat continues to be passed into the system the vapour temperature will rise and the kinetic energy of its molecules will increase. The temperature at which bubbles begin to appear is known as the *bubble point*. For a *pure* substance both dew point and bubble point are at the same temperature which is referred to generally as the *boiling point*.

---

* At the same temperature.

8.3.2 At the temperature T_c both water and water vapour can exist together and if the flow of heat to or from the system is stopped then the relative amounts of the two phases will remain constant. This is obvious since heat must flow into the system in order to provide energy to overcome the van der Waals forces and to separate one molecule from another. Conversely, to condense more vapour the kinetic energy of molecules must be reduced, by extracting heat energy, in order that van der Waals forces can overcome the reduced kinetic energy and so operate between molecules to bind them together.

Although the relative amounts of the two phases remain constant as long as energy neither enters nor leaves the system, a state of dynamic equilibrium exists. Those vapour molecules of higher energy give up their kinetic energy and condense if they strike the surface of the liquid. Since this energy cannot escape from the system it increases the energy level of the neighbouring molecules in the liquid phase so that they are likely to overcome the van der Waals forces operating between them. Hence as they acquire sufficient kinetic energy they join the vapour phase. Thus a small-scale dynamic equilibrium exists between the condensing molecules from the vapour and the vaporising molecules from the liquid. Since the proportions of the phases remain constant the rates of condensation and vaporisation must remain equal. This state of affairs constitutes the basis of *phase equilibrium*. The vapour phase is in equilibrium with the liquid phase so that thermodynamically the free energies of the two phases are equal. The dynamic balance corresponds to a balance between changes in enthalpy and entropy.

In the above it has been assumed that pressure remained constant (at 760 mmHg) and that only temperature was altered. If the pressure also were altered then the equilibrium of the system would be upset. Since enthalpy would change the relative quantities of the two phases present would also change until equilibrium was once more established.

8.3.3 A state of equilibrium will exist between the liquid and vapour phases at temperatures other than the normal* boiling point. At a temperature below the boiling point the liquid exerts a vapour pressure in the space above it because the vapour molecules are in dynamic equilibrium with it. If the *overall* pressure to which the liquid phase is subjected is reduced to the vapour pressure at any given temperature, the liquid will begin to boil. For this reason it is impossible to make a decent cup of tea on Mount Everest. At high altitudes the atmospheric pressure is so low that water boils at temperatures well below 100°C and cannot of course be raised to a higher temperature.

Phase Diagrams

8.4 The relationship between temperature, pressure and the phases in which pure water exists is shown in the simple phase diagram, Fig. 8.3.

* The boiling point at 'atmospheric pressure' (760 mmHg).

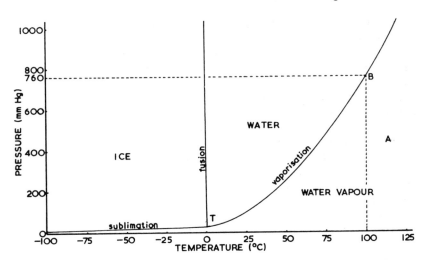

Fig. 8.3 The P/T phase diagram for pure water.

Using this diagram we can, for any given combination of temperature and pressure, find in what phase state 'H_2O' exists, and also determine what effect any alterations of temperature and pressure will have on the stability of a given phase. For example the 'fusion curve' is an almost vertical line indicating that pressure variations have little effect on the freezing temperature of water. The same is true of the freezing points of metals, for which reason we can generally neglect the effects of pressure in our studies of alloy phase equilibrium.

8.4.1 The 'vaporisation curve', however, indicates that the boiling point of water is very dependent upon pressure. Water boils at 100°C at a pressure of 760 mmHg (point B). Increase in pressure above 760 mmHg will raise the boiling point, whilst reduction below 760 mmHg will considerably lower it. The 'sublimation curve' indicates that at low pressures solid ice can vaporise—or *sublime*—without passing through the liquid phase. A number of substances, notably carbon dioxide ('dry ice') and solid iodine, sublime at ordinary atmospheric pressures.

Q | What will be the approximate boiling point of water at the Mount Everest Base Camp if the barometric pressure there is equivalent to 380 mm of mercury?

A | By reference to the 'water vaporisation' curve (Fig. 8.3) we see that at a pressure of 380 mmHg the corresponding water vaporisation temperature (i.e. the boiling point) is approximately <u>63°C</u>.

8.4.2 Possibly the most interesting feature of this simple phase diagram is the *triple point*, T, at which water can exist in the solid, liquid and vapour forms simultaneously. This point T represents a pressure of 4.579 mmHg and a temperature of 0.0098°C and only under these precise conditions of temperature and pressure can the three phases co-exist. Any variation in either temperature or pressure will cause at least one of the phases to disappear.

The triple point in phase equilibrium is of course not confined to water only. For example, solid liquid and gaseous hydrogen can co-exist at −259.2°C and 54.1 mmHg.

The Phase Rule

8.5 First it will be necessary to define more precisely some of the terms used in connection with phase equilibrium. Thus the whole aggregate of different bodies with which we are dealing is described as a *system*. If this system is completely uniform throughout, having the same composition and properties at any point, it is said to be *homogeneous*. Either ice, water or water vapour existing separately would represent such a system. If, however, the system consists of two or more different parts, each of which is homogeneous in itself but whose properties differ from those of the other parts, then the system is said to be *heterogeneous*. Hence, at the triple point, T (Fig. 8.3), ice, water and water vapour existing together constitute a *heterogeneous three-phase system*.

At any point on the curve TB two phases can co-exist, viz. water and water vapour, so that the complete system is heterogeneous and two-phase. Point A represents a single homogeneous phase, water vapour.

8.5.1 A general law dealing with phase equilibrium was developed in 1876–78 by Josiah Willard Gibbs, professor of mathematical physics at Yale. Known as the *Phase Rule* it can be of considerable use in solving some of the more difficult problems of phase equilibrium.

Suppose we again consider point A in Fig. 8.3. This indicates that at a temperature of approximately 110°C and a pressure of 450 mmHg the chemical compound, H_2O, exists as a homogeneous single phase, water vapour. We can alter *both temperature and pressure quite independently* (within limits) without altering the nature of the system which will remain a homogeneous single phase. Thus a system of this type can be said to possess *two degrees of freedom* or to be *bivariant*. The same conditions of course apply to ice and water when they exist as single phases by themselves. The term 'degree of freedom' (or *variance*) describes any variable property such as temperature, pressure, volume or other values which are interrelated in the system.

Point B on Fig. 8.3 represents the temperature (100°C) at which water boils under normal atmospheric pressure (760 mmHg). Thus at that point two phases can co-exist, viz. water and water vapour. However, the number of degrees of freedom have now been reduced to one since we can only vary either temperature *or* pressure. That is, the new temperature (or pressure) must be represented by a point on the vaporisation curve otherwise one of the phases will disappear (either the vapour phase will condense if we move to a point on the left of the curve; or the liquid phase will vaporise if we move to a point on the right of the curve). Consequently if, for the sake of argument, we raise or lower the temperature then the pressure must be raised or lowered *in sympathy* with it. Hence we have only one degree of freedom since in this case pressure is tied to temperature and cannot be varied independently without the disappearance of one or other of the two phases. Such a system is *univariant*.

At point T on Fig. 8.3 all three phases, ice, water and water vapour, can exist together but clearly if we alter either temperature or pressure then at least one of the phases will disappear. Consequently in this three-phase system there are no degrees of freedom and it is said to be *nonvariant*.

8.5.2   If the number of degrees of freedom are represented by f and the number of phases by p, then the phenomena describe above can be related by a single rule:

$$f + p = 3 \qquad\qquad (1)$$

Thus in a single-phase system such as a solid pure metal, $f = 2$, i.e. there are two degrees of freedom whilst at the melting point (where $p = 2$) then $f = 1$, and the system is univariant. If three phases—solid, liquid and vapour—are present at the same time then no degrees of freedom remain and at least one of the phases will disappear if either temperature or pressure are altered.

In an alloy containing two metals, melting occurs over a *range* of temperatures rather than at a single, fixed temperature. Thus, two phases (liquid and solid) can be present over a range of temperatures instead of at a single temperature. Although two phases are present we are able to vary temperature and pressure *independently* without a loss of equilibrium so that there are two degrees of freedom.

We can now make appropriate alterations to equation (1):

$$f + p = 4 \qquad\qquad (2)$$

This equation can be written in the general form:

$$f + p = n + 2 \qquad\qquad (3)$$

where n is the number of *components* in the system. As far as metallic alloy systems are concerned the term 'component' signifies a pure metal. Thus an alloy containing two metals is a two-component system. In practice we often deal with components which are not elements. For example, in the system, ice-salt-solution, the components are, effectively, salt and water both of which are chemical compounds.

8.5.3   Although Gibb's Phase Rule as represented above by (3) deals with the general case in which pressure constitutes one of the degrees of

freedom, a study of metallic alloys does not normally involve a consideration of the effects of pressure. We are normally dealing only with liquid
and solid phases and pressure variation then has a negligible effect on their
equilibrium temperature, generally even less in fact than it has on the slope
of the 'fusion curve' in the ice–water–vapour system (Fig. 8.3).

If pressure is assumed to be kept constant the number of degrees of
freedom falls to one since *one* possibility of variance has been removed
from the system. At *constant pressure*, therefore, the Phase Rule can be
expressed:

$$f + p = n + 1 \tag{4}$$

8.5.4 This modification of the Phase Rule will be generally useful to us
when we are studying alloy systems at a constant atmospheric pressure. A
single-phase molten alloy containing two metals—that is, two components—in a uniform liquid solution in each other, can now be examined
using the modified Rule:

$$f + 1 = 2 + 1$$
$$\therefore \ f = 2$$

Thus there are two degrees of freedom—one being temperature and the
other composition—so that, within certain limits, both temperature and
composition can be varied independently without introducing a second
phase into the system.

Phase Equilibrium in Two-component Systems

8.6 This constitutes by far the most important group of systems to be
studied since it includes all binary metallic alloys. We shall begin, however,
by considering a system involving two liquids which are only partially
*miscible** in each other since a number of important features can be
discussed in terms of this system.

**8.6.1. Two Liquids which are Incompletely Miscible in Each
Other** Figure 8.4 represents the phase-equilibrium conditions for the
system lead–zinc in the liquid state above 500°C. Diagrams of this type are
commonly known as *thermal equilibrium diagrams*, and indicate the
relationships which exist between temperature, composition and phase
constitution in a two-component system. The base line—or *abscissa*—of
the diagram always shows the composition varying between 100% of one
component at one side of the diagram and 100% of the other component at
the opposite end. In most diagrams the percentages are by mass but in
some cases it is more relevant to show the percentages in terms of the
numbers of atoms. The *ordinate* represents temperature. Thus, point P

---

* Literally 'mixable'—or in this case 'soluble'.

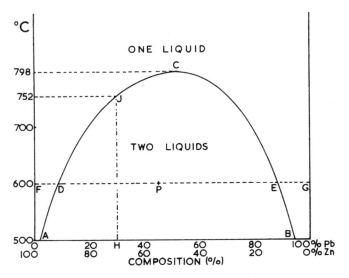

Fig. 8.4   The equilibrium diagram for lead and zinc above 500°C.

(Fig. 8.4) represents a mixture consisting of 45% by mass of lead and 55% by mass of zinc at a temperature of 600°C.

At 500°C molten zinc will dissolve about 2% lead (point A) giving a single solution which becomes saturated at this amount of 2% lead. At the same temperature molten lead will dissolve up to about 5% zinc (point B). Between these two limits given by A and B two separate molten layers will co-exist in the container. The upper layer will consist of molten zinc *saturated* with lead (composition A) and the lower layer will consist of molten lead saturated with zinc (composition B).

The slope of the line ADJC indicates that, as the temperature rises, the solubility of lead in molten zinc increases. Similarly the slope of BEC indicates that the solubility of zinc in molten lead also increases under similar conditions. The two solubility curves meet in C at a temperature of 798°C indicating that above that temperature the two components, lead and zinc, become completely soluble in each other so that only one layer (phase) will be present whatever the composition of the mixture as long as the temperature remains above 798°C.

Suppose we have a molten mixture consisting of 30% lead and 70% zinc (composition H). At, say, 500°C this will consist of a saturated solution of lead in zinc (composition A—2% lead/98% zinc) together with a saturated solution of zinc in lead (composition B—95% lead/5% zinc). Thus the mixture will consist of two phases, i.e. two separate saturated solutions. If the temperature is now raised the amount of lead which dissolves in molten zinc will increase (as will the amount of zinc which dissolves in molten lead), until at J (752°C) the molten zinc will have absorbed the whole 30% lead present and so one phase disappears and a single homogeneous solution remains.

Now let us consider a mixture containing 45% lead/55% zinc at 600°C, represented by point P (Fig. 8.4). Assuming that the molten mixture has had time to reach equilibrium at this temperature—stirring will have assisted this process—then by drawing a 'temperature horizontal' through P we are able to obtain information from the diagram. This temperature horizontal FDPEG, cuts the boundaries between the single- and double-phase fields at D and E respectively. Since the temperatures of the phases in equilibrium must be the same, then D and E must represent the compositions of the two phases in equilibrium at that temperature (600°C). Thus at 600°C the liquid phase, D, contains 8.5% lead/91.5% zinc (from Fig. 8.4) whilst the other liquid phase, E, contains 88.5% lead 11.5% zinc.

From Fig. 8.4 we can also determine the precise proportions in which each phase (compositions D and E respectively) is present:

Let d, e and p be the proportions of lead in the mixtures of compositions D, E and P respectively.

Then:

$$d = \frac{FD}{FG}; \; e = \frac{FE}{FG}; \; \text{and } p = \frac{FP}{FG}$$

If we let the mixture contain a fraction x of the phase of composition D, then it will contain a fraction $(1-x)$ of the other phase (composition E). The proportion of lead in the total molten mixture is p and:

$$p = dx + e(1-x) = dx + e - ex = e + x(d-e)$$

Therefore

$$x = \frac{p-e}{d-e} = \frac{-1.(p-e)}{-1.(d-e)} = \frac{e-p}{e-d}$$

$$x = \frac{\dfrac{FE}{FG} - \dfrac{FP}{FG}}{\dfrac{FE}{FG} - \dfrac{FD}{FG}} = \frac{FE - FP}{FE - FD} = \frac{PE}{DE} \qquad \text{(from Fig. 8.4)}$$

Similarly

$$1 - x = \frac{DP}{DE}$$

Therefore the ratio:

$$\frac{\text{Weight of phase of composition D}}{\text{Weight of phase of composition E}} = \frac{x}{1-x} = \frac{\dfrac{PE}{DE}}{\dfrac{DP}{DE}} = \frac{PE}{DE} \cdot \frac{DE}{DP}$$

$$\frac{\text{Weight of phase of composition D}}{\text{Weight of phase of composition E}} = \frac{PE}{DP} \qquad\qquad (1)$$

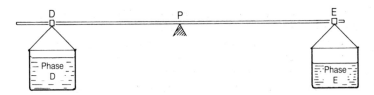

Fig. 8.5   The 'lever rule' as applied to phase equilibrium.

This is commonly known as the *Lever Rule* for fairly obvious reasons (Fig. 8.5). If we imagine the *tie line** DE (Fig. 8.4) to be a lever with its fulcrum at P (which is the composition of the mixture as a whole), from equation (1):

(Weight of phase of composition D).DP
$$= \text{(Weight of phase of composition E)}.PE \quad (2)$$

which is equivalent to 'taking moments' of the weights of the two phases about a fulcrum representing the overall composition of the mixture.

8.6.2 Some General Rules for Interpreting Equilibrium Diagrams   For the general interpretation of thermal equilibrium diagrams, particularly those dealing with metallic alloy systems, a number of rules were derived from Gibb's Phase Rule by the French metallurgist, A. Portevin. Some of these rules and definitions may be found helpful.

(i)   The areas of the diagram are usually referred to as 'phase fields' and on crossing any *sloping* boundary line from one field to the next the number of phases will always change by one. Thus in Fig. 8.6 two

Fig. 8.6

* A 'temperature horizontal' terminating at the boundaries of a phase field.

single-phase fields are separated by a two-phase field containing each of the phases.

(ii)   In a binary system no phase *field* can contain more than two phases. Three phases can co-exist only at a single point such as a eutectic point (Figs 8.12 and 8.13).

(iii) At point P (Fig. 8.6) two phases can exist together. If a temperature horizontal is drawn through P the composition of one phase (in this instance a liquid) is given by L and the composition of the other phase (in this case a solid) which is in equilibrium with it, by S. P represents the composition of the alloy mixture as a whole.

(iv)   The relative amounts of the two phases present at the temperature and composition represented by P are given by the lengths LP and PS respectively:

$$\frac{\text{Weight of phase of composition L}}{\text{Weight of phase of composition S}} = \frac{\text{PS}}{\text{LP}}$$

This is of course the Lever Rule derived above.

(v)   A phase which does not occupy a field by itself but appears only in a two-phase field is either a pure metal (Fig. 8.13) or an intermetallic compound of invariable composition (Fig. 8.18). (In each case the 'phase field' is in fact infinitely narrow, being a vertical line representing the appropriate fixed composition.)

(vi)   If a vertical line representing the overall composition, C, of some alloy (Fig. 8.7) crosses a sloping line of the diagram it means that a

Fig. 8.7

change in the number of phases will occur at that point (O), i.e. a phase will be precipitated or absorbed. Thus in Fig. 8.7 on raising the temperature from T_1 to T_2 the phase X will be absorbed, whilst on cooling from T_2 to T_1 the phase X would be precipitated.

8.6.3 Two Components which are Soluble in Each Other in All Proportions in Both the Liquid and Solid States   Since this case is of particular importance in a study of metallic alloys we shall deal here with a system in which both components are metals and which form a series of disordered substitutional solid solutions in all proportions (7.2.4). Many pairs of metals form solid solutions which are basically of this type (7.3.1). Since the copper–nickel alloys are of commercial interest this system is used as the example.

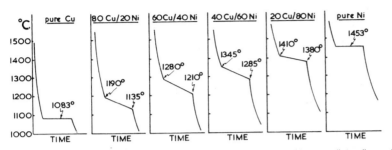

Fig. 8.8 Cooling curves for pure copper, pure nickel and a series of intermediate alloys of the two metals.

8.6.3.1 Whilst a pure metal freezes and melts at a constant single temperature an alloy, with certain exceptions (8.6.4.1), freezes and melts over a range of temperatures. Suppose we make up a number of copper/nickel alloys of different compositions. We shall find that they solidify over different temperature ranges and by plotting the cooling curves (Fig. 8.8) we can determine the temperatures at which freezing begins and ends for each alloy. This information can then be used to plot the phase equilibrium diagram for the complete system (Fig. 8.9), assuming, of course, that each

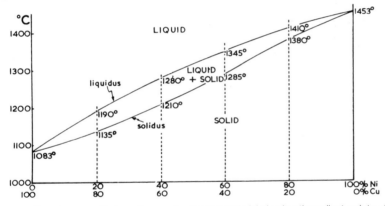

Fig. 8.9 The phase equilibrium diagram for copper and nickel using the salient points of the cooling curves from Fig. 8.8.

alloy cooled slowly enough for it to attain something approaching equilibrium at each stage of the cooling process.

The resultant phase diagram is one of the most simple we encounter in a study of the equilibrium systems of metals. It consists of two lines only, the upper—or *liquidus*—above which any point will represent in temperature and composition a completely liquid phase, and the lower—or *solidus*—below which any point will represent in temperature and composition a completely solid phase. Between these two lines is a phase field in which both liquid and solid phases co-exist.

Applying the Phase Rule ($f + p = n + 1$) to the single-phase fields, both above the liquidus and below the solidus, we have:

$$f + 1 = 2 + 1$$
or $$f = 2$$

so that the system, being bivariant, both temperature and composition can be varied independently within limits without any alteration in the structure taking place. In the two-phase field, between liquidus and solidus:

$$f + 2 = 2 + 1$$
or $$f = 1$$

so that the system is univariant and temperature and composition can no longer be altered independently without a loss of equilibrium. Any change in temperature must be accompanied by an appropriate change in composition (and vice versa) or balance between the proportions of the phases present will be upset.

Suppose we have made up an alloy containing 70% nickel and 30% copper (X in Fig. 8.10). At temperatures above T this will exist as a homogeneous liquid but as the temperature falls to T, the liquidus temperature of this alloy, solidification will begin. If we draw a temperature horizontal through X and T it cuts the solidus at Y. Y is therefore the composition of the solid phase which can exist in equilibrium with the liquid phase of composition X at the temperature T. Consequently dendrites of composition Y begin to solidify.

Since these dendrites are much richer in nickel (approximately 91% Ni as read from the diagram) than was the original liquid, the remaining liquid will have become depleted in nickel and correspondingly richer in copper as a result. Thus the composition of the remaining liquid moves to the left of the diagram—say, to X_1. No further solidification occurs until the temperature has fallen to T_1 and at this temperature we find that the solid phase which is in equilibrium with the liquid X_1 is now of composition Y_1, i.e. rather less rich in nickel than was the original dendrite. Solid of composition Y_1 therefore forms a coating over the original dendrite.

However, we will assume that the system is *cooling extremely slowly so that phase equilibrium is attained at every stage of solidification.* Clearly, (at T_1), crystals of solid of *uniform* composition Y_1 throughout constitute the equilibrium phase under these conditions and this uniformity is achieved by diffusion which proceeds according to Fick's Laws (7.4.2), nickel atoms moving outwards from the cores of the dendrites, to be replaced by copper atoms which move inwards.

Assuming that a state of equilibrium has been thus achieved at the temperature T_1 we can use the 'lever rule' to obtain complete information about the system at this temperature:

Weight of liquid (composition X_1).X_1O
$$= \text{Weight of solid (composition } Y_1\text{)}.OY_1$$

or

$$\frac{\text{Weight of liquid (composition } X_1)}{\text{Weight of solid (composition } Y_1)} = \frac{OY_1}{X_1O}$$

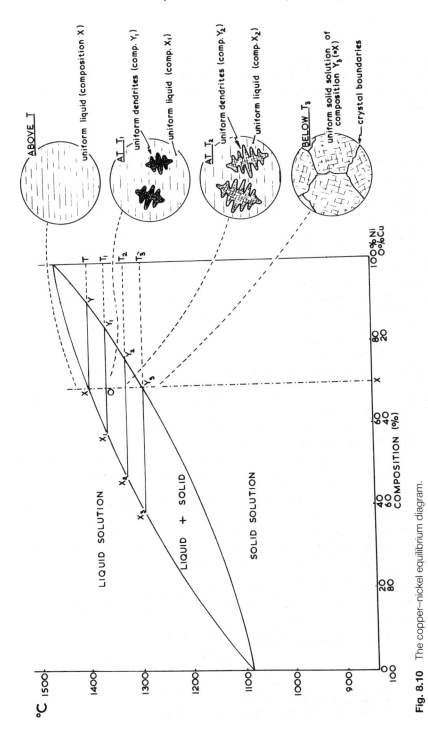

Fig. 8.10 The copper–nickel equilibrium diagram.

In the case of the alloy chosen—70% Ni/30% Cu—we have (reading from Fig. 8.10):

$$\frac{\text{Weight of liquid (59\% Ni)}}{\text{Weight of solid (85\% Ni)}} = \frac{(85-70)}{(70-59)} = \frac{15}{11}$$

We have therefore been able to determine the precise compositions of each phase and the proportions in which they co-exist in equilibrium at 1365°C (T_1).

Solidification continues in this way as the temperature falls slowly, the composition of the liquid changing (due to mixing caused by convection) along $X_1 X_2 X_3$ whilst the composition of the solid changes (due to diffusion) along $Y_1 Y_2 Y_3$. Thus at temperature T_2 the composition of the growing dendrites will have changed to Y_2 whilst the composition of the uniform liquid in equilibrium with them will be X_2. Meanwhile the proportion of liquid to solid will also have changed to OY_2/OX_2.

Finally, at temperature T_3 the last trace of liquid (of composition X_3) solidifies at the crystal boundaries and is immediately absorbed by diffusion so that the solid becomes uniform and of composition Y_3 throughout—which is, of course, the same as that of the original liquid (X). It could not be otherwise since matter has been conserved and the initial liquid phase and the final solid phase are of uniform composition.

8.6.3.2 In the above exposition we assumed that the rate of cooling was 'infinitely slow' so that complete equilibrium by means of convection (in the liquid) and diffusion (in the solid) could be achieved at each stage of

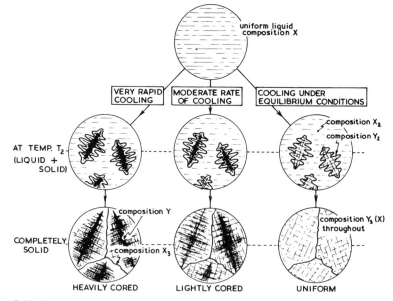

Fig. 8.11 Illustrating the effects of the rate of cooling on the extent of coring in the 70–30 nickel–copper alloy dealt with here.

the process. Under prevailing conditions of industrial casting, however, the rate of solidification outstrips the rate of diffusion and the solid structure is unable to attain uniformity and equilibrium. Thus the composition of the core of each dendrite is in the region of Y, that is it contains more nickel than the equilibrium amount, so that the outer fringes must contain correspondingly more copper than the equilibrium amount. This is the mechanism of coring and, clearly, the greater the rates of solidification and cooling the greater will be the extent of coring in the final structure since there will be less chance for diffusion to take place (Fig. 8.11).

If the cored structure is heated to a sufficiently high temperature diffusion will take place according to Fick's Laws and after an adequate period of time the concentration gradient will approach zero as nickel atoms diffuse outwards from the core and are replaced by copper atoms moving inwards towards the core. Diffusion ceases when a uniform composition prevails. As the degree of coring approaches zero during annealing, the *rate* of any further diffusion becomes extremely slow so that detectable coring is still present in most industrially annealed materials.

Q   A copper–nickel alloy contains 80% Cu/20% Ni. By reference to Fig. 8.10 determine as accurately as is possible:

   (i)   the temperature at which such a molten alloy begins to solidify;
  (ii)   the composition of the initial dendrites which form;
 (iii)   the temperature at which solidification will be complete. Assume an extremely slow rate of cooling.

A   (i)   Draw a vertical line on the diagram representing the composition (20% Ni/80% Cu). This cuts the liquidus line at a point representing a temperature of 1205°C (approximately). This is the temperature at which freezing begins.

  (ii)   The temperature 'horizontal' from 1205°C cuts the solidus line at a point which represents an alloy composition, 47% Ni/53% Cu, which will be the composition of the first dendrites to form.

 (iii)   The vertical line representing the overall composition of the alloy (20% Ni/80% Cu) cuts the solidus line at 1125°C (approximately). This is the temperature at which solidification of the 20% Ni/80% Cu alloy is complete—assuming that solidification takes place very slowly so that equilibrium is reached at every stage of solidification.

8.6.4 Two Substances, Completely Miscible as Liquids, Become Totally Insoluble in Each Other as Solids   Most readers will know that the easiest way to remove a coating of ice from the garden path is to sprinkle a small amount of 'common salt' (sodium chloride) on it. The ice soon begins to melt and a glance at the relevant part of the phase diagram (Fig. 8.12) explains why this happens. Although we may use only a small proportion of salt (say C) we see that both ice and solution are in equilibrium at temperatures between about −4°C and −23°C. Therefore 'slush' will soon

Fig. 8.12   Part of the ice–sodium chloride phase equilibrium diagram.

form and the slipperiness of the ice be eliminated due to the formation of some liquid. In the unlikely event of the temperature being below −23°C scattering salt would be unfruitful since solid ice and solid salt are then the phases in equilibrium with each other and ice would not melt. When salting our roads the Local Authority uses substances which are cheaper than sodium chloride but equally effective both in de-frosting the tarmac and in causing corrosion of our motor cars.*

8.6.4.1   The diagram is a typical example of what is commonly called a *eutectiferous series*. The term *eutectic* is derived from the Greek, 'eutektos' or 'capable of being melted easily', a reference to the fact that a mixture of precisely eutectic composition (23.6% sodium chloride/76.4% water in the above example) has the lowest freezing temperature range in the series. In a system of this type the components are mutually soluble as liquids but during solidification separate out either partially or completely. In this section we are dealing with cases of complete insolubility in the solid state. The phenomenon of *depression of freezing point* is involved here. Thus the addition of sodium chloride to water progressively depresses its freezing point. Conversely the addition of water to sodium chloride will progressively depress its freezing point.† The two curves meet in a minimum—or eutectic—point at −23°C and 23.6% (by mass) of sodium chloride. Whichever component is present in excess of this eutectic composition will crystallise out first as the mixture cools so that the final liquid will always be of eutectic composition. For this reason any mixture in the series—other than one of eutectic composition—will solidify over a range of temperature. A mixture of exactly eutectic composition, however, will solidify at a single temperature in the same manner as pure water.

8.6.4.2   It is doubtful whether two substances can ever be classed as being *completely* insoluble in each other. For example, we normally speak of, say, silver chloride as being 'insoluble in water' whereas in fact it has a

*Calcium chloride is used principally. It is a low-value by-product of soda manufacture. The eutectic mixture (with water) freezes at −51°C making it effective at very low temperatures.
† The melting point of sodium chloride (801°C) is of course above the boiling point of water but the relationship holds true for water-rich mixtures below 100°C.

measurable slight solubility. The same is true of solids, and in particular, metals. Total insolubility in the solid state is extremely rare and probably non-existent when those two metals were initially completely soluble in each other as liquids. Thus, although we shall discuss the bismuth/cadmium system here as an example of *complete insolubility* in the solid state we now believe that a small degree of solid solubility exists between the metals. Nevertheless we shall ignore this slight solid solubility for the sake of our argument.

Bismuth, one of the 'less-metallic metals' in a chemical sense, forms a fairly complex type of crystal structure. Whilst this is basically rhombic the metallic bond is partly replaced by covalent bonding with the result that atoms arrange themselves in layers with van der Waals forces operating between these layers. It is not surprising therefore that bismuth is extremely brittle and that it is reluctant to form mixed crystals with cadmium which crystallises in a normal CPH pattern. Again the phase diagram is of simple form consisting of two lines only, the liquidus AEB and the solidus CED, or, more properly ACEDB.

8.6.4.3 We will consider the solidification of a bismuth/cadmium alloy of composition X (Fig. 8.13) that is, containing 25% bismuth/75% cadmium. At any point above the liquidus line AEB this will consist of a completely homogeneous liquid and solidification will commence when the temperature falls to T. At this temperature the solid in equilibrium with the liquid X is pure cadmium since the relevant solidus phase boundary in this case is the 100% cadmium ordinate. Hence dendrites of pure cadmium begin to form. This causes the remaining liquid to become correspondingly richer in bismuth so that its composition moves to the left, say to X_1. Further solidification occurs as the temperature falls to T_1 when more pure cadmium will crystallise. This process continues until the temperature has fallen to 144°C by which time the remaining liquid contains 60% bismuth/40% cadmium, that is, the temperature and composition are represented by the point E—the eutectic point.

Applying the 'lever rule':

Weight of liquid (composition E) . EP
$$= \text{weight of pure cadmium} . \text{DP}$$

or

$$\frac{\text{Weight of liquid (composition E)}}{\text{Weight of pure cadmium}} = \frac{\text{DP}}{\text{EP}}$$

$$(\text{Substituting values from Fig. 8.13}) = \frac{100-75}{75-40} = \frac{25}{35} = \frac{5}{7}$$

Thus there are seven parts by mass of solid cadmium to five parts by mass of liquid.

At this stage the two metals are in equilibrium with each other in the remaining liquid. It is reasonable to suppose, however, that due to the

Fig. 8.13   The bismuth–cadmium equilibrium diagram.

momentum of crystallisation of the primary cadmium, a little too much cadmium deposits throwing the composition of the liquid slightly to the left of E. Equilibrium is almost immediately restored as a film of bismuth deposits and in this manner the remaining liquid solidifies by depositing alternate layers of cadmium and bismuth as the liquid composition swings to and fro about E. The temperature remains constant at 144°C until the solidification process is complete.

The final structure therefore consists of *primary* crystals of cadmium in a matrix of eutectic which in turn consists of alternate layers of pure cadmium and pure bismuth in a laminated structure which is typical of metallic eutectics. The proportion of primary cadmium to eutectic will obviously be the same as indicated above since the liquid referred to then has now solidified as eutectic. Thus there will be 7 parts by mass of solid cadmium to 5 parts by mass of eutectic which in turn contains 60% bismuth and 40% cadmium.

8.6.4.4   A homogeneous liquid containing, say, 80% bismuth/20% cadmium will begin to solidify by depositing primary dendrites of bismuth. These do not resemble the dendrites of other metals but are roughly cubic in shape (Fig. 8.13). Since bismuth has been rejected from solution the latter has become correspondingly richer in cadmium so that its composition moves to the right towards the eutectic point. Ultimately the liquid will contain 60% bismuth and 40% cadmium at 144°C when it will solidify as eutectic as in the case of the 25% bismuth/75% cadmium alloy. In fact any bismuth/cadmium alloy will contain eutectic which will always be of fixed composition, namely, 60% bismuth/40% cadmium. Any liquid solution of composition either side of the eutectic amount will first deposit which ever metal is in excess of the eutectic composition in the form of primary crystals until the final liquid is of the eutectic composition at 144°C (Fig. 8.14). It must be emphasised that in the final structure there is no question of solid solubility since even in the eutectic portion the alternate layers of the two pure metals are readily resolvable using quite a low-power microscope.

At the eutectic point three phases can co-exist in equilibrium, i.e. liquid,

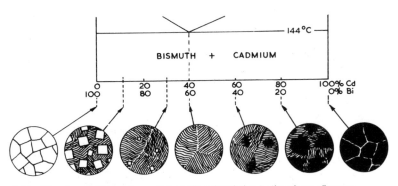

Fig. 8.14   Structures of bismuth–cadmium alloys in relation to the phase diagram.

solid cadmium and solid bismuth. The solid cadmium and solid bismuth of course constitute the eutectic. Hence there are two components (cadmium and bismuth) and three phases. Applying the modified Phase Rule:

$$f + p = n + 1$$
$$f + 3 = 2 + 1$$
$$\therefore f = 0$$

This means that the system is invariant so that if either composition or temperature are altered at least one of the phases will disappear.

8.6.4.5 We can summarise the general properties of the eutectic as follows:

(i)   in a binary system it consists of a heterogeneous *mixture* of two phases, i.e. it must not be regarded as a single phase itself;

(ii) it solidifies at a single fixed temperature—the lowest temperature in the series;

(iii) it will be of fixed composition whatever the overall composition of the alloy in which it exists;

(iv) it solidifies in typical lamellar manner, though in some cases the lamellae of one of the phases degenerate to globular form due to surface-tension effects.

Fig. 8.15 A typical lamellar eutectoid structure (9.2.1). This micrograph is of a cast 11.8% aluminium-in-copper alloy annealed at 700°C. It shows lamellae of solid solution α and the intermediate phase γ_2 (11.2.5). (Etched in Acid ferric chloride solution.) ×500.

> **Q** Given that the relative densities of cadmium and bismuth are 8.65 and 9.8 respectively calculate the relative thicknesses of the cadmium and bismuth laminations in the Cd/Bi eutectic (Fig. 8.13).
>
> **A** From Fig. 8.13 the Cd/Bi eutectic contains (by *weight*) 40% Cd and 60% Bi.
>
> $$\text{Relative volume of Cd} = \frac{40}{8.65} = 4.62$$
>
> and
>
> $$\text{Relative volume of Bi} = \frac{60}{9.8} = 6.12$$
>
> Thus if we assume that the surface areas of the Cd and Bi layers are equal then the Bi layers in the eutectic will be 6.12/4.62 = 1.32 times thicker than the Cd layers.

8.6.5 Two Substances Completely Miscible as Liquids but Only Partially Soluble in Each Other as Solids

In the two preceding sections we have considered cases both of complete solid solubility (Cu/Ni) and total insolubility (Bi/Cd) in the solid state. These were, in a sense, special cases of which this one is the general, since here we shall deal with the case of partial solid solubility. Since the two special cases referred to permit of simpler interpretation they were considered first, so that the reader could progress from the simple to the more complex.

8.6.5.1 The silver–copper system is a typical example of partial solid solubility. The principal features not encountered during our study of previous phase systems are the phase boundaries CRC_1R_1F and DUD_1U_1G in Fig. 8.16. These boundaries occur *below* the solidus line and consequently separate phase fields containing only solid phases. This means that phase changes can take place in an alloy after solidification is complete. A line such as CF or DG is generally termed a *solvus*. As the diagram indicates we have in the system two solid solutions α and β* respectively since:

(i)   silver will dissolve up to a maximum of 8.8% copper at the eutectic temperature forming the solid solution α;

(ii) copper will dissolve up to a maximum of 8.0% silver at the eutectic temperature forming the solid solution β.

The slope of these boundary lines CF and DG indicates that both the solubility of copper in silver (α) and the solubility of silver in copper (β)

---

*In phase diagrams of metallic alloys it is the custom to label phases from left to right using letters of the Greek alphabet as appropriate.

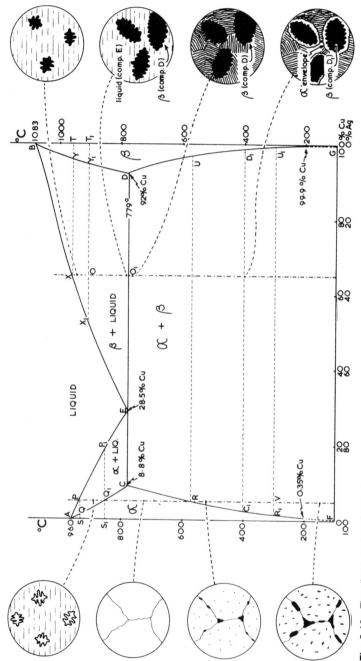

Fig. 8.16 The silver–copper phase diagram.

decrease with fall in temperature. This is a phenomenon mentioned earlier (7.5.2) and is also a common feature of partial solubility in the liquid state as instanced by the zinc/lead system (8.6.1).

8.6.5.2 We will now examine the solidification of two representative alloys in the series. First, we will consider an alloy of composition X (Fig. 8.16) that is, containing 35% Ag/65% Cu. This will begin to solidify as the liquidus is reached at temperature T, when dendrites of the solid solution β of composition Y will commence to form. Since the β dendrites are relatively rich in copper then the remaining liquid will be left correspondingly rich in silver and its composition will move to the left, say to X_1. Further deposition of β, this time of composition Y_1 will occur as the temperature falls to T_1 and, if we assume *slow cooling*, then the original dendrites wil also have changed to this composition (Y_1) by means of diffusion. Thus, at temperature T_1, we have:

$$\text{Weight of uniform liquid (composition } X_1 \text{)} . X_1 O$$
$$= \text{Weight of uniform solid solution } \beta \text{ (composition } Y_1 \text{)} . OY_1$$

As the temperature continues to fall slowly, β continues to deposit and at the same time change in composition along $Y_1 D$ whilst the remaining liquid changes in composition along $X_1 E$. At 779°C, the eutectic temperature, we ultimately have liquid of composition E and solid solution β of composition D in the ratio:

$$\text{Weight of uniform liquid (composition E)} . EO_1$$
$$= \text{Weight of uniform solid solution } \beta \text{ (composition D)} . O_1 D$$

As the temperature begins to fall below 779°C the remaining liquid solidifies as a eutectic by depositing alternate layers of α (composition C) *and* β (composition D) until the liquid is used up. Obviously the ratio:

$$\frac{\text{Weight of primary } \beta}{\text{Weight of eutectic}} \text{ will be the same as that indicated above, i.e.}$$

$EO_1/O_1 D$.

If we assume a very slow rate of cooling then further changes will occur in the structure of the solid alloy as the temperature falls and the solid solubilities of each metal in the other decrease along CF and DG respectively. Thus copper will be rejected from the α layers of the eutectic in the form of particles of the other solid solution β. These particles will join the layers of β adjacent to those of α. Similarly silver will be precipitated from the β layers in the form of a small amount of α which will join the α layers.

The crystals of primary β will also reject some silver in the form of α and this will sometimes be seen as a film surrounding the primary β crystals (Fig. 8.16) in cases like this. At, say, 400°C the composition of α will be given by C_1 and the composition of β (both eutectic and primary) by D_1. Because of the extremely slow rate of ionic diffusion at low temperatures little change in structure can be expected in this case below approximately 300°C.

8.6.5.3 We will now consider the cooling of an alloy of composition P, that is, containing 95% Ag/5% Cu. This will, of course, begin to solidify at temperature S (Fig. 8.16) by depositing dendrites of α of composition Q. Solidification will be complete at temperature S_1 when the final traces of liquid will be of composition P_1 but if we assume slow cooling such that equilibrium is maintained throughout, then liquid P_1 will be absorbed by diffusion so that the overall composition of the solid solution α is Q_1 (which, of course, is the same as that of the original liquid P).

No further changes take place in the uniform α crystals as they cool slowly towards R. On reaching R, however, the solid solution α becomes fully saturated with copper and as the temperature falls below that corresponding to R, copper will be rejected from solution, not of course as pure copper but as solid solution β of composition U, which is the phase in equilibrium with α at that temperature. As the temperature falls slowly β continues to precipitate from α and both solid solutions in fact change in compositions along RF and UG respectively until at, say 300°C, α contains little more than 0.5% copper (R_1) and β no more than 0.2% silver (U_1). The relative amounts of α and β at this temperature will be given by:

$$\text{Weight of } \alpha \text{ (composition } R_1). R_1V$$
$$= \text{Weight of } \beta \text{ (composition } U_1). VU_1$$

Up to this point we have been dealing with the solidification and cooling of alloys on the assumption that cooling has been slow enough for equilibrium to have been maintained at each stage. In practice this is rarely so and in a system such as this, rapid cooling is likely to have the following principal effects:

(i)   primary crystals, whether of α or β, will be heavily cored;
(ii) α is likely to be supersaturated with copper, and β with silver since cooling will have been too rapid to permit diffusion and precipitation. This applies both to primary α and β and also to eutectic α and β.

Q Two 'mythological metals' fantasium (Fa) and hilarium (Hi) have melting points of 1250°C and 800°C respectively. They are completely soluble as liquids but only partially soluble as solids, forming two solid solutions α and β. These solid solutions form a eutectic at 620°C containing 62% Hi and 38% Fa. The following table shows the maximum solubilities of each metal in the other under equilibrium conditions over a range of temperatures:

Temperature (°C)	0	100	200	300	400	500	600	620
Max. solubility of Hi in Fa (α) (% wt)	5.0	6.5	8.3	10.4	13.0	16.4	22.3	24.0
Max. solubility of Fa in Hi (β) (% wt)	1.0	1.4	1.8	2.4	3.3	4.6	6.4	7.0

   (i)   **Draw the thermal equilibrium diagram for the system on squared paper as accurately as the above figures allow and label all phase fields.**

  (ii)   **Describe step by step the solidification and cooling under equilibrium conditions from 1000°C to 0°C of an alloy containing 70% Fa/30% Hi. Sketch the expected structures at each stage.**

 (iii)   **Calculate the proportions of phases in this 70% Fa/30% Hi alloy at temperatures of (a) 800°C; (b) 400°C.**

 (iv)   **Describe step by step the solidification and cooling under equilibrium conditions from 1150°C to 0°C of an alloy containing 85% Fa/15% Hi. Sketch structures at each stage.**

  (v)   **Sketch and describe the structure which would most probably be obtained if the 85% Fa/15% Hi alloy were cooled slowly to 700°C and then quenched in cold water.**

A   (i)

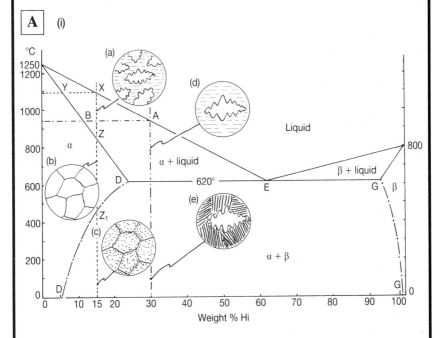

(ii)   The 70% Fa/30% Hi alloy begins to solidify at A (935°C) by depositing nuclei of α of composition B (88.2% Fa/11.8% Hi) causing the composition of the remaining liquid to change along AE as the alloy cools slowly. Diffusion occurs and by the time the temperature has fallen to 620°C the composition of the growing α dendrites has changed along BD so that they are uniform and of composition D (76% Fa/24% Hi). The remaining liquid (composition E) now crystallises as a eutectic by depositing alternate layers of α (composition D) and β (composition G).

     As the alloy cools slowly to 0°C the composition of α changes along DD_1 whilst β will change in composition along GG_1, i.e. particles of β will precipitate from α (both primary and eutectic) whilst particles of α will precipitate from the eutectic β. The precipitated particles will fuse into the appropriate layers of the eutectic, i.e. α particles will join the α layers whilst β particles join the β layers.

(iii) (a) At 800°C:

Weight of solid α (83% Fa/17% Hi) $\times$ (30 – 17)
= Weight of liquid (56% Fa/44% Hi) $\times$ (44 – 30)

or:

$$\frac{\text{Weight of solid } \alpha \text{ (83\% Fa/17\% Hi)}}{\text{Weight of liquid (56\% Fa/44\% Hi)}} = \frac{44-30}{30-17} = 14/13$$

(b) At 400°C:

Weight of solid α (87% Fa/13% Hi) $\times$ (30 – 13)
= Weight of eutectic (38% Fa/62% Hi) $\times$ (62 – 30)

or:

$$\frac{\text{Weight of solid } \alpha \text{ (87\% Fa/13\% Hi)}}{\text{Weight of eutectic (38\% Fa/62\% Hi)}} = \frac{62-30}{30-13} = 32/17$$

but in terms of actual *phases* present (since the *eutectic* is a *mixture* of phases):

$$\frac{\text{Weight of solid } \alpha \text{ (87\% Fa/13\% Hi)}}{\text{Weight of solid } \beta \text{ (3.3\% Fa/96.7\% Hi)}} = \frac{96.7-30}{30-13} = 66.7/17$$

(iv) An alloy containing 85% Fa/15% Hi will begin to solidify at X (1090°C) by depositing α of composition Y (94% Fa/6% Hi). As solidification proceeds under equilibrium conditions it will be complete at 850°C (Z) when the composition of α wll be 85% Fa/15% Hi (and the last trace of liquid was of composition 61.5% Fa/38.5% Hi—absorbed by diffusion into the uniform α).

Uniform α continues to cool until at 462°C (Z_1) it becomes saturated with Hi so that further cooling below 462°C causes progressive precipitation of particles of β as the composition of α changes along DD_1 (at the same time the precipitated β will change in composition along GG_1 as the temperature continues to fall, i.e. migration of both Fa and Hi atoms between α and β is continuous).

(v) The expected structure obtained by quenching an 85% Fa/15% Hi alloy from 700°C will be similar to (b) in the diagram. Quenching (rapid cooling) will suppress any precipitation of β from the α so that a *metastable* solid solution of α will ensue. It will be supersaturated with Hi at 0°C and therefore unstable. In some alloy systems precipitation of minute particles of β would indeed take place with passage of time depending upon the relative thermal activation at ambient temperature.

8.6.6 A System in Which a Peritectic Reaction Occurs So far we have been dealing with alloy systems in which solidification is either of the solid solution or eutectic type. A further mode of solidification involves the *peritectic reaction*. Here two phases which are already present together in a heterogeneous mixture react to produce a third phase, one or both of the original phases disappearing in the process. Frequently one of the reacting phases is a liquid though this is not a pre-condition and quite often two solids will take part—albeit very slowly—in a peritectic reaction. When a liquid reacts with a solid in this way it invariably coats it with an envelope of the new phase. This film or envelope effectively insulates the initial solid phase from further reaction, except by the very slow process of diffusion. The term 'peritectic' is derived from the Greek, 'peri' which means 'around' (as in 'perimeter'), a reference here to this surrounding envelope of the new phase.

8.6.6.1 We will consider the platinum/silver system which is one of the few based almost solely on a peritectic reaction. In this system any alloy containing between 12% and 69% silver will begin to solidify by forming dendrites of the solid solution α but, when the temperature falls to 1185°C (the peritectic temperature), α reacts with the remaining liquid to produce a new solid solution δ, i.e.

$$\alpha + \text{liquid} \rightleftharpoons \delta$$

First let us consider the solidification of an alloy of composition X (75% Pt/25% Ag) (Fig. 8.17). This begins to solidify in the normal manner of a solid solution as the temperature falls to T by depositing dendrites of α of composition A. Assuming that cooling is taking place very slowly so that equilibrium is attained at each stage of the process, then by the time the peritectic temperature of 1185°C has been reached the structure consists of uniform α of composition P and liquid of composition R in the ratio:

Weight of solid solution α (composition P) . PX_1
= Weight of liquid (composition R) . X_1R

Just below 1185°C the peritectic reaction takes place between the solid solution α and the remaining liquid:

$$\alpha + \text{liquid} \xrightarrow{1185°C} \delta$$

The composition of the new solid solution δ at this temperature is given by Q, i.e. 55% Pt/45% Ag. However the overall composition of the alloy is X, i.e. 75% Pt/25% Ag and it follows that not all of the α will be used up since it was obviously in excess of that required to produce a structure solely of δ. Therefore when the reaction is complete some α will remain. In fact the composition of the alloy, when solidification is just complete is given by:

Weight of solid solution α (composition P) . PX_1
= Weight of solid solution δ (composition Q) . X_1Q

Substituting appropriate values in the above equation:

Weight of solid solution α (88% Pt/12% Ag)(25 − 12)
= Weight of solid solution δ (55% Pt/45% Ag)(45 − 25)

or

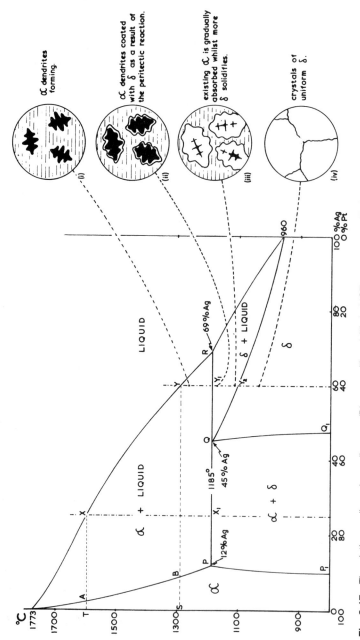

Fig. 8.17 The platinum–silver phase diagram. (Diagrams (ii) and (iii) do NOT represent equilibrium structures but are typical of the sort of situation which would be likely to occur with a moderate rate of cooling.)

$$\frac{\text{Weight of solid solution } \alpha \text{ (88\% Pt/12\% Ag)}}{\text{Weight of solid solution } \delta \text{ (55\% Pt/45\% Ag)}} = \frac{45-25}{25-12} = \frac{20}{13}$$

As the alloy continues to cool slowly α will change in composition along PP_1 and δ along QQ_1.

An alloy of composition Y (40% Pt/60% Ag) will also begin to solidify by depositing solid solution α. In this case solidification begins at temperature S as dendrites of α of composition B form. Again assuming very slow cooling we shall have a solid solution α of composition P and remaining liquid of composition R by the time the temperature has fallen to 1185°C. However, the proportions of α and liquid are different from what they were in the previous case since this alloy is richer in silver. Thus:

Weight of solid solution α (composition P). PY_1
      = Weight of liquid (composition R). Y_1R

Consequently the liquid is present in such excess that when the peritectic reaction takes place all of the α is used up and some liquid remains. Just below the peritectic temperature we have:

Weight of solid solution δ (composition Q). QY_1
      = Weight of liquid (composition R). Y_1R

As the temperature continues to fall the remaining liquid solidifies as δ, the crystals of which change in composition along QY_2. Solidification is complete at Y_2 when uniform crystals of δ of this composition (which of course is the same as Y) remain.

8.6.6.2 Mention has already been made of the tendency for a phase produced as a result of a peritectic reaction to coat a solid reactant. This will most certainly happen during any but extremely slow rates of cooling. The effect is illustrated in Fig. 8.17. Here the α dendrites (shown in black), on reacting with the remaining liquid, have become coated with the new solid solution δ (light) (ii). This will inevitably 'seal off' the α from the remaining liquid and so stop further direct reaction between the two. As the temperature falls the remaining liquid solidifies as δ.

However, due to diffusion, platinum atoms will migrate outwards from the α cores so that these regions will change progressively in composition to that of δ (iii). If cooling is slow enough, or if the solid alloy is subsequently annealed at a high temperature, diffusion will proceed to completion and uniform δ ultimately remain (iv).

8.6.6.3 A peritectic reaction may also result in the formation of an intermetallic compound of fixed composition. Such a situation is illustrated in Fig. 8.18 which consists of the relevant section of the gold/lead phase diagram. Here an alloy of composition X (80% Au/20% Pb) will begin to solidify at temperature T by depositing dendrites of pure gold. This process continues until at 418°C, the peritectic temperature, we have crystals of pure gold along with the remaining liquid of composition L (55% Au/45% Pb). A peritectic reaction then occurs between the gold dendrites and the

Fig. 8.18 Part of the gold–lead phase diagram.

remaining liquid to produce the new phase Au_2Pb, an intermetallic compound of fixed composition as is indicated by the fact that its phase 'field' is the ordinate through 65.6% Au/34.4% Pb. When the alloy is completely solid its phase make-up will be given by:

$$\text{Weight of gold} . SX_1 = \text{Weight of } Au_2Pb . X_1P$$

or $$\frac{\text{Weight of gold}}{\text{Weight of } Au_2Pb} = \frac{X_1P}{SX_1} = \frac{34.4 - 20}{20} = \frac{14.4}{20}$$

An alloy containing, say, 60% Au/40% Pb (Y) will begin to solidify by depositing dendrites of gold as before but when the peritectic reaction occurs (at Y_1) all of the solid gold will be used up in forming Au_2Pb. (Some liquid will remain and as the temperature falls the structure will undergo further changes not indicated in this limited section of the gold/lead phase diagram.)

8.6.7 A System Which Contains One or More Intermediate Phases
So far we have been dealing with alloy systems in which the metals have had little chemical affinity for each other. Consequently, apart from the gold/lead system just mentioned, no intermetallic compounds have been formed in these alloy series and the resultant phase diagrams have been of a relatively simple nature.

If the reader refers to Dr Max Hansen's comprehensive work Constitution of Binary Alloys,* he will find that many phase diagrams are of complex construction and often contain twenty or more phase fields. This state of affairs is due to the formation of a number of intermediate phases in the series. Indeed the complete gold/lead system referred to above is of this type since it includes other intermetallic compounds as well as Au_2Pb.

_____
* Since Dr Hansen's very useful book was published many years ago and has not recently been revised many of the phase diagrams are now out of date in the light of more recent investigational work.

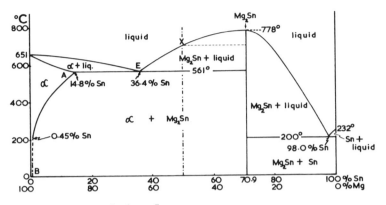

Fig. 8.19   The magnesium–tin phase diagram.

8.6.7.1   The magnesium/tin system (Fig. 8.19) is one which includes a single intermediate phase. This phase is an intermetallic compound of the 'valency' type, Mg_2Sn, and its phase 'field' is represented by the ordinate through 70.9% tin. This indicates that the phase is of invariable composition and that Mg_2Sn dissolves neither magnesium nor tin. It will also be noted that this stable intermetallic compound has a melting point (778°C) which is higher than that of either magnesium or tin.

The system can be interpreted as though we were dealing with two separate eutectic series joined together at the Mg_2Sn ordinate. For example a 50/50 alloy will begin to solidify at X by depositing crystals of pure Mg_2Sn (the temperature horizontal through X cuts the other boundary of the phase field at the ordinate representing pure Mg_2Sn). Mg_2Sn continues to crystallise out until at 561°C, the eutectic temperature of the Mg/Mg_2Sn section of the diagram, the remaining liquid is of composition E. This liquid then solidifies as a eutectic consisting of alternate layers of α (composition A) and Mg_2Sn. On further slow cooling, α changes in composition along AB by rejecting particles of Mg_2Sn (which will join the Mg_2Sn layers in the eutectic). Since Mg_2Sn is of fixed composition no change will take place either in its primary crystals or in its eutectic layers. It will be noted that whilst interpreting the solidification of this alloy X we have not crossed the Mg_2Sn ordinate to the other half of the diagram.

8.6.7.2   Quite often an intermetallic compound will take into solution a limited amount of one or both of its constituent components. Then the intermediate phase field occupies 'width' on the diagram instead of being represented by an ordinate through the composition of the intermetallic compound. Part of the nickel/tantalum phase diagram (Fig. 8.20) is used here to illustrate this case. The intermetallic compound $TaNi_3$ contains 50.67% Ta/49.33% Ni when pure. However, the diagram indicates that $TaNi_3$ will take into solid solution both tantalum and nickel giving rise to the phase field labelled 'β'.

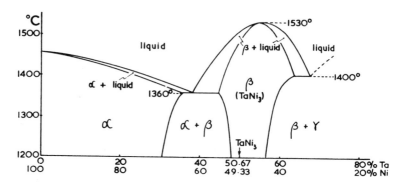

Fig. 8.20 Part of the nickel–tantalum phase diagram.

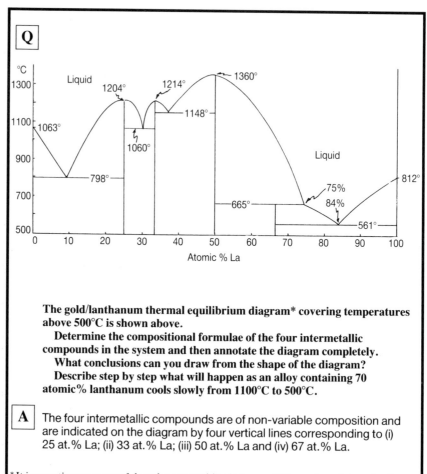

The gold/lanthanum thermal equilibrium diagram* covering temperatures above 500°C is shown above.

Determine the compositional formulae of the four intermetallic compounds in the system and then annotate the diagram completely.

What conclusions can you draw from the shape of the diagram?

Describe step by step what will happen as an alloy containing 70 atomic% lanthanum cools slowly from 1100°C to 500°C.

The four intermetallic compounds are of non-variable composition and are indicated on the diagram by four vertical lines corresponding to (i) 25 at.% La; (ii) 33 at.% La; (iii) 50 at.% La and (iv) 67 at.% La.

---

*It is sometimes more useful to plot composition in terms of '% number of atoms' rather than the more usual '% weight'.

Atomic % Lanthanum	Ratio: $\dfrac{\% \text{ Au atoms}}{\% \text{ La atoms}}$	Chemical formula
(i) 25	$\dfrac{75}{25} = 3:1$	Au_3La
(ii) 33	$\dfrac{67}{33} = 2:1$	Au_2La
(iii) 50	$\dfrac{50}{50} = 1:1$	$AuLa$
(iv) 67	$\dfrac{33}{67} = 1:2$	$AuLa_2$

The equilibrium diagram can now be annotated as below:

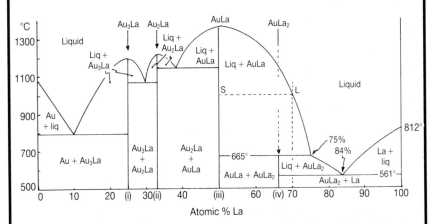

Here we follow the rules that a single-phase field (in this case the vertical lines represent intermetallic compounds of *non-variant* composition) always separates two-phase fields and that on crossing a sloping phase boundary the number of phases always changes by one.

This phase diagram indicates that gold and lanthanum are completely soluble in all proportions in the liquid state but are completely insoluble in each other in the solid state. The four intermetallic compounds form a series of eutectic systems, over appropriate ranges of composition, with gold, lanthanum and with each other. At 665°C and 67% La a peritectic reaction takes place so that with alloys containing more then 67% La a eutectic series $AuLa_2$/La is present.

When an alloy containing 70% La cools to the liquidus temperature (L) (about 1000°C) solidification commences by depositing crystal nuclei of composition S (pure AuLa). This process continues until at 665°C the system consists of solid AuLa and a remaining liquid containing 75% La in the ratio:

$$\text{Weight of AuLa} \times (70 - 50) = \text{Weight of liquid} \times (75 - 70)$$

or

$$\frac{\text{Weight of AuLa}}{\text{Weight of liquid}} = \frac{(75 - 70)}{(70 - 50)} = \frac{5}{20} = \frac{1}{4}$$

A peritectic reaction then takes place at 665°C between solid AuLa and the remaining liquid giving rise to the intermetallic compound $AuLa_2$. The original AuLa crystals are used up first in this reaction and the excess liquid continues to solidify as $AuLa_2$ until at 561°C the remaining liquid contains 84% La. This is a eutectic point at which the remaining liquid solidifies by depositing alternate layers of $AuLa_2$ and pure lanthanum as a eutectic. Thus the final structure consists of uniform crystals of $AuLa_2$ in a eutectic matrix of $AuLa_2$ and La, in a ratio given by:

$$\text{Weight of primary } AuLa_2 \times (70 - 67)$$
$$= \text{Weight of eutectic} \times (84 - 70)$$

or

$$\frac{\text{Weight of primary } AuLa_2}{\text{Weight of eutectic}} = \frac{(84 - 70)}{(70 - 67)} = \frac{14}{3}$$

Ternary Alloy Systems

8.7   So far we have been dealing only with two component systems. Since, in studying metallic systems in particular, we have neglected the effects of pressure variation, only two variables have required consideration. One of these has been temperature and the other, variation of the amount of one component with respect to the other.* Hence only two graphical dimensions are required to represent such a system and flat paper suffices for this purpose.

In the case of a ternary alloy, however, the amounts of any two of the three metals can be varied independently and the extra variable involved requires an extra dimension for its representation, i.e. we must use a three-dimensional phase diagram. The base of this will be in the form of an equilateral triangle, each apex of which represents 100% of one of the metals (Fig. 8.21). Thus an alloy containing 30% Bi/60% Sn/10% Pb will be represented by point P. (This can be demonstrated mathematically.)

* Clearly, in a binary system the proportions of each metal must add up to 100%. Thus if the amount of one is changed the amount of the other changes in sympathy and there is only one independent variable involved.

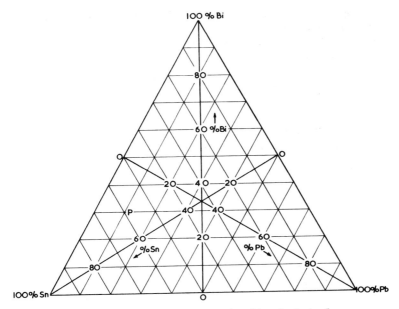

Fig. 8.21   The base plane for a three-component system (bismuth–tin–lead).

Temperature will be represented by ordinates normal to the triangular base.

8.7.1   Figure 8.22 illustrates (within its limitations) the ternary alloy system for bismuth/tin/lead. The binary systems tin/lead, lead/bismuth and bismuth/tin are fundamentally of the eutectic type and are represented by

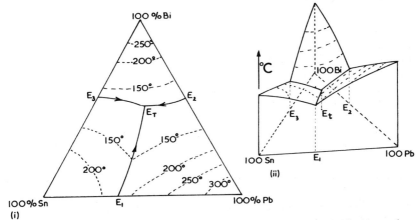

Fig. 8.22   The bismuth–tin–lead phase system. (i) This is a projection of the liquidus surfaces on to a horizontal plane in the manner of a contoured ordnance survey map. (ii) Is a space model showing the liquidus surfaces. E_1 is the eutectic point of the Sn–Pb system (183°C); E_2 the eutectic point of the Pb–Bi system at 124°C and E_3 the eutectic point of the Bi–Sn system at 135°C. The *ternary* eutectic point E_t is at 100°C.

the 'vertical' end faces of the space model (Fig. 8.22(ii)). The binary liquidus lines develop to liquidus surfaces in the three-dimensional model and these surfaces intersect in three 'valleys' which drain down to a point of minimum temperature vertically above E_t. This is the ternary eutectic point of the system, its temperature being lower than that of either of the binary eutectic points E_1, E_2, E_3. For this system the ternary eutectic contains 50% Bi/28% Pb/22% Sn and freezes at 100°C.

At this ternary eutectic point four phases can co-exist, i.e. solid bismuth, solid tin, solid lead and liquid. The components of the sytem are bismuth, tin and lead. Applying the modified Phase Rule:

$$f + p = n + 1$$
$$f + 4 = 3 + 1$$
$$f = 0$$

Thus at this eutectic point, as at a binary eutectic point, the system is invariant.

8.7.2   In considering this ternary system we have dealt only with the liquidus 'surface'. Generally we shall be more involved with changes which take place in *solid* alloys at lower temperatures, i.e. 'inside' the space model. Clearly such three-dimensional phase 'volumes' are rather inconvenient to illustrate, though for demonstration purposes three-dimensional space models may be constructed using bits of differently coloured electrical wire to indicate phase boundaries. Many years ago such a project used to be a favourite exercise leading to the award of a Ph.D. in many university metallurgy departments. Much painstaking work was involved including the casting of alloys of different compositions within the series,

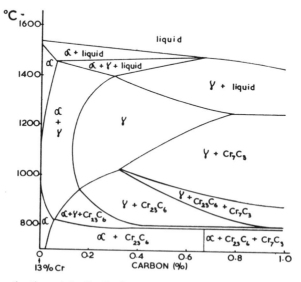

Fig. 8.23   A section through the Fe–Cr–C space model at 13% chromium.

followed by careful chemical analysis, prolonged annealing of samples at different temperatures and then quenching. The phase structure was then examined, generally using the microscope.

It is generally more useful to use either horizontal 'slices' (that is, isothermal sections) at the required temperatures; or vertical slices at a fixed amount of one of the components. Figure 8.23 represents a vertical slice from the ternary iron/chromium/carbon system. This has been taken at the 13% Cr base line and the section shown will be parallel to the iron/carbon end face of the space model.

The same sort of arrangement can be used for multi-component alloys. Then, all but one of the components are fixed and a two-dimensional diagram drawn showing temperature and *one* component as variables. These diagrams—usually called 'pseudo-binary' diagrams—are very useful for studying alloy steels which may contain up to six component elements.

9

Steel and its Heat-treatment

9.1 Iron was known to the Ancients as 'Metal from Heaven' or 'Metal from the Stars' from which we may conclude that such iron as was used in those times was of meteoric origin. The deliberate smelting of iron ores seems to have originated some 3500 years ago and by that time the Iron Age can be said to have arrived. Presumably historians would agree that we are still in the Iron Age for, despite the developments of aluminium alloys and plastics materials during the twentieth century, steel undoubtedly remains our most important engineering material. It is unlikely that the present state of technological development could have been achieved in any field without the help of steel as our main structural, constructional and tool material—and all this at a relatively low financial outlay.

Almost all ferrous materials—carbon steels, alloy steels and cast irons—are derived from pig iron produced in the blast furnace. Pig iron is a relatively impure material containing up to a total 10% of carbon, silicon, manganese, phosphorus and sulphur. The manufacture of steel from pig iron involves the progressive oxidation of these impurities which are lost as slag or gas. Such processes were, until the latter half of the twentieth century, carried out either in a Bessemer converter or an open-hearth furnace which, whilst removing the bulk of the impurities present in the original pig iron, left behind a significant amount of these impurities in the metal and, in the case of the now obsolete Bessemer process, introduced a new impurity, atmospheric nitrogen. Whilst amounts of these residual impurities were small they were sufficiently variable in quantity to make precise alterations in properties (by alloying) difficult. In the years following the Second World War, however, new steel-making processes were developed—the new *basic oxygen processes* where low priced 'tonnage' oxygen replaces air and so eliminates nitrogen contamination—and electric processes where totally controlled and chemically clean conditions lead to a high-purity product of predictable properties. The high ductility of such a material has allowed steel—or more properly, *iron*—to replace many of the more expensive non-ferrous metals for deep-drawing and other forming processes.

9.1.1   Steel is essentially an alloy of iron and carbon containing up to roughly 2.0% carbon. By varying this carbon content and the heat-treatment of the resultant alloy we can obtain an enormous range of mechanical properties, whilst the addition of alloying elements such as nickel, chromium and molybdenum extends the properties still further. Nevertheless it is a sobering thought that, but for the *polymorphic* changes which occur in iron (3.6.4), the heat treatment of steel as we practise it would be impossible and that iron/carbon alloys would then have extremely limited properties. Certainly tool steels and high-strength constructional steels would not be available to the engineer. It is possible that we would still be in the Bronze Age.

9.1.2   The relatively small carbon atoms dissolve interstitially in iron. However, whilst the FCC form of iron will dissolve up to 2.0% C, the maximum solubility of carbon in BCC iron is only 0.02% at 723°C and this falls almost to zero (probably of the order of $2.0 \times 10^{-7}\%$) at 0°C. This is because the interatomic spaces in BCC iron are very small so that carbon atoms cannot readily be accommodated. In fact carbon atoms are too large for extensive interstitial solid solution but too small to form a stable substitutional solid solution with iron. Whilst the interatomic spaces are larger in FCC iron they are not large enough to accommodate carbon atoms freely. Some crowding therefore occurs producing strain in the structure so that not all spaces can be occupied.

The maximum solubility of carbon in the FCC form of iron is 2.0% and this sets the limit of the carbon content of a steel. As the FCC form changes to BCC any carbon in solution in excess of 0.02% will be precipitated from solid solution assuming that it is cooling slowly enough to permit this. Under the slowest rate of cooling which is possible carbon does not precipitate as graphite but as iron carbide, Fe_3C. Like most metallic carbides this is a hard substance. Consequently as the carbon content of a steel increases so does its hardness even though it has not been heat-treated but is only in the slowly cooled condition.

9.1.3   Since steel is such an important alloy and has been studied since the very early days of metallography* names have been adopted for the three phases mentioned above. The interstitial solid solution of carbon in FCC iron is known as *Austenite*. It can contain up to 2.0% C and is a stable phase between 723 and 1493°C. The interstitial solid solution of carbon in BCC iron is known as *Ferrite*. This is stable below 910°C and can contain a maximum of only 0.02% C (at 723°C). Nevertheless although ferrite contains so little dissolved carbon this carbon exerts considerable influence on the mechanical properties of ferrite by opposing the movement of dislocations (7.5.1.2). For this reason very high purity iron (99.9999%) is extremely soft and ductile as compared with the low-carbon ferrite of dead-mild steel. Iron carbide, Fe_3C, is generally known as *Cementite*.

---

*Professor Henry Sorby (of Sheffield) was conducting the systematic microscopical metallography of steels in the early 1860s.

Other phases, and mixtures of phases, are also given special names as we shall see subsequently.

9.1.4 What is popularly known as the 'iron/carbon diagram' is strictly speaking not a true equilibrium diagram since cementite (iron carbide) is not a completely stable phase. Graphite is in fact more stable than cementite but nucleation of cementite takes place in advance of precipitation of graphite mainly because at that stage the FCC iron is saturated with carbon. Once cementite has formed it is quite stable and for most practical considerations can be treated as an equilibrium phase. However, whilst Fig. 9.1 represents what is generally known as the 'iron/carbon equilibrium diagram' we should bear in mind that it is more properly labelled the 'iron/*iron carbide* metastable system'.

9.1.5 Phase changes within the structures of solid alloys are often sluggish for reasons such as those described in 7.4.1. Once an alloy has solidified atoms move with greater difficulty into positions of equilibrium and often require considerable thermal activation to effect this state. This also applies to the polymorphic transformation $(\alpha \rightleftarrows \gamma)$ which takes place in iron—and consequently in carbon steels. Figure 9.3 indicates that for a 0.4% carbon steel the transformation, austenite→ferrite, begins at U_1 (approximately 820°C). In fact on *heating* such a steel the ferrite→austenite change would probably not be complete until 850°C whilst on *cooling* the reverse transformation would not be complete until about 790°C. Thus the (austenite$\rightleftarrows$austenite + ferrite) change point for this 0.4% carbon steel is taken as 820°C—the equilibrium temperature midway between 790°C and 850°C.

Similarly the austenite$\rightleftarrows$pearlite transformation may not occur until about 750°C on heating whilst it may be delayed until 690°C on cooling— 723°C is an average value in this case for the 'equilibrium' temperature. These measured temperatures depend largely on the rates of heating (or cooling) which prevail during the plotting of the cooling curves.

The temperatures measured on *heating* were designated A_{c1} and A_{c3} whilst those measured during *cooling* were designated A_{r1} and A_{r3} when these transformation points were investigated more than a century ago by the French scientist F. Osmond. His nomenclature is still used, i.e. 'c' stands for 'chauffage' (heating) whilst 'r' stands for 'refroidissement' (cooling). 'A' represents 'arret' (arrest) since these temperatures were measured as arrest points on corresponding heating or cooling curves.

Incidentally the A_2 temperature refers to the Curie point (768°C for pure iron) above which iron loses its ferromagnetism. The temperature does not appear on the phase diagram since no crystal change is involved and solubilities of added elements are not affected.

9.1.6 Whilst methods for determining the positions of phase boundaries become more sophisticated metallurgists nevertheless rely to a great extent on visual analysis of microstructures using simple optical microscopes. For reasons such as this it is difficult for example, to determine whether or not

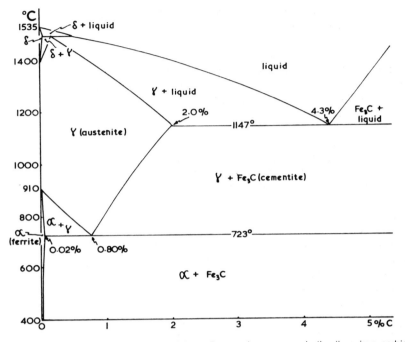

Fig. 9.1 The iron–carbon phase equilibrium diagram (more properly the 'iron–iron carbide metastable system').

a steel is of precise eutectoid structure when viewed through the microscope and that the observed structure does not contain a very small excess of either ferrite or cementite. During my lifetime as a metallurgist the eutectoid composition of carbon steel has successively been established between 0.89% and 0.77% carbon, values of 0.81; 0.83; 0.85 and 0.80% being in fashion during different periods the while. For this reason I have adopted the value 0.8% carbon and have retained this over the years. Other values in the iron/carbon system show similar variations but are of little consequence in practice. I make these points since readers may be puzzled by variations in phase diagrams as between different text books.

9.1.7   If iron did not undergo polymorphic changes at 1400 and 910°C but retained its FCC structure at all temperatures it would presumably form a simple eutectic system with a eutectic point at 4.3% C and 1147°C, and the solid solubility of carbon in austenite falling from 2.0% at 1147°C to a very low value at 0°C. The resultant alloy would have limited use and would not be amenable to the extensive heat-treatments applied to carbon steels as we know them, though some limited precipitation-hardening treatment might be possible.

The very limited solid solubility of carbon in the BCC forms (α and δ) introduces modifications to the diagram as indicated. Thus a steel containing less than 0.51% C will begin to solidify as dendrites of δ which, at

1493°C will undergo a peritectic reaction with the remaining liquid to produce γ (austenite). As the steel cools the FCC structure will ultimately begin to transform to one which is BCC and carbon will be thrown out of solid solution as cementite.

Since carbon dissolves interstitially in iron the diffusion of carbon atoms is relatively rapid and normally there is no coring of carbon with respect to iron. Other alloying elements, and particularly impurities, may, however, be cored especially if they dissolve substitutionally in iron.

The Transformation of Austenite Under Near Equilibrium Conditions

9.2 We will now study the solidification and transformations which occur as a 0.4% C steel cools slowly from the liquid state to room temperature. Such a steel will begin to solidify when its temperature falls to T (Fig. 9.2)

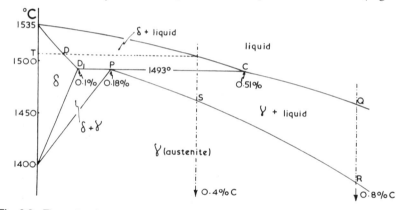

Fig. 9.2   The peritectic region of the iron–carbon diagram.

and dendrites of BCC δ (of composition D) begin to form. This phase, δ, continues to crystallise and at the same time change in composition along DD_1 until, at 1493°C, its composition is given by D_1 (0.1% C). The remaining liquid in equilibrium with δ at this temperature is of composition C (0.51% C). A peritectic reaction then occurs between δ and this remaining liquid producing a new phase—the solid solution γ (austenite). Since the overall composition of the steel lies between P and C all of the δ will be used up in the reaction and the surplus liquid will then solidify out as γ whilst the temperature falls to S and the γ changes in composition along PS by diffusion of carbon.

No further phase change takes place until the temperature has fallen to T_1 (Fig. 9.3). This is known as the *upper critical*—or A_3—temperature for this particular steel.*

*The *upper critical temperature* of a steel is given by the appropriate point corresponding with its composition on the line UEV (Fig. 9.3).

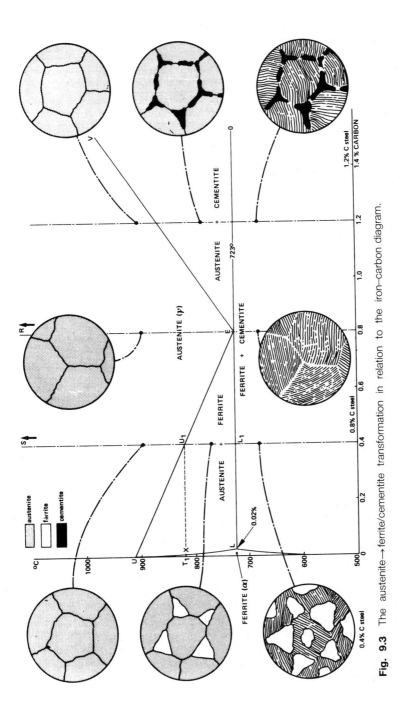

Fig. 9.3 The austenite→ferrite/cementite transformation in relation to the iron–carbon diagram.

As the temperature falls just below T_1 crystals of BCC α (ferrite) of composition X begin to grow within the austenite which of course they gradually replace. Since the ferrite contains less than 0.02% C it follows that the remaining austenite becomes richer in carbon so that its composition moves to the right towards E. By the time the temperature has fallen to 723°C we shall have ferrite of composition L (0.02% C) and remnant austenite of composition E (0.80% C) in the ratio:

$$\frac{\text{Weight of ferrite (0.02\% C)}}{\text{Weight of austenite (0.8\% C)}} = \frac{L_1E}{LL_1} \tag{1}$$

i.e. there are approximately equal masses of each phase present at this instant.

9.2.1 The *lower critical*—or A_1—temperature is given by the line LO (Fig. 9.3) and is of course constant at 723°C for all carbon steels. The point E will be recognised as being associated with a eutectic type of transformation. Previously we have been dealing with the formation of eutectics from liquid solutions. Here, however, the transformation of a *solid* solution (austenite) is involved and so the resultant structure in this case is termed a *eutectoid*.

As the temperature falls just below 723°C the remaining austenite, now of composition E, transforms to eutectoid by precipitating alternate layers of ferrite (composition L) and cementite (Fe_3C). This particular eutectoid is known as *pearlite*, a reference to the fact that, like mother-of-pearl, it behaves as a diffraction grating in dispersing white light. It is important to realise that pearlite is *not* a phase but a mixture of two phases, ferrite and cementite. Since the austenite present in the structure at just above 723°C has now transformed to pearlite the ratio:

$$\frac{\text{Weight of primary ferrite}}{\text{Weight of pearlite}} = \frac{L_1E}{LL_1}$$

i.e. as in (1) above. All *hypo*-eutectoid steels, that is, those containing *less* than 0.8% C consist of varying quantities of pearlite and primary ferrite, assuming that they are in an equilibrium state.

Further fall in temperature will have no significant effects on the microstructure. The carbon contents of both primary and eutectoid ferrite will decrease from the already low value of 0.02% to virtually zero. This carbon will precipitate as tiny particles of cementite mainly at the ferrite grain boundaries.

9.2.2 A steel containing exactly 0.8% C will solidify directly as austenite (between Q and R—Fig. 9.2) and will undergo no further changes until it reaches the eutectoid point E when the whole of the structure will transform from austenite to pearlite. *Hyper*-eutectoid steels, i.e. those containing *more* than 0.8% C, will also solidify directly as austenite. On reaching the upper critical temperature (a point on EV), primary cementite will begin to separate out—generally as embrittling films at the austenite grain boundaries—so that the composition of the remaining austenite moves to the left towards E. At 723°C this remaining austenite contains exactly 0.8% C and will transform to pearlite as before.

The Formation of Pearlite

9.3 When austenite transforms isothermally just below 723°C the reaction product is pearlite, the eutectoid of ferrite and cementite indicated by the iron/carbon diagram. As emphasised already pearlite is not a phase but a mixture of two phases. These two phases, ferrite and cementite, are present in a definite ratio as indicated by the lever rule:

$$\text{Weight of ferrite} . (0.8) = \text{Weight of cementite} . (6.67 - 0.8)$$

Since cementite, Fe_3C, contains 6.67% by weight of carbon.
    Therefore:

$$\frac{\text{Weight of ferrite}}{\text{Weight of cementite}} = \frac{(6.67 - 0.8)}{0.8} = \frac{5.87}{0.8} = \frac{7.3}{1}$$

As the relative densities of ferrite and cementite are roughly equal the lamellae of ferrite are approximately seven times the width of those of cementite. As a result ferrite forms a continuous phase with the plates of cementite surrounded by this ferrite matrix. Since ferrite is tough and ductile this makes pearlite tougher and stronger than it would be if the brittle cementite were continuous.

9.3.1 The mechanism of the austenite→pearlite transformation is typical of many reactions within solid alloys in that it begins at grain boundaries in the main and continues into the existing grains. This is predictable since the atoms at grain boundaries have higher energy states than atoms within the grains. Hence these atoms at grain boundaries require less extra energy in order to break away from their neighbours to form a new structure.
    It is thought that carbon atoms collect together and so form a cementite nucleus (Fig. 9.4(ii)). As this nucleus grows inwards from the boundary the surrounding austenite becomes successively depleted in carbon and, since its composition moves to the left of the eutectoid point, so layers of ferrite separate on either side of the cementite nucleus (iii). Beyond these ferrite layers an increase of carbon occurs so that more cementite nucleates (iv) and so on, the structure building up as alternate layers of cementite and ferrite (vi).

Uses of Steel

9.4 Ordinary carbon steels can be classified conveniently into four main groups in terms of their specific uses:

9.4.1 'Dead-mild' Steels containing up to 0.15% C, which are used for

Fig. 9.4 The nucleation and growth of pearlite (it is assumed that the steel in this instance contains 0.8% carbon).

general presswork and other applications where high ductility is necessary in forming, e.g. motor-car bodies, tin cans, nails and wire. Since 1950 there has been a considerable improvement in the chemical purity of mild steel generally. This has resulted in improved ductility so that dead-mild steel has replaced some of the more expensive deep-drawing alloys such as 70–30 ('cartridge') brass for many traditional uses.

9.4.2 Mild Steels containing 0.15–0.3% C. Wrought forms are used as RSJ and other structural members, shafting, levers and various forgings. Steels used for sand castings usually contain 0.3–0.35% C.

9.4.3 Medium-carbon Constructional Steels containing 0.4–0.6% C, which are widely used for components in the engineering industries, e.g. axles, connecting rods, gears, wire ropes, rails, etc.

9.4.4 Tool Steels containing 0.6–1.5% C.

9.4.5 Steels in the first two of these groups, i.e. containing less than 0.3% C, do not harden sufficiently when heat treated to make this generally worth while. This is not due specifically to lack of carbon as we shall see later (9.6.6). Such steels are therefore used in the cast, hot-rolled, cold-worked or normalised forms. The third and fourth groups contain steels with more than 0.4% C and can be hardened or strengthened by an appropriate heat treatment programme.

   In Britain the motor-car industry is the largest single user of steel, followed by various branches of the building and plant construction industries. Thus, despite the attractive properties offered by steel in its heat-treated forms we must not forget that the bulk of ferrous material is of the low-carbon type which is not heat-treated. Some of the many uses of carbon steels related to carbon content and mechanical properties are illustrated in Fig. 9.5.

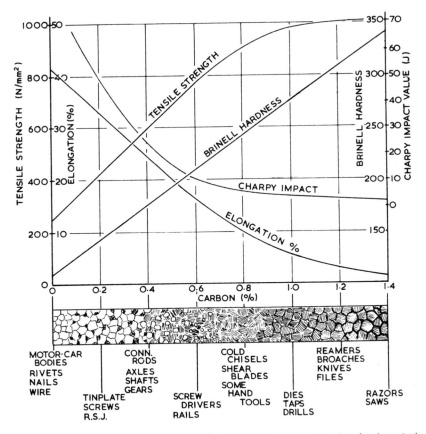

Fig. 9.5 The relationship between % carbon, uses and mechanical properties of carbon steels in the hot-rolled condition.

Heat-treatment Processes which Produce Equilibrium Structures in Steels

9.5 Several heat-treatment processes applied to steels result in the formation of pearlitic* structures. In the first two of these processes mentioned here no phase change occurs since the treatment is carried out *below* the A_1—or lower critical—temperature.

9.5.1 Process Annealing This is an inter-stage annealing operation such as might be applied to any heavily cold-worked metal or alloy in order to achieve recrystallisation (5.8.3) and lead to the recovery of malleability and ductility so that further cold-work may be applied. In this context it is

---

* In this sense the term 'pearlitic' means that the steel has cooled slowly enough for pearlite to form, not that the structure will necessarily be wholly pearlitic, which of course it will not be if less—or more—than 0.8% C is present.

used mainly to soften mild and dead-mild steels which have become work-hardened by cold-rolling or drawing and which must undergo further cold-forming. A typical example is the drawing of low-carbon steel wire.

The presence of small amounts of carbon in the form of scattered patches of pearlite is not significant here. We are involved mainly with the work-hardening of ferrite. The recrystallisation temperature of ferrite is of the order of 500°C but on an industrial scale annealing at about 650°C is generally employed so that recrystallisation will proceed more quickly and a treatment time of about one hour be ample. The structure remains one of ferrite/pearlite throughout and since no phase change occurs the equilibrium diagram is not involved here.

9.5.2 Spheroidising Anneals   The machining of high-carbon steels even in the normalised condition is difficult since the layers of pearlitic cementite provide hard barriers to cutting. The structure can be suitably modified by a prolonged annealing treatment—upwards of twelve hours—at a temperature between 650°C and 700°C, that is, just *below* A_1. Again, no phase change will occur but, due to surface energy effects, the cementite layers of the pearlite gradually break up and assume a globular form (Fig. 9.6).

Fig. 9.6   The spheroidisation of pearlitic cementite on prolonged annealing just below A_1.

Not only is machinability improved but cold-forming processes are possible when the steel is in this condition. After shaping is completed the steel can be hardened or the original pearlitic structure restored by normalising.

9.5.3 Normalising   The general aim of this process is to refine the grain of a steel which, for some reason, has become coarse-grained. Such refinement of grain improves the mechanical properties particularly in respect of ductility and toughness. Normalising is applied to hypoeutectoid steels and involves heating the material to a temperature not more than 50°C above the A_3—or upper critical temperature. Between A_1 and A_3 the initial structure recrystallises to austenite but, and this is the important point, to austenite of *very fine grain*. After a short period of 'soaking' the steel is removed from the furnace and cooled in still air when the fine-grained austenite will transform to fine-grained ferrite and pearlite.

It is important that neither the temperature nor the time of treatment be exceeded or grain growth of the newly formed austenite will take place and give rise to coarse-grained ferrite and pearlite. Normalising also improves the machinability of hypo-eutectoid steels. The resultant surface finish is also superior to that obtained by machining an annealed component, since high ductility in the latter leads to a tendency to local tearing of the surface.

9.5.4 Annealing of Castings Sand castings in steel commonly contain about 0.35% C and, because of the slow rate of solidification and cooling in the sand mould, considerable grain growth takes place in the austenite whilst it is at temperatures above 1000°C. Consequently such a steel is very coarse grained by the time the temperature has fallen to A_3 and the primary ferrite tends to separate out as elongated plates at the austenite grain boundaries and also along the close-packed {111} planes *within* the coarse austenite grains (Fig. 9.10). This is termed a Widmanstätten* structure and is characterised by brittleness since fracture can propagate easily along the boundaries of the coarse primary ferrite plates.

The structure can be refined by heat treatment in a process closely resembling normalising. The casting is heated to some 50°C above the A_3 temperature, soaked for a period depending upon the section of the casting

Fig. 9.7 The replacement of the coarse Widmanstätten pattern by a fine-grained structure during the annealing of a casting.

and then cooled, usually in the furnace. Air cooling would be more likely to lead to the formation of internal stresses particularly in castings of variable section which would therefore cool unevenly.

As the casting is heated to the A_1 temperature small austenite grains

---

* A structure of this type was first observed on the surface of a meteorite by Aloys Beck von Widmanstätten in 1804—possibly one of the first excursions into practical metallography.

replace the coarse pearlite and as the temperature rises to A_3, the coarse Widmanstätten ferrite plates gradually transform to *fine-grained* austenite also. On cooling again this fine-grained austenite is replaced by fine-grained ferrite/pearlite. Whilst the tensile strength is little affected by this treatment, ductility is almost doubled and impact toughness increased considerably.

9.5.5 'Full Annealing' is sometimes used for carbon steels of all composi-tions. Hypo-eutectoid steels are heated to approximately 50°C above the A_3 temperature followed by slow cooling, generally in the furnace. Such a treatment applied to *hyper*-eutectoid steels would give rise to the precipita-tion of coarse brittle intercrystalline networks of primary cementite (9.2.2) during the subsequent cooling process between A_3 and A_1. This could only be dispersed by hot forging—impracticable if the work piece is already 'to size'. Hence hyper-eutectoid steels are heated to 30°C above A_1 followed by slow cooling. Under these conditions *globular* primary cementite (formed by the original rolling (or forging) process just above A_1) is not dissolved and so cannot precipitate as needles on cooling.

In either case a soft coarse-grained structure free of stress and easily machinable is formed. Subsequently the finished product may be normal-ised or hardened (and tempered) as required.

The Non-equilibrium Transformation of Austenite

9.6 Phase transformations in solid alloys do not occur instantaneously but are very time-dependent. Many such transformations rely on diffusion which in substitutional solid solutions is inevitably a slow process (7.4). For this reason, particularly in many industrial processes, the rate of tempera-ture fall far outstrips the ability of the alloy to maintain phase equilibrium during cooling. Often when ambient temperature is reached the thermal energy of the system is too low to allow further transformation, and a non-equilibrium structure is trapped, sometimes permanently unless furth-er heat-treatment is applied.

9.6.1 Since carbon dissolves interstitially in γ-iron it is able to diffuse comparatively quickly so that, for example, the resultant coring of carbon with respect to iron is negligible. Nevertheless when austenite is cooled rapidly below the A_1 temperature—to about 200°C or so—the driving force tending to cause transformation from FCC to BCC is so great that it occurs *before* carbon has an opportunity to form separate crystals of carbide. Consequently supersaturation of the new BCC structure with carbon is bound to result. This causes considerable distortion of the structure which in turn makes slip virtually impossible, a fact which manifests itself in the great hardness produced by quenching a steel in cold water. Such quenching of a steel from the austenitic condition gives rise to a solid solution of carbon in BCC iron which is of the order of at least one

%C

0·05

0·15

0·45

0·65

0·80

Fig. 9.8 Illustrates the effect of increase in carbon content on the structures of normalised plain carbon steels. Ferrite (light) with increasing amounts of pearlite (dark). The pearlite is not generally resolved at this magnification. (Etched in 2% Nital.) ×150.

Fig. 9.9   1.2% carbon tool steel, annealed. Pearlite and primary cementite (Fe₃C). Here the high magnification is sufficient to resolve the pearlite showing lamellae of cementite in a ferrite matrix. (Etched in 2% Nital.) ×1000.

thousand times supersaturated. It appears under the microscope as acicular (needle-shaped) crystals though what is being observed is in reality a cross-section through roughly discus-shaped plates. This structure, known as *Martensite*, is single-phased since all of the carbon has been retained in solid solution.

If the degree of undercooling is less severe allowing transformation to proceed isothermally at some temperature between 250°C and 550°C then carbide will precipiate though not in the lamellar form found in pearlite. In this case the carbide tends to form as small particles too small to be visible except at high magnifications, and the structure is known as *Bainite*.

Whereas pearlite formation is initiated by the nucleation of carbide, bainite formation appears to be initiated by ferrite growth which in turn is followed by the precipitation of carbide particles adjacent to the ferrite. Transformation which occurs at the lower end of the range (around 250°C) gives rise to very small carbide particles and the structure, known as *lower bainite*, resembles most closely that of *martensite* in shape. Transformations occurring in the region of 500°C on the other hand, allow much larger carbide particles to form and the structure, which has a different appearance to that formed around 250°C, is usually known as *upper bainite*.

Transformations which occur at temperatures just below 723°C and down to about 550°C will give rise to pearlite. The lower the transformation temperature within these limits the finer the structure of pearlite produced.

Fig. 9.10 0.5% carbon steel overheated to 1100°C and furnace cooled. Ferrite (light) has precipitated along the original austenite grain boundaries and also as Widmanstätten plates. (Etched in 2% Nital.) ×150.

Fig. 9.11 0.5% carbon steel annealed at 700°C for 12 hours to give spheroidisation of the pearlitic cementite lamellae. (Etched in 2% Nital.) ×1000.

Fig. 9.12   0.5% carbon steel water quenched from 850°C. Martensite. (Etched in 2% Nital.) ×1000.

Fig. 9.13   0.5% carbon steel, water quenched from 850°C and tempered at 400°C. Tempered martensite. (Etched in 2% Nital.) ×1000.

Fig. 9.14   0.5% carbon steel oil quenched from 850°C, showing a mixture of martensite (light) and bainite (dark). (Etched in 2% Nital.) ×150.

Fig. 9.15   The thermal treatment sequence for the derivation of a TTT diagram. The specimens are of roughly the dimensions of a British 1p piece or an American dime coin.

9.6.2   The relationship between the temperature at which transformation of austenite occurs and the structure and properties produced can be studied by reference to a time–temperature–transformation (TTT) diagram. Such a diagram (Fig. 9.17) relates the temperature at which the transformation of austenite is taking place with the time necessary for this to occur and the type of structure produced as outlined above. This form of diagram is produced for a steel of single composition and we shall consider one of eutectoid (0.8% C) composition.

As Fig. 9.3 indicates this steel becomes completely austenitic when heated just above 723°C so a sufficient number of specimens, all identical in size and composition (0.8% C) are heated in a bath of molten salt at 730°C so that they attain this state. About ten of these specimens are then removed simultaneously from the austenitising bath and quickly transferred to another salt bath which will be maintained at some predetermined

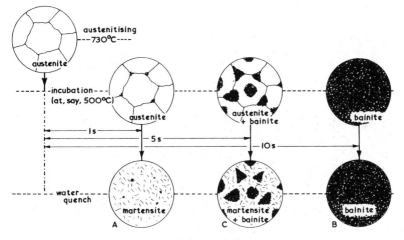

Fig. 9.16   The extent of transformation during incubation.

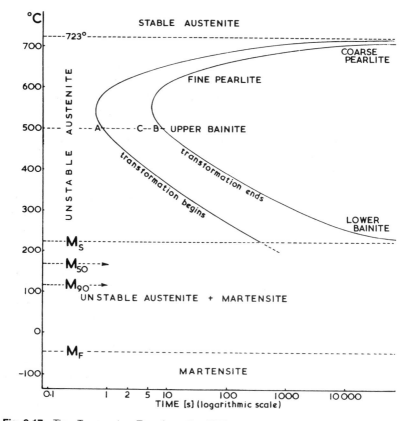

Fig. 9.17 Time Temperature Transformation (TTT) curves for a 0.8% plain carbon steel.

temperature *below* 723°C. Individual specimens are then removed from this second—or *incubation*—bath after increasing time intervals and immediately quenched in cold water. This final quench effectively halts any transformation which was taking place at the incubation temperature in the second bath and any austenite which remained will be converted to martensite by the final quench.

Microscopical examination will reveal the extent to which transformation had occurred during incubation in the second salt bath (Fig. 9.16). Thus a specimen which was almost completely martensitic (A) had only just begun to transform after being held for one second at the incubation temperature (500°C in this case); whilst one which contained no martensite but which was almost completely bainite (B) had already transformed to bainite after ten seconds at the incubation temperature. The final water quench could then have no further effect on the structure since the transformation of austenite was already complete. Any specimen incubated at 500°C for a time intermediate between these two extremes would, on quenching in the final water bath, contain a mixture of bainite and martensite (C) indicating that transformation was incomplete after that holding time (in this case five seconds).

The same sequence of operations is then repeated with other batches of the same austenitised steel but using different incubation temperatures. At each incubation temperature the time for transformation to begin and to be completed can then be ascertained and the complete TTT curves plotted.

In practice small *thin* specimens of approximately the dimensions of a British 1p piece or an American dime coin are used in order that they shall attain very quickly the temperature of the incubation bath to which they are transferred. A length of temperature-resistant wire is generally fixed to the specimen to facilitate rapid transfer and quenching.

9.6.3 The resultant TTT curves plotted from information obtained by studying the *isothermal* transformations of a 0.8% C steel are shown in Fig. 9.17. The horizontal broken line representing the temperature of 723°C is of course the A_1 temperature above which the structure of the 0.8% C steel in question consists entirely of stable austenite. At temperatures below this line austenite becomes increasingly unstable and the two approximately C-shaped curves indicate the time intervals necessary for the austenite $\rightarrow$ ferrite + carbide transformation to begin and to reach completion when a specimen is cooled rapidly to the incubation temperature.

As might be expected transformation is sluggish at temperatures immediately below 723°C since austenite is then only slightly short of its lowest equilibrium temperature, but as the temperature falls towards 550°C both the delay in starting and the time required for completion of transformation decrease rapidly. Again this is to be expected since the greater the degree of undercooling of the austenite the greater will be its urge to transform. Both the interval before transformation begins and the period of transformation itself are at a minimum at about 550°C.* Any transformation which takes place between 723°C and 550°C is initiated by cementite precipitation since in this range the diffusion of carbon in FCC iron is rapid and the transformation product is pearlite (9.3)—coarse pearlite just below 723°C where transformation is sluggish and broad lamellae have time to form; but fine pearlite when rapid transformation occurs near 550°C.

At temperatures between 550°C and 220°C transformation becomes increasingly sluggish with decrease in temperature, for, although the FCC structure becomes increasingly unstable the further the temperature falls below 723°C, the thermal diffusion of carbon at lower temperatures becomes much slower and influences transformation. In this temperature range transformation is initiated by *ferrite* precipitation (instead of cementite as in pearlite formation) and the product, bainite, varies in appearance from a dark feathery mass for that formed in the region of 550°C (upper bainite) to an acicular structure resembling martensite for that formed in the region of 220°C (lower bainite).

---

* Note that a logarithmic scale is used.

9.6.4 The horizontal lines at the foot of the diagram are, strictly speaking, not part of the TTT diagram but represent the temperature range over which the transformation: austenite→martensite begins (M_s) and is about 90% complete (M_{90}). The M_f line at which transformation is 100% complete is at approximately $-50°C$ so that most steels quenched to room temperature exhibit some 'retained austenite' in the microstructure.

It must be emphasised that these TTT curves represent *isothermal* transformations, that is, transformations which have taken place at a series of single incubation temperatures, and it would be inadmissible to superimpose on them cooling curves which represent more or less continuous rates of cooling. However, it is possible to derive modified TTT curves which are related to continuous rates of cooling. These are similar in shape to true TTT curves but are displaced somewhat to the right as indicated in Fig. 9.18.

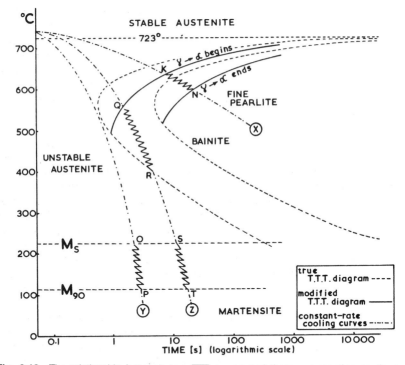

Fig. 9.18 The relationship between true TTT curves and those representing continuous-cooling conditions.

9.6.5 On this diagram curves X, Y and Z, representing different continuous cooling rates, are superimposed. Curve X indicates a rate of cooling such as might prevail during a normalising process. Transformation of the unstable austenite begins at K and is complete at N and the product is of course fine pearlite. A very rapid fall in temperature is represented by curve Y. Here the 'nose' of the modified transformation-begins curve is missed so that unstable austenite persists until at O (on the M_s line) transformation direct to martensite begins. The rapid cooling necessary in order to avoid the nose of the transformation-begins curve explains why it is not possible to harden uniformly very heavy sections in ordinary carbon steel. Whilst the *surface* of a thick bar might cool sufficiently rapidly to be represented by curve Y and so become martensitic, the core of the same bar will cool more slowly and be represented, say, by curve Z. Thus, in the core, transformation to bainite will begin at Q, be arrested again somewhere in the region of R and the austenite which remains will begin to transform to martensite at S. Consequently whilst the outer skin of the bar will be totally martensitic the core will contain a mixture of martensite and softer bainite.

9.6.6 The TTT curves we have been considering are those for a 0.8% C steel. If the carbon content is either above or below this amount the curves are displaced to the left so that the critical cooling rate* required to produce a totally martensitic structure will be even greater. Hence the need for drastic water quenches in order to harden plain carbon steels becomes apparent. With a steel of less than 0.3% C the transformation-begins curve has moved so far to the left (Fig. 9.19(i)) that it is impossible to cool a steel such that the formation of considerable amounts of ferrite in the upper temperature ranges is avoided. The resultant structure therefore contains martensite interspersed with soft ferrite.

9.6.7 An unfortunate result of this need to cool plain carbon steels so rapidly, if they are to be completely hardened, is that internal quenching strains are set up. These strains lead to distortion and, particularly with components of complex shape, to quench cracking. With rapid cooling it is inevitable that the outer skin and the inner core of a piece of steel will pass through the FCC→BCC transformation after *different time intervals*. Since a volume change accompanies this transformation (3.6.3) it follows that the skin, as it changes to BCC, will tend to expand away from the core which at that instant is still FCC in structure. Internal strain is set up and this is very likely to cause cracking of the skin which has just transformed to hard, brittle martensite.

Fortunately the addition of alloying elements like nickel, chromium or molybdenum to a steel reduces considerably the rate of cooling necessary to produce a martensitic structure. Thus the TTT curves are displaced to the right (Fig. 9.19(ii)). This makes it possible to oil quench—or even air harden—a suitable steel and still obtain a hard martensitic structure throughout.

It will be noted (Fig. 9.19(ii)) that for a low-alloy steel separate C-shaped TTT curves represent the austenite→ferrite + pearlite and austenite→bainite transformations. A similar situation arises with the TTT diagram for a eutectoid plain-carbon steel but the two sets of curves

*Represented by a cooling curve which just grazes the nose of the relevant transformation-begins curve.

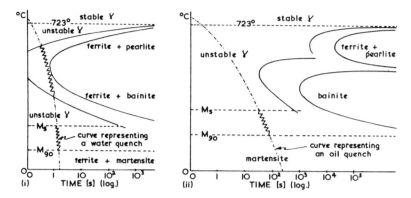

Fig. 9.19 (i) TTT curves for a 0.3% carbon steel. (ii) TTT curves for a low-alloy steel.

overlap to the extent that single continuous curves (Fig. 9.17) can be used to represent the transformation phenomena.

Mass Effects and Hardenability

9.7 It will have become obvious from the foregoing account that there is a maximum cross-section for carbon steels which, if exceeded, will not be uniformly martensitic even when quenched at the greatest possible rate from the austenitic state. Thus whilst a small-section bar (Fig. 9.20) will have cooled rapidly enough for martensite to have been formed throughout, thicker bars will contain bainite or even fine pearlite in those regions where austenite did not cool fast enough to avoid cutting into the nose of

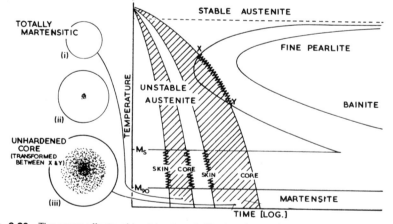

Fig. 9.20 The mass effects of heat-treatment. The heavier sections (ii) and (iii) will cool too slowly for martensite to form throughout; (ii) is of slightly greater diameter than the ruling section, but (i) has hardened completely throughout. The core of (iii) has cooled so slowly that transformation to bainite and pearlite has taken place between X and Y.

the appropriate transformation-begins curve. The displacement of TTT curves to the right due to the addition of alloying elements to a carbon steel makes it possible to oil- or air-harden quite heavy sections of suitably alloyed steels. However, to prevent the misuse of their steels by over-optimistic customers who may assume incorrectly that any alloy steel can be oil-hardened in heavy sections, manufacturers generally specify a *ruling section* for a particular alloy steel. This ruling section is the maximum diameter which can be hardened (using the recommended heat-treatment programme) if the stated mechanical properties are to be obtained. If this diameter is exceeded then the mechanical properties may vary across the section since the core will not have cooled rapidly enough to have transformed to martensite during quenching. Any subsequent tempering will not produce the desired structure or properties.

9.7.1   The *Jominy end-quench test* is of practical value in assessing the 'depth of hardening' of a steel. Here a standard test piece (Fig. 9.21) is

Fig. 9.21   The Jominy End-quench Test. (i) Details of the test-piece. (ii) Schematic diagram of the test apparatus.

heated to its austenitic range, dropped into position in a suitable holding jig as shown and then quenched at its end by a jet of water at standard temperature and pressure. In this way different rates of cooling are achieved along the length of the test piece. When it is cool, a 'flat' about 0.4 mm deep is ground along the side of the test piece and hardness determinations made every millimetre along the length from the quenched end. The results are then plotted (Fig. 9.22). Whilst there is no simple relationship between the results of this test and the ruling section of a steel, it is of value in making a preliminary estimation of the latter.

Fig. 9.22 Depth of hardening as indicated by the Jominy Test. These curves indicate that low-alloy steels harden to a greater depth than do plain carbon steels of similar carbon content.

Heat-treatments Based on the Non-equilibrium Transformation of Austenite

9.8 Heat-treatments most popularly associated with steels are those which lead to the development of great hardness or strength. All of these processes result in the formation of structures which are not in equilibrium at room temperature and involve cooling from the range of stable austenite at such a rate that pearlite formation is prevented.

9.8.1 Hardening A hardening process seeks to produce a uniform martensitic structure throughout the steel and if this is to be achieved the whole of the component must be cooled at a rate which exceeds the critical cooling rate for the steel. That is, the cooling curve for the core must pass to the left of the transformation-begins curve (Fig. 9.20). To attain this objective a plain carbon steel must generally be quenched in cold water and heavier sections may need to be quenched in brine or caustic soda solution which will promote even quicker rates of cooling. Very heavy sections in carbon steel cannot be completely hardened and then alloy steels must be used instead.

A major difficulty associated with water quenching plain carbon steels is that distortion and cracking are always likely to occur (9.6.7). Less drastic quenching in oil can be used for thin sections in carbon steel and for heavier sections in low-alloy steels and a martensitic structure still obtained. In all quenched steels varying amounts of austenite may be retained since the M_f line is at sub-zero temperatures. Such retained austenite will give rise to soft spots but can be made to transform to martensite by further treatment in solid carbon dioxide or liquid oxygen as required.

Quenching oils are of animal, vegetable or mineral origin but traditionally whale oil was the most commonly used. This has now been replaced by mineral oils and various synthetic organic compounds based on polyalkane glycols (in water solution). Most will agree this was excellent news for the whale which is now only hunted to satisfy the depraved appetites of a few gourmands who apparently wish to *eat* these noble docile creatures.

A hypo-eutectoid steel is generally hardened by heating it to a temperature just above A_3 so that it becomes completely austenitic, and then quenching it in the appropriate medium. Symmetrical components like axles should be quenched 'end-on' and agitated vigorously so that they cool as uniformly as is possible thus minimising distortion. Hyper-eutectoid steels are usually hardened by quenching from just above the A_1 temperature to give a structure of primary carbides in martensite. In fact the hot-rolling of hyper-eutectoid steels is finished just above the A_1 temperature so that the primary cementite is rendered globular in form by the working process which is taking place simultaneously with the precipitation of primary cementite. If working were finished above the A_{cm} temperature* very brittle networks of primary cementite would form around the austenite grains as the steel cooled between the A_{cm} and A_1 temperatures. For a similar reason a hyper-eutectoid steel (supplied as hot-rolled bar) should never be heated very far above A_1 as once the globular primary cementite has been re-absorbed by the austenite it is likely to be deposited as brittle films along the austenite grain boundaries on cooling.

9.8.2 Tempering   Martensite is a non-equilibrium phase for which reason its name does not appear on the iron-carbon diagram. If, therefore, martensite is heated the structure is enabled to proceed in some degree towards equilibrium depending upon the temperature used. Essentially tempering causes the dispersed dissolved carbon atoms to precipitate as carbide particles which increase in size as the temperature increases.

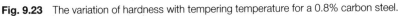

Fig. 9.23   The variation of hardness with tempering temperature for a 0.8% carbon steel.

At low temperatures (between 100°C and 200°C) extremely thin platelets of ε-carbide, which is of different composition to Fe_3C, are formed and the carbon content of the remaining martensite falls to about 0.3% as a result. At first the hardness increases slightly (Fig. 9.23) as carbide precipitation

*The temperature at which primary cementite begins to separate from the austenite, in a hyper-eutectoid steel.

begins but then begins to fall again as the amount of precipitated carbide increases. At about 400°C the ε-carbide begins to change to ordinary Fe_3C and low-carbon martensite starts to revert to ferrite. These changes are accompanied by a fall in hardness but a corresponding increase in ductility and toughness.

During low-temperature tempering the original martensite becomes dark in appearance due to the precipitation of the ε-carbide particles. These particles are too small to be visible using an optical microscope. When higher tempering temperatures are used the structure loses its original martensitic appearance and globular carbide particles in a ferrite matrix can be seen at high magnifications using an optical instrument. The former terms of 'troostite' and 'sorbite' to describe these respective structures are now generally obsolete and metallurgists prefer to speak in terms of 'tempered martensite'.

At tempering temperatures approaching A_1 precipitation of Fe_3C is such that the structure resembles that produced by a spheroidising anneal (9.5.2). In fact this treatment is often used to spheroidise a steel following a suitable quenching operation since it is quicker and gives a more regular dispersion of cementite.

Plain-carbon (and also low-alloy) steels are usually tempered below 250°C where hardness and wear-resistance are of paramount importance. When greater ductility is required, as in stress-bearing parts in aircraft and automobile engineering, then higher tempering temperatures up to 650°C are employed.

9.8.3 Isothermal Treatments   These treatments are based on information obtained about the isothermal transformations of steels as embodied in the TTT curves (9.6.2), though it must be admitted that many old industrial processes such as the 'patenting' of steel wire predated the *theoretical* consideration of such matters by many years. The principal advantage of isothermal transformation is that it reduces the possibility of internal stresses being generated due to rapid cooling through the M_s–M_f (or more properly M_s–M_{90}) range.

9.8.3.1 Martempering   This is rather an unfortunate title for this treatment since it is, in no sense, a tempering process. For that matter it is not an isothermal treatment either since transformation to martensite occurs on a falling temperature gradient. The object of the treatment is of course to minimise warping, distortion and cracking which are likely to occur during customary quench-hardening which might otherwise be employed.

The treatment involves heating the steel to its austenitising temperature and then quenching it *as rapidly as possible* in a molten salt bath which is maintained just above the M_s temperature. The steel is held in the salt bath for long enough to permit its structure to attain the same temperature throughout (Fig. 9.24(i)). It is then withdrawn from the salt bath and allowed to cool *quite slowly* to room temperature (remember the TTT diagram employs a logarithmic time axis). As the diagram indicates both the skin and the core of the component will then pass through the M_s–M_{90}

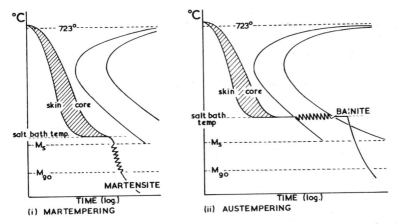

Fig. 9.24 Isothermal treatments (i) martempering and (ii) austempering. In each case the steel remains in the salt bath for long enough for a *uniform temperature to be attained throughout* before transformation of the unstable austenite begins. This is the most important feature of these processes.

range more or less simultaneously so that distortion arising from the generation of internal stresses will be minimised. A uniform martensitic structure will have been produced.

9.8.3.2 Austempering   The object of this treatment is again to eliminate the need for a drastic water quench and the 'thermal shock' associated with it. In this case a true isothermal transformation to bainite is involved. In so far as *mechanical* properties are concerned bainite produced by isothermal transformation is as satisfactory as the parallel structure produced by tempering martensite (9.8.2). The former is initiated by ferrite separation, the latter by carbide precipitation. Thus, by using isothermal transformation the need for an extra tempering process is eliminated but, more important still, the customary severe quench needed to precipitate the birth of martensite becomes unnecessary. Moreover the toughness and ductility of austempered steels is often greater than that of similar steels hardened and tempered in the orthodox ways to similar hardness values.

Again as Fig. 9.24(ii) shows, it is necessary to cool the steel rapidly past the nose of the transformation-begins curve. However, the quenching-bath temperature, which is always of course above M_s, is adjusted to give bainite of the desired form. The steel is allowed to remain in the quenching bath long enough for transformation to bainite to be complete, when it can be cooled to room temperature at any convenient rate.

In view of the obvious advantages of these isothermal treatments the reader may well ask: 'Why are the old methods of quenching/tempering still used?' The answer lies in the fact that successful isothermal treatments are dependent upon cooling the steel rapidly enough to miss the 'nose' of the transformation-begins curve, otherwise fine pearlite will form in the range 500°–600°C. Thus, as far as *plain-carbon* steels are concerned these processes are limited to the treatment of materials not more than 10 mm thick which can be cooled sufficiently rapidly. For example, austempering is a useful treatment for springs, pins, needles, screws, garden spades and similar thin components. High-carbon steel wire too is similarly treated.

With alloy steels the scope is greater because of the displacement of the TTT curves which indicate the lower critical cooling rates prevailing. In such cases an oil quench will be very roughly equivalent to martempering since the cooling curve representing an oil quench levels out near to the M_s line because of the higher boiling point of oil as compared with that of water.

Surface Hardening of Steels

9.9. Frequently components like gears and axles need to be tough and ductile but, at the same time, have a hard, wear-resistant surface. Ductile materials are usually soft whilst hard materials are generally brittle, so the most effective way to achieve this difficult combination of properties is to use a steel which has a surface of different composition from that of the core; as is contrived by carburising or nitriding. Alternatively a steel of uniform carbon content can be used and its surface heat-treated in a different manner from the core.

9.9.1 Case-hardening This is a title generally given to those processes which involve carburising the surface of a tough, ductile low-carbon steel. Since carbon dissolves interstitially in FCC iron it follows that the low-carbon steel employed must be heated to above A_3 (about 900°–950°C) in an atmosphere which will release free carbon atoms. Treatment time must be sufficient to produce a 'case' of the required depth.

Solid, liquid and gaseous carburising media are used. Solid media contain some form of carbon such as charcoal along with an activator like barium carbonate. The components are packed into steel boxes with this medium and heated for up to five hours. Liquid media are based on molten sodium cyanide/sodium carbonate mixtures. The poisonous nature of these mixtures and the disposal of process waste poses problems in the modern world. Most large-scale industrial carburising now employs gaseous media, such as town gas, natural gas or other hydrocarbons like propane. Gaseous methods are clean, rapid and efficient.

During carburising a concentration gradient of carbon in iron is set up as indicated in Fig. 9.25. Often specifications quote a hardness value as determining the limit of the *useful* case. In Fig. 9.25 a hardness of 550 H_V is judged to coincide with this limit which in this instance is at a depth of 1.2 mm where the carbon content will probably be about 0.3%.

Following carburising heat-treatment of the component is necessary. In many instances quenching direct from the carburising medium is employed but the counsel of perfection is to use a double heat-treatment process. The core structure is first refined by normalising the component just above the

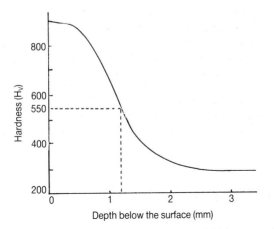

Fig. 9.25 The relationship between case depth and hardness (H_V) for a case-hardened steel.

A_3 temperature *for the core*. The component is then reheated to about 760°C which is above the A_3 temperature for the *case*, and then quenched.

Q Axle shafts produced from a carbon steel containing 0.15% carbon are gas carburised at 950°C. Calculate the time required to achieve a carbon content of 0.4% at a distance of 1.0 mm below the surface. Assume that a uniform carbon content of 1.0% is maintained at the surface during carburisation and that the diffusion coefficient (D) under these conditions is 3.0×10^{-11} m²/s.

A The Grube solution (7.4.3) of Fick's Second Law can be used:

$$\frac{C_s - C_x}{C_s - C_0} = \text{erf.}\left(\frac{x}{2\sqrt{(D.t)}}\right)$$

so

$$\frac{1.0 - 0.4}{1.0 - 0.15} = \text{erf.}\left(\frac{1.0 \times 10^{-3}}{2\sqrt{(3.0 \times 10^{-11}.t)}}\right)$$

and hence:

$$\therefore 0.706 = \text{erf.}\left(\frac{0.5 \times 10^{-3}}{5.477 \times 10^{-6}.\sqrt{t}}\right)$$

But when erf. $y = 0.706$; $y = 0.742$ (Appendix I). Hence:

$$0.742 = \frac{0.5 \times 10^{-3}}{5.477 \times 10^{-6}\sqrt{t}} = \frac{500}{5.477\sqrt{t}}$$

and

$$\sqrt{t} = \frac{500}{5.477 \times 0.742} = 123.03$$

Therefore:

$$t = (123.03)^2 = 15\,136.38 \text{ s} = \underline{4.2 \text{ hours}^*}$$

9.9.2 Nitriding This process involves the interstitial solid solution of nitrogen atoms to form a very hard skin on the surface of the component. Special steels containing either 1.0% aluminium or small amounts of chromium, molybdenum and vanadium are used so that hard nitrides of these metals will form in the surface layers. The charge is heated at 500°C for upwards of forty hours in an atmosphere of ammonia. This dissociates:

$$NH_3 \rightleftharpoons 3H + N$$

and the *atomic* nitrogen released dissolves interstitially in the surface of the BCC iron.

The main advantages of nitriding as compared with carburising are:

(i)   Since no quenching is required after nitriding, cracking and distortion are unlikely.
(ii) High surface hardness up to 1150 *VPN* is obtainable with some of the special 'Nitralloy' steels.
(iii) Resistance to corrosion and resistance to fatigue are good.
(iv) Hardness is retained up to 500°C whereas a hardened carburised component begins to soften at 200°C due to normal effects of tempering.

The principal disadvantage of nitriding is that the plant outlay is high whereas components can be carburised in solid media quite cheaply. As with many processes the question of unit cost is dependent largely upon the scale of 'throughput'.

9.9.3 Induction and Flame Hardening In these processes a steel of uniform carbon content—usually about 0.4% C—is used. The component is first normalised to give a tough, ductile, pearlitic core. Its surface is then heated, either by a gas torch or by high-frequency induction methods, and, almost immediately, quenched by means of water jets. This produces a hard martensitic skin on top of the tough pearlitic core. The two zones are generally separated by a layer of bainite which reduces the probability that the case will flake away. Symmetrical components such as gears and axles can be spun between centres during heating and quenching, thus ensuring a more uniform structure.

---

*It is extremely unlikely that such information would be obtained in the real world of industry by calculations of this type—which are devised mainly to terrorise students during examinations. Instead a series of pieces of the axle material in question would be carburised for different periods of time (all at the same predetermined temperature). The case depth, measured by simple metallographic inspection, would then be plotted *vs* carburising time. In any event an experiment of this type would need to be performed to derive *D* for the carburising conditions if the mathematical approach were to be used—a 'chicken-and-egg' situation!

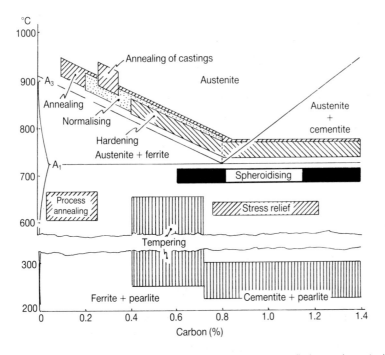

Fig. 9.26 A summary of the various heat-treatment processes applied to carbon steels in relation to carbon content and treatment temperature.

10

Alloy Steels and Engineering Cast Irons

10.1 Man's early ferrous implements were of wrought origin, forged from meteoric iron and, later, from the spongy mass produced by the low-temperature reduction of iron ore. Development of furnaces which could reduce and *melt* iron and so produce *cast* iron began in about 1000 B.C. in China whence the technology spread to India before the start of the Christian era. The process was independently, if somewhat belatedly, discovered in Europe in the fourteenth century A.D. The World's first metal bridge—that at Ironbridge in Shropshire—was built of cast iron in 1779. The use of cast iron was widespread in Victorian times for ornamental (?) as well as for engineering purposes, whereas it is now used solely as a sophisticated engineering material with a highly developed technology.

10.1.1 Though the first alloy steel was a high-manganese alloy originated by Hadfield in 1882, commercial alloy steels are of the twentieth century. Thus, high-speed steel was first produced in the USA by Taylor and White in 1900 and the first stainless steel by Brearley of England in 1913, though it did not attain commercial status until much later.

In the foregoing chapter the structures, properties and heat-treatments of ordinary carbon steels were outlined. Here we shall deal with some of the more complex iron-base alloys which depend for their specific properties on the presence of other elements in addition to carbon.

Alloy Steels

10.2 Alloying elements are added to carbon steels to improve their existing properties but, in some cases, also to introduce new properties such as corrosion resistance. Small amounts of manganese (0.3–0.8%) are always present in so-called plain-carbon steels as a residue from deoxidation and desulphurisation. Such manganese has a measurable effect in reducing the critical cooling rate for a steel. However, it is not regarded as an alloying element unless it exceeds 2.0%, whilst it has been traditional to regard 0.1% Mo or V; 0.3% W, Co or Cr; or 0.4% Ni as being the minimum quantities of these elements to qualify for 'alloy status'. The introduction of the modern generation of *micro-alloyed steels* (10.3.4) has rendered such a definition spurious since amounts as small as 0.0005% of some elements made to a high-purity mild steel effectively improve properties.

The principal effects which the main alloying elements have on the structure and properties of steels can be classified as follows:

10.2.1 Increase in Tensile Strength   Most of the elements used in alloy steels dissolve substitutionally to some extent and, predictably, increase the tensile strength as a result. This of course is a general effect of alloying (7.5). In low-alloy steels the ferrite is strengthened in this way whilst high ductility and toughness are maintained. Though metallurgically unspectacular, this is one of the most important features of alloying in steels.

10.2.2 Changes in the Polymorphic Transformation Temperatures   The polymorphic transformation temperatures involved here are the A_3, at 910°C in pure iron, where the $\alpha \rightleftharpoons \gamma$ transformation occurs, and the A_4, at 1400°C, where the $\gamma \rightleftharpoons \delta$ change occurs.

Some elements, notably nickel, manganese, copper and cobalt, raise the A_4 temperature and depress the A_3 as indicated in Fig. 10.1(i). Consequently, when added to a carbon steel these elements will tend to stabilise austenite (γ) by increasing the temperature range over which it can persist as a stable phase. Many of these elements are FCC in structure at room temperature thus favouring the retention of the FCC austenite in which

Fig. 10.1   The effects of alloying additions on the A_3 and A_4 temperatures in iron. (i) Elements which tend to stabilise γ, i.e. Ni, Mn, Cu and Co. (ii) Elements which tend to stabilise α, i.e. Cr, W, Mo, V, Al and Si.

they are substitutionally dissolved and retarding its transformation to BCC ferrite. Moreover since most of these elements do not form carbides, the carbon tends to remain in solid solution in the austenite. When austenite is stabilised at room temperature by amounts of the alloying element in excess of X (Fig. 10.1(i)) then the steel will also lose its ferromagnetism.

Other elements, the most important of which are chromium, molybdenum, tungsten, vanadium, aluminium and silicon, have the opposite effect in that they stabilise ferrite (α) by raising the A_3 temperature and depressing the A_4. These elements restrict the field over which austenite exists, forming what is generally termed a 'γ-loop' (Fig. 10.1(ii)) in the appropriate diagram. Most of these elements are BCC in structure at room temperature so it is not surprising that they stabilise BCC ferrite. Since some of them (the Cr–Mo–W group of the transition metals) form very stable carbides they favour the precipitation of carbon in this form and further aid the $\gamma \rightarrow \alpha$ transformation.

10.2.3 Relative Stability of Carbides   Some elements form very stable carbides when added to a carbon steel. Many of these carbides are harder than iron carbide. Since these elements will increase the hardness of a steel they are generally used in tool steels of superior quality. Such elements include chromium, molybdenum, tungsten, vanadium and manganese; whilst titanium and niobium also have very strong carbide-forming tendencies (Fig. 10.2). Complex carbides are often present when one or more of these elements are added to a steel.

Proportion dissolved in ferrite (◄—)	Element	Proportion present as carbide (—►)
	Nickel	
	Silicon	
	Aluminium	
	Manganese	
	Chromium	
	Tungsten	
	Molybdenum	
	Vanadium	
	Titanium	
	Niobium	
Solubility–0.3% max	Copper	

Fig. 10.2   The relative carbide-forming tendencies of alloying elements in steel.

Not all of the alloying elements used in steel have a chemical affinity for carbon when in the presence of iron. Thus nickel, aluminium and silicon do not form carbides in the presence of iron and in fact cause instability of Fe_3C so that carbon is precipitated in the form of graphite. Therefore if it is necessary to add appreciable amounts of these elements to steel it can only be done in the presence of one or more of the carbide-stabilising elements already mentioned. For this reason there are no high-carbon steels containing nickel as the sole alloying element.

10.2.4 Influence on Grain Growth   During heat-treatment, particularly at high temperatures, grain growth of austenite will inevitably take place. The presence of some elements will accelerate this grain growth and so increase the resultant brittleness of the component. The most important element which produces this unwelcome effect is chromium. Consequently care must be taken in heat-treating chromium steels to avoid undue over-heating or using excessive treatment times.

Fortunately grain growth is rendered more sluggish by the presence of some elements, notably nickel and vanadium. Hence these elements are generally classed as 'grain refiners'.

10.2.5 Displacement of the Eutectoid Point   The addition of any of the important alloying elements to a carbon steel will cause displacement of the eutectoid point such that the structure will be totally pearlitic even though the steel contains less than 0.8% carbon. For example, the addition of 2.5% manganese to a steel containing 0.65% carbon will produce a completely eutectoid structure (Fig. 10.3), whilst a stainless cutlery steel

Fig. 10.3   The effects of manganese and titanium on the displacement of the eutectoid point in steel.

containing 13% chromium and 0.35% carbon is also eutectoid. The eutectoid temperature may be raised or lowered (Fig. 10.3), the A_1 temperature being displaced in sympathy with A_3 as mentioned above (10.2.2). Hence alloy steels generally contain less carbon than plain carbon steels used for similar purposes.

10.2.6 Retardation of Transformation Rates   We have seen (9.6.7) that the addition of alloying elements tends to reduce the critical cooling rates necessary to obtain a completely martensitic structure in steels. This is indicated by the displacement of the TTT curves to the right in Fig. 9.19(ii). All alloying elements, with the exception of cobalt, have this effect and it is one of the most useful features imparted by alloying since it renders possible oil quenching or even air hardening as alternatives to a drastic water quench. For example, a steel containing 4.25% nickel and 1.25% chromium can be air hardened in thin sections or oil hardened in sections up to 150 mm diameter. Atoms of these elements which are dissolved substitutionally impede the $\gamma \rightarrow \alpha$ transformation.

Extremely small amounts of boron (up to 0.005%) may be added to low-carbon/low-alloy steels in order to improve depth of hardening. The atomic size of boron renders it too large for easy interstitial solid solution but too small for ready substitutional solid solution. It is thought that the boron atoms therefore migrate to lattice imperfections which exist at grain boundaries thus decreasing strain energy in these regions. This in turn delays the incidence of nucleation of ferrite and so retards the transformation of austenite.

10.2.7 Improvements in Corrosion Resistance   Those metals which become coated with a thin but dense and adherent film of oxide (19.2) are often protected from further corrosion by this useful barrier. Thus aluminium, a relatively reactive metal, is stable as an engineering material because of the protection afforded by the dense and closely adherent film of Al_2O_3. Aluminium, together with silicon and chromium, impart this property to steel when used as alloying elements, but when resultant mechanical properties have to be considered, only chromium of the elements mentioned can be used. Chromium forms a dense but adherent film on steel and protects it effectively when a minimum of 13% chromium is used.

Constructional Alloy Steels

10.3   The addition of alloying elements to medium-carbon constructional steels has the following advantageous results:

(i)   mechanical properties, but in particular yield strength, tensile strength, toughness and ductility are improved;

(ii)   critical transformation rates are considerably reduced so that a steel can be hardened by oil quenching or air cooling.

10.3.1   Nickel is useful in promoting these properties since it strengthens ferrite by solid solution and improves toughness by its tendency to impede grain growth. Unfortunately nickel also favours graphitisation of any cementite present so that, in a medium-carbon steel, it is generally accompanied by a carbide stabiliser such as chromium, which will also assist it in strengthening ferrite and reducing transformation rates. Such steels contain rather more than two parts of nickel to one of chromium, maximum quantities being 4.25% Ni and 1.25% Cr, giving an air-hardening steel. These nickel–chromium steels regrettably suffer a fault known as 'temper brittleness'. This manifests itself as a very low impact toughness value in a nickel–chromium steel which has been tempered in the range 250–400°C. Tensile properties, however, are not substantially impaired. Fortunately this defect can be minimised by adding about 0.3% molybdenum, thus giving rise to the popular range of 'nickel–chrome-moly' steels. Constructional steels of this type are used for the manufacture of axles, shafts, gears, connecting rods, bolts, brackets, etc.

10.3.2 Of recent years greater use has been made of small additions of the relatively cheap alloying element manganese to replace some nickel and chromium in the cheaper grades of low-alloy steels. These steels are used quite widely in the automobile industry. Manganese promotes increased depth of hardening by reducing critical cooling rates. In low-carbon steels it effectively reduces the temperature at which brittle fracture (16.3) will take place. Compositions, heat-treatments, properties and uses of some popular low-alloy constructional steels are given in Table 10.2.

10.3.3 Although most of the constructional alloy steels are in the low-alloy class, a few high-strength/high-alloy steels are produced. These are the 'maraging' steels, typical of which is one containing 18% Ni, 8% Co, 4% Mo and up to 0.8% Ti. Since such a high nickel content will have a considerable graphitising influence carbon must be kept low—about 0.03%. Such a steel is first solution treated (11.3.2) to absorb precipitated compounds. On air-cooling the structure will become martensitic because of the reduction in the transformation rates incurred by the high-alloy content. Unlike ordinary tetragonal martensite this is of BCC form and hence much softer and tougher. However, the steel can now be *precipitation hardened* (11.3.2) at 500°C for two hours so that coherent precipitates of compounds such as $TiNi_3$ are formed and tensile strengths of over $2000\,MPa^2$ developed. The term 'maraging' refers of course to the 'age-hardening'—or 'ageing'—of martensite. Such steels are naturally very expensive and their use is generally limited to sophisticated technology such as rocketry.

Table 10.1 *Relative Effects of Alloying Elements on Steel*

Element	Influence on the properties of steel	Uses in steel
Nickel	Stabilises γ by raising A_4 and depressing A_3. It is the universal grain refiner in alloy steels (and many non-ferrous alloys). Strengthens ferrite by solid solution. Unfortunately a powerful graphitiser	In amounts up to 5% as a grain refiner in case-hardening steels. Along with chromium and molybdenum in low-alloy constructional steels. In larger amounts in stainless and heat-resisting steels
Manganese	Like nickel it stabilises γ but unlike nickel it forms stable carbides	Low-manganese steels are not widely used though in recent years it has been used to replace small amounts of more expensive alloying elements, e.g. nickel. The high-manganese (Hadfield) steel contains 12.5% Mn and is austenitic but hardens on abrasion
Chromium	Stabilises α by raising A_3 and depressing A_4. Forms hard stable carbides. Strengthens ferrite by solid solution. In amounts above 13% imparts stainless properties. Unfortunately increases grain growth	In small amounts in constructional and tool steels. Also in ball bearings. In larger amounts in stainless and heat-resisting steels
Molybdenum	Strong carbide-stabilising influence. Raises the high-temperature creep strength of suitable alloys. Imparts some sluggishness to tempering influences	Reduces 'temper brittleness' in nickel–chromium steels. Increases red-hardness of tool steels. Now used to replace some tungsten in high-speed steels

Table 10.1 *Cont.*

Element	Influence on the properties of steel	Uses in steel
Vanadium	Strong carbide-forming tendency. Stabilises martensite and increases hardenability. Like nickel it restrains grain growth. Induces resistance to softening at high temperatures once the steel is hardened	Used in steels required to retain hardness at high temperatures, e.g. hot-forging dies, extrusion dies, die-casting dies. Also increasingly in high-speed steels
Tungsten	Has similar effects to chromium in stabilising α. Also forms very hard carbides. Renders transformations very sluggish—hence, once hardened, a steel resists tempering influences	Used mainly in high-speed steels and other tool and die steels, particularly for use at high temperatures
Cobalt	Induces sluggishness to the transformation of martensite. Hence increases 'red-hardness'	Super high-speed steels and 'maraging' steels. Permanent magnet steels and alloys
Silicon	A strong graphitising influence—hence not used in high-carbon steels. Imparts casting fluidity. Improves oxidation resistance at high temperatures	Up to 0.3% in steels for sandcastings where it improves fluidity. In some heat-resisting steels (up to 1.0%)
Boron	Very small amounts displace the TTT curves to the right, thus increasing hardenability	0.0005 to 0.002% added to some structural, constructional and automobile engineering steels

10.3.4 High-strength Low-alloy (HSLA) Steels   One of the most significant developments in the steel industry in recent decades has been the steady improvement in the chemical purity of dead-mild steels. This has meant that 'micro' additions of suitable alloying elements can be made to produce effective and *consistent* increases in yield strength whilst retaining high ductility and toughness. Such consistency of properties could not be attained using a dead-mild steel of irregular quality.

Quite small quantities of Nb, V, Ti and Al are now used in these HSLA and other *micro-alloyed steels*. HSLA steels are currently employed in the structural work of buildings, bridges and pipelines. The principal function of the added elements is to combine with the small amounts of carbon and nitrogen present to form very small carbo-nitride precipitates (TiC; TiN or NbCN) which severely restrict crystal growth during hot-forming and annealing processes. Strength of HSLA steels increases as grain-size decreases seemingly due to dispersion hardening (14.1.5) associated with these precipitates. Niobium in amounts between 0.06 and 0.1% will produce yield strengths up to 350 MPa in cold-worked annealed forms and 550 MPa in some hot-rolled HSLA steels. A typical off-shore pipeline material may contain C–0.09; Si–0.3; Al–0.03; Mn–1.35; Nb–0.035 and V–0.07%. Such an alloy cannot be referred to as a 'steel' in the normal sense but rather as a 'dispersion-strengthened iron'. Nevertheless the term 'steel' is generally used industrially.

In the automotive industries cold-formed annealed *micro-alloyed steels* are not confined to high-strength materials. Thus auto-body pressings in 'difficult' shapes require a *low-strength* steel with *high ductility*. Vacuum degassed steel with small amounts of titanium and/or niobium contain minute separate precipitates of Ti or Nb carbides and nitrides but *no* carbon or nitrogen in interstitial solid solution. Such a 'steel' has a very low yield stress but a very high ductility—ideal for cold forming.

Alloy Tool Steels

10.4   Alloying elements are added to tool steels:

(i)   to increase hardness and wear-resistance by the formation of carbides which are much harder than Fe_3C;
(ii) to impart resistance to the tempering effects associated with high-temperature working. This applies not only to exposure to high temperatures when these steels are used in hot-working dies, but also to frictional heat generated in cutting tools during high-speed machining processes. A correctly hardened high-speed steel will resist tempering up to dull-red heat.

The alloying elements involved here are the refractory metals chromium, tungsten, molybdenum and vanadium, all of which form very hard stable carbides. Moreover since they are all BCC metals they limit the range over which austenite may exist (as indicated by the γ-loop formed in the binary diagram when each is added to iron) and so stabilise ferrite and martensite.

10.4.1   Low-alloy tool steels generally contain small amounts of chromium and are oil-hardened from temperatures slightly above those used for equivalent plain carbon steels. Such steels are used for ball-bearings, chisels and hot-working dies.

The high-alloy tool steels, however, contain considerable amounts of tungsten, molybdenum and vanadium as well as chromium (Table 10.3). Of recent years the relative scarcity of tungsten has led to the wider use of molybdenum and vanadium, particularly in high-speed steels.

In order that these steels shall not lose their hardness at high working temperatures the maximum amount of tungsten (or molybdenum) must be present in *solid solution* in the original austenite before it is quenched. Figure 10.4—a pseudo-binary diagram (8.7.2) representing a high-speed steel—indicates that at a temperature of say 900°C, little more than 0.2% C *along with its associated tungsten* is in solid solution in the austenite and that even at the solidus temperature of about 1300°C only an amount, S, is in solid solution. For this reason, in order to retain as much carbon and tungsten in solid solution in the quenched structure as is possible and so obtain both hardness and *temper-resistance*, very high quenching temperatures are necessary. At these high temperatures decarburisation and grain growth are rapid unless special precautions are adopted. Thus a two-chamber furnace enables preliminary pre-heating to be used and so limit grain growth by reducing the time of contact with the high quenching temperature; whilst the use of inert furnace atmospheres prevents decarburisation of the tool surface.

Table 10.2 Carbon and Low-alloy Constructional Steels

Type of steel	Composition (%)	Heat-treatment	Mechanical properties				Uses
			Yield stress (MPa)	Tensile strength (MPa)	Elonga-tion (%)	Izod (J)	
Carbon steel	0.40 C 0.80 Mn	Harden by quenching from 830/860°C. Temper between 550 and 660°C	500	700	20	55	Axles, crankshafts, spindles, etc. under medium stress
Carbon steel	0.55 C 0.70 Mn	Harden by quenching from 810/840°C. Temper between 550 and 660°C	550	750	14	—	Gears, cylinders and machine-tool parts requiring resistance to wear
Low manganese	0.28 C 1.5 Mn	Oil-quench 860°C (water-quench for sections over 38 mm diameter). Temper as required	355	590	20	—	Automobile axles, crankshafts, connecting rods, etc., where a relatively cheap steel is required
Manganese–molybdenum	0.36 C 1.5 Mn 0.26 Mo	Oil-quench from 830°–850°C; temper between 550° and 650°C and cool in oil or air	700	1000	15	55	A substitute for the more highly alloyed nickel–chrome–molybdenum steels
Nickel–chromium–molybdenum	0.40 C 0.55 Mn 1.50 Ni 1.20 Cr 0.30 Mo	Oil quench from 830/850°C: 'light temper'—180/200°C 'full temper'—550/650°C	— 988	2010 1080	14 22	27 69	Differential shafts, crankshafts and other highly-stressed parts where fatigue and shock resistance are important. In the 'light tempered' condition it is suitable for automobile gears. Can be surface hardened by nitriding
Manganese–nickel–chromium–molybdenum	0.38 C 1.40 Mn 0.75 Ni 0.50 Cr 0.20 Mo	Oil-quench from 830°–850°C; temper at 550°–660°C, and cool in air	958	1040	21	85	Automobile and general engineering components requiring a tensile strength of 700 to 1000 N/mm^2

Table 10.3 *Alloy Tool and Die Steels*

Type of steel	Composition (%)	Heat-treatment	Uses
High carbon–high chromium	2.10 C 0.30 Mn 12.50 Cr	Heat slowly to 750°–800°C and then raise to 960°–990°C. Oil-quench (small sections can be air cooled). Temper at 150°–400°C for 30–60 minutes	Blanking punches, dies and shear blades for hard, thin materials. Dies for moulding abrasive powders, e.g. ceramics. Master gauges. Thread-rolling dies
Hot-working die steel	0.35 C 1.00 Si 5.00 Cr 1.50 Mo 0.40 V 1.35 W	Pre-heat to 800°C, soak and then heat quickly to 1020°C and air cool. Temper at 540°–620°C for 1½ hours	Extrusion dies, mandrels and noses for aluminium and copper alloys. Hot-forming, piercing, gripping and heading tools. Brass-forging and hot-pressing dies
High-speed steels 18% Tungsten	0.75 C 4.25 Cr 18.00 W 1.20 V	Quench in oil or air blast from 1290°–1310°C. Double temper at 565°C for 1 hour	Lathe, planar and shaping tools; millers and gear cutters. Reamers; broaches; taps; dies; drills and hacksaw blades. Bandsaws. Roller-bearings at high temperatures (gas turbines)
12% Cobalt	0.80 C 4.75 Cr 22.00 W 1.50 V 0.50 Mo 12.00 Co	Quench in oil or air blast from 1300°–1320°C. Double temper at 565°C for 1 hour	Lathe, planar and shaping tools, milling cutters, twist drills, etc. for exceptionally hard materials. Has maximum red-hardness and toughness. Suitable for severest machining duties, e.g. manganese steels and high-tensile steels, close-grained cast irons
Molybdenum '562'	0.83 C 4.25 Cr 6.50 W 1.90 V 5.00 Mo	Quench in oil or air blast from 1250°C. Double temper at 565°C for 1 hour	Roughly equivalent to the standard 18–4–1 tungsten high-speed steel but tougher. Drills, reamers, taps, milling cutters, punches, threading dies, cold-forging dies
9% Mo–8% Co	1.00 C 3.75 Cr 1.65 W 1.10 V 9.50 Mo 8.25 Co	Quench in oil or air blast from 1180°–1210°C. Triple temper at 530°C for 1 hour	Similar uses to the 12% Co–22% W high-speed steel

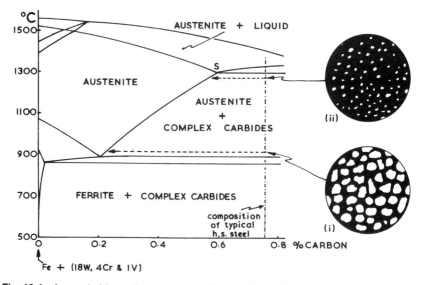

Fig. 10.4 A pseudo-binary diagram representing an 18–4–1 high-speed steel. If such a steel were quenched from 900°C the structure (i) would consist of mixed carbides (light) in a matrix of martensite (dark) which would be deficient in both tungsten and carbon (0.2%). When quenched from almost 1300°C much of the carbide has been dissolved (ii) and the martensite matrix now contains much more dissolved carbon (0.6%) along with its previously combined tungsten.

10.4.2   The large amounts of alloying elements present slow down transformation considerably and so air-hardening (or oil-quenching of heavy sections) can generally be used. Considerable amounts of austenite are retained on quenching and a 'secondary hardening' treatment is necessary. This involves heating the tool to about 550°C to promote the transformation of retained austenite to martensite. Two or three such treatments are often necessary before the austenite→martensite transformation approaches completion. Once formed, the martensite is similarly very sluggish to tempering at temperatures up to 700°C, wherein lies the most important feature of high-speed steels.

Stainless and Heat-resisting Steels

10.5   The role of chromium in imparting corrosion resistance to steels has been mentioned (10.2.7) and amounts between 13% and 21% are commonly used for this purpose. Those steels containing only chromium as an alloying element are of course ferritic since chromium stabilises BCC iron and quenching those chromium steels which contain sufficient carbon will produce a hard martensitic structure. Thus, high-chromium steels containing 0.3% C are used for the manufacture of table knives and other rust-resistant edged tools; whilst those containing little or no carbon are used for press-work and deep-drawing since they can be hardened only by cold-work.

10.5.1 Chromium steels tend to become brittle during heat treatment because chromium promotes grain growth. Stainless steels containing both nickel and chromium are therefore less likely to suffer from brittleness because of the grain-refining influence of nickel. The well-known '18–8' stainless steels are in this class. These are non-hardening alloys in which the carbon content is normally less than 0.1%. Even so the final treatment is usually a quench from 1050°C in order that carbon is retained in solid solution and to prevent it from precipitating as particles of chromium carbide which would impair the corrosion resistance of the alloy (19.7.1). Alternatively the addition of up to 1.0% of either niobium or titanium (which have high affinities for carbon) effectively 'ties up' any carbon present so that it is no longer available to precipitate as chromium carbides. This is necessary for stainless steels which are to be welded and would be prone to 'weld decay' (17.6.2) unless so 'proofed'.

10.5.2 Heat-resisting steels must resist oxidation at high temperatures and, in some cases, attack by sulphurous gases (19.2.2). For this reason they may contain as much as 30% chromium together with small amounts of silicon. The general tendency for chromium to induce grain growth, with its attendant brittleness, is offset by adding nickel.

These steels must be reasonably strong at high temperatures so they are 'stiffened' by adding small amounts of two or more of the following: carbon, tungsten, molybdenum, titanium and aluminium, so that small particles of carbides or intermetallic compounds are formed. In this manner the limiting creep stress is raised due to *particle hardening* (14.1.5). Metallurgical principles involved in the heat-resisting steels also apply to the 'Nimonic' series of alloys. These are not steels but contain basically 75% nickel–20% chromium, stiffened with small amounts of carbon, titanium, aluminium, cobalt, molybdenum or niobium.

Engineering Cast Irons

10.6 In terms of cost per unit mass cast iron is the cheapest metallurgical material available to the engineer. It is in fact re-melted pig iron, the composition of which has undergone some adjustments during the melting process. Apart from its low cost other commendable properties of cast iron include good rigidity and *compressive* strength; excellent fluidity so that it makes good casting impressions; and good machinability. Whilst the ductility and *tensile* strength of ordinary grey cast iron are not very high both of these properties can be considerably improved by treatments which modify the microstructures of suitable irons.

During the nineteenth century cast iron was used both for engineering and domestic purposes. Today the Victorian fireplaces, lamp standards and water fountains have largely disappeared—except as collector's items—

Table 10.4 Stainless and Heat-resisting Steels

Type of steel	Composition (%)	Condition	Yield stress (MPa)	Tensile strength (MPa)	Elongation (%)	Hardness (Brinell)	Heat-treatment	Uses
			Typical mechanical properties					
Stainless iron	0.04 C 0.45 Mn 14.00 Cr	Soft	340	510	31	—	Non-hardenable except by cold work	Wide range of domestic articles—forks, spoons. Can be spun, drawn and pressed
Cutlery steel	0.30 C 0.50 Mn 13.00 Cr	Cutlery temper / Spring temper	— / —	1670 / 1470	— / —	534 / 450	Water- or oil-quench (or air-cool) from 950°–1000°C. Temper: (for cutlery)—at 150°–180°C; (for springs)—at 400°–450°C	Cutlery and sharp-edged tools requiring corrosion resistance. Circlips, etc. Approximately pearlitic in structure when normalised
18/8 stainless	0.05 C 0.80 Mn 8.50 Ni 18.00 Cr	Softened / Cold-rolled	278 / 803	618 / 896	50 / 30	170 / —	Non-hardening except by cold-work. (Cool quickly from 1050°C to retain carbon in solid solution)	Particularly suitable for domestic and decorative purposes. An austenitic steel
18/8 stainless (weld-decay proofed)	0.05 C 0.80 Mn 8.50 Ni 18.00 Cr 0.80 Ti	Softened / Cold-rolled	278 / 402	649 / 803	45 / 30	180 / 225	Non-hardening except by cold-work. (Cool quickly from 1050°C to retain carbon in solid solution)	A weld-decay proofed steel (fabrication by welding can be safely employed). Used extensively in nitric acid plant and similar chemical processes

Type of steel	Composition (%)	Condition	Testing temperature (°C)	0.1% P.S. (MPa)	Tensile strength (MPa)	Elongation (%)	Maximum working temperature (°C)	Uses
Heat resisting steel	0.15 C 20.00 Cr 25.00 Ni	Forged or rolled	—	—	—	—	820	Conveyor chairs and skids, heat-treatment boxes, recuperator valves, valves and other furnace parts
Heat resisting steel	0.10 C 1.50 S 1.00 Mn 19.00 Cr 11.00 Ni	Air-cooled from 1100°C	20, 400, 600, 800, 900	347, 248, 217, 139, 85	698, 527, 465, 248, 155	55, 40, 37, 45, 50	1000 (air) / 950 (flue gases)	A fairly cheap grade of heat-resisting steel with a good combination of properties

Fig. 10.5 18–8 stainless steel, furnace-cooled from 1100°C. This treatment resulted in the precipitation of some chromium carbide as dark films along the austenite grain boundaries. Some austenite crystals show annealing twins in common with many other FCC structures. (Etched in 'Mixed acids with glycerol'.) ×100.

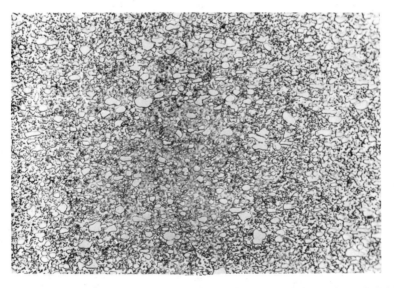

Fig. 10.6 18–4–1 high-speed steel, annealed at 900°C. Rounded particles of carbide (light) in a ferritic-type matrix. (Etched in 2% Nital.) ×600.

and cast iron is now used solely as a reasonably sophisticated engineering material. Alloy—and other special cast irons are now used for the manufacture of such components as crankshafts, axles and connecting rods, hitherto the province of forged steels.

10.6.1 The Relationship Between Composition and Structure of Cast Iron Most engineering cast irons contain 2.0–3.5% C. Reference to the right-hand half of the iron–carbon equilibrium diagram (Fig. 9.1) indicates that an alloy in this range begins to solidify by depositing dendrites of austenite and that by the time that the eutectic temperature (1147°C) has been reached the structure will consist of austenite (γ) crystals containing 2.0% C and a remaining liquid containing 4.3% C. This liquid then solidifies at 1147°C by forming a eutectic, known at one time as *ledeburite*, consisting of austenite (2.0% C) and cementite (6.68% C). On further cooling both primary and eutectic austenite will change in composition along VE (Fig. 9.3) by depositing particles of cementite until at 723°C the remaining austenite contains 0.8% C where it transforms to pearlite. Consequently a final structure of cementite and pearlite would result (Fig. 10.9). In practice of course we are not dealing with simple iron–carbon alloys since cast iron contains, in addition to carbon, up to 2.5% Si; 0.8% Mn; 0.8% P and 0.1% S and the form in which carbon exists depends to some extent on the amounts of these other elements present. We have seen (10.2.3) that the presence of some elements in steel renders carbides unstable so that they tend to decompose forming graphite. Silicon is one of these elements and its presence in varying amounts in cast iron has a similar influence on the state in which carbon exists. Thus a high-silicon iron—containing as much as 2.5% Si—will be almost completely graphitic and will contain little or no cementite. The presence of large amounts of graphite in the structure will lead to a fractured surface having a grey appearance. That is, it will be a 'grey iron'.

If, on the other hand, the silicon content is low—in the region of 0.5%—then cementite will tend to be stable and little or no graphite will form. Since cementite is extremely brittle such an iron will be mechanically weak and a fractured surface will be silvery white since cementite is a white compound. For this reason it is termed a 'white iron'.

Quite small amounts of sulphur will have the opposite effect to that of silicon, that is, sulphur will tend to stabilise cementite and so favour the formation of a white iron. However, the direct effect of sulphur in iron, as in steel, is to increase the brittleness excessively so the foundryman does *not* use sulphur as a means of controlling the structure of cast iron. Generally he will attempt to keep the sulphur content to a minimum, and instead use a varying silicon content as a means of controlling the state of the carbon present.

10.6.2 The Influence of Cooling Rates on the Structure of Cast Iron Even though silicon is present in a cast iron in sufficient quantities to graphitise any cementite present such decomposition does not occur instantaneously but is time dependent. Consequently if a cast iron solidifies rapidly its

Fig. 10.7 The relationship between section thickness—and hence rate of cooling—and structure in a cast iron.

fracture may well be white because cementite has had insufficient opportunity to decompose, under the influence of silicon, to form graphite.

Clearly the sectional thickness of a casting influences its rate of solidification and cooling and this, in turn, will control the condition of the carbon. An iron which, in heavy sections, will produce a grey iron may, when 'chilled' by casting it as a thin section, give a brittle white structure. The extent to which a particular iron will 'chill' in this way is generally examined by casting a sample in the form of a stepped bar (Fig. 10.7).

The Structure and Properties of Cast Iron

10.7.   A white cast iron contains both *primary* and *pearlitic* cementite. The addition of silicon to the composition causes first the primary cementite and then the pearlitic cementite to decompose to graphite and ferrite. The best combination of mechanical properties in an ordinary grey iron is obtained when the structure consists of small flakes of graphite in a pearlitic matrix. Such a structure will be achieved by the foundryman by adjusting the silicon content of the iron to suit the sectional thickness of the casting and hence the rate of cooling which will prevail during its solidification.

Fig. 10.8   The influence of silicon and cooling rate on the structure and properties of cast iron.

Under the microscope graphite flakes in cast iron appear as irregular strands with pointed ends. In fact what we see are sections through graphite flakes which, in reality, are shaped something like the corn flakes sold as a breakfast cereal. Nevertheless the sharp rims, which appear in cross-section as pointed ends, act as stress-raisers in the structure and are responsible for considerable reductions in strength and toughness of the resultant casting. In the design of stress-bearing components generally, the engineer makes every attempt to avoid in-cut sharp corners which can act as stress-raisers. Similarly, on the microscopic scale, the metallurgist can provide an iron in which the stress-raising flakes of graphite are replaced by rounded globules of carbon.

Special Cast Irons

10.8   These are cast irons in which the mechanical properties have been improved either by alloying or by some treatment which has caused graphite to precipitate in a spheroidal form and so eliminate the weakening graphite flakes.

10.8.1 Spheroidal-graphite (S.G.) Cast Iron   Also known as 'ductile iron' in the USA, this is one such iron where flake graphite is replaced by spherical particles of graphite. This is achieved by adding small amounts of magnesium—usually as a magnesium/nickel alloy—to give a resultant magnesium content of no more than 0.1% in the iron. This is sufficient to 'nucleate' the graphite so that it forms as spherical globules (Fig. 10.12).

Such an iron may have a tensile strength of 775 MPa. Tensile strengths of the order of 1600 MPa (elongation 1.0%) can be attained by austempering such an iron at 250°C (following initial austenitising at 900°C).

10.8.2 Compacted-graphite (C.G.) Irons   These are intermediate in structure—and hence in mechanical properties—between ordinary grey flake-graphite irons and SG irons. Molten iron of near-eutectic composition is treated with a single alloy addition containing small amounts of Mg, Ti and Ce. This causes short graphite flakes with *rounded* edges to be formed, leading to the term 'vermicular iron' being adopted in the USA. These rounded edges are less hostile as stress raisers than the acute edges of ordinary graphite flakes.

10.8.3 Malleable Cast Irons   These are cast to shape in the normal way using a low-silicon iron which will produce a brittle white structure. Both primary and pearlitic cementite are subsequently eliminated by suitable heat treatment. In the *Blackheart Process* the castings are heated for upwards of forty hours at about 900°C so that cementite decomposes to form spherical 'rosettes' of 'temper carbon' in a tough ferritic matrix. Blackheart malleable iron is used widely in the motor-car industry for such components as rear-axle housings, wheel hubs, brake shoes and door hinges.

In the *Whiteheart Process* the heating process is carried out under oxidising conditions so that carbon is progressively eliminated from the structure. As carbon in the surface layers oxidises and is lost as carbon

Table 10.5 *Compositions, Properties and Uses of Some Cast Irons*

| Type of iron | Composition (%) | Representative mechanical properties | | Uses |
		Tensile strength (MPa)	Hardness (Brinell)	
Grey iron	3.30 C 1.90 Si 0.65 Mn 0.10 S 0.15 P	Strengths vary with sectional thickness but are generally in the range 150–350 MPa	—	Motor brake drums
Grey iron	3.25 C 2.25 Si 0.65 Mn 0.10 S 0.15 P		—	Motor cylinders and pistons
Grey iron	3.25 C 1.25 Si 0.50 Mn 0.10 S 0.35 P		—	Heavy machine castings
'Chromidium'	3.20 C 2.10 Si 0.80 Mn 0.05 S 0.17 P 0.32 Cr	275	230	Cylinder blocks, brake drums, clutch casings, etc.
Wear- and shock-resistant	2.90 C 2.10 Si 0.70 Mn 0.05 S 0.10 P 1.75 Ni 0.10 Cr 0.80 Mo 0.15 Cu	450	300	Crankshafts for diesel and petrol engines. (Good strength, shock-resistance and vibration-damping capacity)
'Ni-resist'	2.90 C 2.10 Si 1.00 Mn 0.05 S 0.10 P 15.00 Ni 2.00 Cr 6.00 Cu	215	130	Pump castings handling concentrated chloride solutions. An austenitic corrosion-resistant alloy

dioxide, more carbon diffuses outwards from the core and is in turn eliminated leaving a totally ferritic structure though a central core of pearlite invariably remains. The surface often exhibits oxide penetration and malleable iron produced by this method is widely used in fittings for gas, water and steam pipes; bicycle and motor-cycle frame fittings and switch-gear equipment.

In *Pearlitic Malleable Iron* heat treatment is applied to cause decomposition of all primary cementite. The work is then heated to 900°C so that some of the temper carbon is dissolved in the austenite then present. Subsequent cooling leads to the formation of a strong, tough pearlitic matrix.

Fig. 10.9   A 'white' cast iron consisting essentially of pearlite (dark) and cementite (white). The pearlite 'dendrites' were derived from primary austenite dendrites. On close examination the light areas are seen to be lamellae of cementite and pearlite, arising from a eutectic of cementite and austenite formed at 1147°C. (Si—0.35%.) (Etched in 2% Nital.) ×150.

Fig. 10.10   A 'grey' cast iron. The structure consists of graphite flakes (black) in a matrix of pearlite. Typical of a good-quality engineering cast iron. (Si—1.85%.) (Etched in 2% Nital.) ×150.

Fig. 10.11 Blackheart malleable iron, produced by annealing the appropriate white iron at 900°C for 70 hours. The structure consists of 'rosettes' of graphite in a matrix of ferrite. (Etched in 2% Nital.) ×90.

Fig. 10.12 Spheroidal-graphite iron, in the cast condition. The structure consists of spheroidal graphite (black) in a matrix of pearlite and cementite (white). Annealing would lead to decomposition of the brittle cementite giving a matrix of pearlite and ferrite. (Etched in 2% Nital.) ×90.

10.8.4 Alloy Cast Irons   These make use of the same elements as are employed in alloy steels. *Nickel* is the most common element used. It promotes graphitisation and has a useful grain-refining—and hence, toughening—effect. *Chromium* on the other hand increases hardness and wear-resistance by stabilising carbides. *Vanadium* promotes heat-resistance in cast iron by rendering carbides very stable so that they are less likely to decompose at high temperatures. *Copper*, which is only very sparingly soluble, improves the resistance to atmospheric corrosion.

11

Non-ferrous Metals and Alloys

11.1 Limited amounts of copper, like gold, occur uncombined in the Earth's crust and it was for this reason that both metals were used in ancient times. It is thought that a short 'Copper Age' preceded the well-documented 'Bronze Age'. Lead was used by the Romans—one can still see the remnants of their plumbing in Pompeii—whilst Phoenicians came from the East to trade for tin with the Ancient Britons of Cornwall. The production of other important metals like aluminium, zinc, nickel and magnesium is of more recent origin and the industrial history of others such as titanium spans decades rather than centuries.

Copper and its Alloys

The bulk of the world output of copper is used unalloyed mainly in the electrical industries. Though it is second to silver in terms of conductivity relative costs are of course much in favour of copper. Since the presence of impurities seriously impairs electrical conductivity (18.4) either electrolytically refined (OFHC—'oxygen-free high-conductivity') copper or fire-refined ('tough pitch') copper is used for electrical purposes.

11.2.1 Pure copper is a very ductile metal, being FCC in structure. In the 'soft' condition it has a % elongation of 60 but is relatively weak with a tensile strength of no more than 215 MPa. Cold-drawn copper has a strength of only 340 MPa and for use as telephone wires the addition of 1% cadmium will increase the strength to 460 MPa whilst reducing specific conductivity by only 5% (Fig. 18.5). Because of its high ductility and good corrosion resistance much copper is also used in domestic plumbing, particularly since the availability of standard pipe connectors (containing pre-set solder) has made joining a simple procedure.

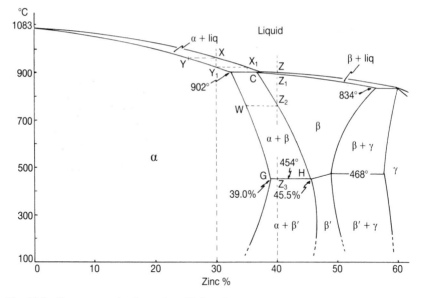

Fig. 11.1   The copper–zinc thermal equilibrium diagram.

11.2.2   α-phase alloys in the Cu–Zn; Cu–Sn and Cu–Al systems are all ductile. They were therefore used widely for deep drawing and presswork but vast improvements in the ductility of mild steel in recent decades coupled with developments in forming techniques have meant that much cheaper mild steel has largely replaced brasses and bronzes except where the higher corrosion resistance of the latter is a priority.

11.2.3 Brasses   are Cu–Zn alloys containing up to 45% Zn (Fig. 11.1). The α-phase brasses containing up to approximately 37% Zn are noted mainly for high ductility which reaches a maximum at 30% Zn. 70–30 brass is used for the deep-drawing of components requiring good resistance to corrosion. This group of brasses is shaped almost entirely by cold-working processes in the finishing stages.

An alloy containing 30% Zn begins to solidify at X (about 960°C) by depositing α dendrites of composition Y. Since these dendrites are copper-rich (approx. 23% Zn) the remaining liquid becomes less rich in copper. If we assume solidification to proceed *slowly* then the liquid changes in composition along XX_1 whilst the α dendrites change in composition along YY_1 as the temperature falls. At Y_1 solidification is complete and uniform α (70% Cu–30% Zn) is the result.

Rapid cooling under industrial conditions would not allow diffusion to proceed to completion so that some coring would be retained in the growing α crystals. Hence the shrinking volume of liquid would become so enriched with Zn that at 902°C a peritectic reaction would occur resulting in the formation of β of composition C (36.8% Zn) along the α grain

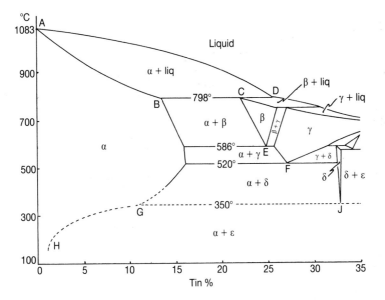

Fig. 11.2 The copper–tin thermal equilibrium diagram up to 35% Sn. Broken lines suggest that indicated phases in that region are unlikely to be formed under normal rates of cooling due to sluggishness of transformations caused by lack of thermal activation at low temperatures.

boundaries. Fortunately this β is generally quickly absorbed during the early stages of hot-working and the structure attains equilibrium showing a uniform α structure as indicated by the equilibrium diagram (see Figs 11.3 and 11.4).

A 60–40 brass begins to solidify at Z as the β phase and solidification is complete at Z_1. Since the range between liquidus and solidus is small there will be little coring in the β crystals. As the temperature falls to about 770°C (Z_2) dendrites of α (composition W) begin to separate out along certain planes within the β structure producing a typical Widmanstätten structure (Fig. 11.5). This process continues as the temperature falls, α changing in composition along WG whilst β changes along Z_2H. Hence at 454°C:

$$\text{Weight of } \alpha \text{ (composition G)} \times \text{GZ}_3 = \text{Weight of } \beta \text{ (composition H)} \times \text{Z}_3\text{H}$$

or $$\frac{\text{Weight of } \alpha}{\text{Weight of } \beta} = \frac{\text{Z}_3\text{H}}{\text{GZ}_3} = \frac{(45.5 - 40)}{(40 - 39)} = \frac{5.5}{1}$$

Below 454°C a disorder–order transformation occurs, i.e.

$$\begin{array}{ccc} \beta & \rightleftarrows & \beta' \\ \text{(disordered)} & & \text{(ordered)} \end{array}$$

Whereas β′ is hard and rather brittle β is soft and malleable. Hence 60–40 (α + β′) brasses are shaped by hot-working (rolling, forging or extrusion) above 770°C but on a *falling temperature gradient* so that the process

Fig. 11.3 70–30 brass in the chill-cast condition. The structure consists of heavily cored crystals of the α phase only. Neighbouring crystals have etched to different intensities because of different orientations of the crystallographic planes. (Etched in ammonia/hydrogen peroxide.) ×60.

Fig. 11.4 70–30 brass, cold-worked and then annealed at 650°C. Uniform α solid solution with annealing twins produced on recrystallisation of the cold-worked structure. (Etched in ammonia/hydrogen peroxide.) ×60.

Fig. 11.5   60–40 brass in the chill-cast condition. This shows a Widmanstätten distribution of α (light) in a matrix of β′ (dark). (Etched in ammonia/hydrogen peroxide.) ×150.

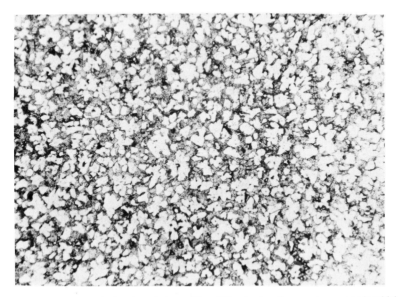

Fig. 11.6   60–40 brass in the extruded condition. Extrusion in a temperature range in which α was being deposited from β has resulted in the former being broken up and the weak Widmanstätten structure prevented from re-forming. (Etched in ammonia/hydrogen peroxide.) ×150.

finishes below 770°C. Thus as α precipitates it is prevented from forming a brittle Widmanstätten structure but is broken up into a granular pattern instead (Fig. 11.6).

Additions of up to 2% Pb which is insoluble in both liquid and solid brass, improves the machinability of brass by initiating chip-cracks in advance of the cutting tool; whilst 1.0% Sn will increase corrosion resistance.

11.2.4 Tin Bronzes   These also fall into two groups. The single-phase α-alloys are tough, ductile, strong and corrosion resistant. They are more heavily cored in the cast condition than are the α brasses and this is indicated by the wide range between the liquidus AD and the solidus AB in the Cu–Sn system (Fig. 11.2) as compared with the narrow liquidus–solidus range in the Cu–Zn system (Fig. 11.1). Moreover diffusion of tin in copper is slow so that cast Cu–Sn alloys with as little as 6% Sn may contain some of the brittle intermetallic compound δ ($Cu_{31}Sn_8$). In industrial practice cooling is never slow enough to permit transformation to ε (Cu_3Sn) to occur; for which reason the lower (broken-line) part of the diagram is of theoretical rather than practical interest. Thus a 5% Sn bronze which has cooled rapidly will be heavily cored (Fig. 11.7) and will contain small 'islands' of sky blue δ ($Cu_{31}Sn_8$) interspersed between the α dendrites. A similar alloy slowly cooled will exhibit a completely α structure with very light coring but it will be most unlikely to contain any of the phase ε which can only occur under strictly controlled laboratory conditions. These α bronzes are cold-rolled (Figs 11.9 and 11.10) to sheet and drawn to wire and other sections.

Bronzes containing in excess of 10% Sn will show considerable amounts of δ due to excessive coring of the α. Such an alloy will undergo a number of phase changes as it cools. At 798°C the α dendrites will react peritecti-cally with the remnant liquid to produce the solid solution β (composition C–22% Sn). This β changes in composition along CE and in turn undergoes a eutectoid transformation at 586°C to give a eutectoid (E) containing 24.6% Sn; i.e.

$$\beta \rightleftharpoons \alpha + \gamma$$

in which γ is another solid solution. At 520°C yet another eutectoid transformation occurs forming a eutectoid of composition F (27% Sn) containing the phases α and δ ($Cu_{31}Sn_8$). At each stage of this series of phase changes some α will remain in excess of the eutectoid composition so that the final structure will consist of cored dendrites of the solid solution α with an infilling of α + δ eutectoid (Fig. 11.8). As mentioned above it is unlikely that with normal cooling rates the eutectoid transformation at approximately 350°C (J) will occur, much less the precipitation of ε (Cu_3Sn) from α as is indicated by the steeply sloping boundary GH.

Due to the presence of considerable quantities of the extremely brittle intermetallic compound δ ($Cu_{31}Sn_8$) in bronzes containing more than 10% Sn these alloys are too brittle to permit cold-working using normal industrial processes. In fact during the Second World War one of my tasks was to produce slabs of *cold-rolled* 12% Sn bronze in connection with early prototypes of Whittle's jet engine. This entailed solution annealing of the cast ingots for 12 hours at 740°C, i.e. just below the solidus temperature, to absorb the brittle δ. Even so the uniform α so obtained work-hardened very quickly and many inter-stage anneals were involved in rolling to size. My team finally produced a cold-worked 12% Sn bronze with a tensile strength in excess of 1000 MPa but it was very expensive to produce!

Bronzes containing more than 10% Sn are used for the manufacture of corrosion-resistant castings and bearings in which the presence of particles of the hard δ phase, cushioned in a touch matrix of α provides a hard-wearing, low-friction surface. Tin bronzes are generally deoxidised with small amounts of phosphorus, leaving a residue of no more than 0.04% phosphorus in the alloy. The addition of phosphorus in excess of this amount and up to 1.0% is made when increased strength, corrosion resistance and casting fluidity are necessary. These alloys are known as *phosphor bronzes*.

In 'gunmetals' some of the expensive tin is replaced by zinc. This not only reduces cost but makes deoxidation with phosphorus unnecessary since zinc itself has a high affinity for oxygen forming insoluble zinc oxide which joins the dross on the surface of the molten alloy.

11.2.5 Aluminium Bronze As in the two previous alloy systems copper forms a ductile α solid solution, here containing a maximum of 9.4% Al. These α alloys are also very corrosion resistant. Since the 5% Al alloy is of the colour of 18 ct gold it was used largely in the manufacture of 'Woolworth's wedding rings'. Good resistance to corrosion ensured that it was unlikely to leave behind an embarrassing green stain on the wearer's finger during its weekend of use. Now that proof of marriage is no longer a social requirement these α-aluminium bronzes have lapsed into obsolescence!

As the phase diagram (Fig. 11.11) indicates alloys containing between 9.5% and 14.95% Al solidify as a solid solution β and as the temperature falls either primary α or primary γ_1/γ_2 will precipitate within the β crystals until at 565°C the remaining β transforms to a eutectoid of $\alpha + \gamma_2$. This eutectoid transformation $(\beta \rightarrow \alpha + \gamma_2)$ is similar to that which occurs in steels at 723°C (Austenite $(\gamma) \rightarrow$ Pearlite $(\alpha + Fe_3C)$). The similarity does not end there as the products of the transformation of β are also influenced by *cooling rates* as are those in steel.

Thus an 11% Al alloy (Fig. 11.11) of β composition which is cooling *extremely slowly* begins to deposit primary α as the temperature falls below X and at 565°C, the eutectoid temperature, the structure consists of primary α (composition Q) and remaining β (composition R) in the ratio:

$$\text{Weight of } \alpha \text{ (composition Q)} \times QX_1$$
$$= \text{Weight of } \beta \text{ (composition R)} \times X_1R$$

or $$\frac{\text{Weight of } \alpha}{\text{Weight of } \beta} = \frac{X_1R}{QX_1} = \frac{(11.8 - 11.0)}{(11.0 - 9.4)} = \frac{0.8}{1.6} = \frac{0.5}{1}$$

Fig. 11.7 95–5 tin bronze, chill cast. Cored dendritic structure of α solid solution. (Etched in ammonia/hydrogen peroxide.) ×100.

Fig. 11.8 90–10 tin bronze, chill cast. Cored dendrites of α (dark) with an infilling of α + δ eutectoid. The δ has coagulated so that the lamellar nature of the eutectoid has disappeared. The large dark areas are shrinkage cavities (they follow the outline of the dendrites). (Etched in ammonia/hydrogen peroxide.) ×150.

Fig. 11.9 95–5 tin bronze, cold-worked, annealed and cold-worked again. Unifom α solid solution exhibiting annealing twins from the previous annealing operation and large numbers of strain bands caused by the last cold-worked operation. (Etched in ammonia/hydrogen peroxide.) ×250.

Fig. 11.10 95–5 tin bronze, cold-worked and annealed. Uniform α solid solution, many of the crystals showing annealing twins. The strain bands are now absent because the alloy has recrystallised, and is now soft. (Etched in ammonia/hydrogen peroxide.) ×250.

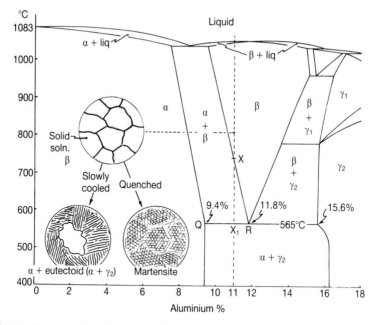

Fig. 11.11   Part of the Cu–Al thermal equilibrium diagram.

Just below 565°C remaining β transforms to a eutectoid consisting of alternate lamellae of α (9.4% Al) and γ_2 (15.6% Al) (Fig. 8.15). Fundamentally γ_2 is a brittle compound Cu_9Al_4 and occupies a similar position here to that of Fe_3C in steel.

Water quenching of the 11% Al alloy from 800°C produces a very hard martensitic type of structure which responds to tempering as does martensite in steel, whilst air cooling from 800°C gives a structure which is analogous to bainite in steels. The response to cooling rates by an aluminium bronze is therefore similar to that of a low-alloy (oil-quenching or air-quenching) steel. Despite these apparently attractive properties little use is now made of such alloys and most $\alpha + \gamma_2$ aluminium bronzes now used are as corrosion resistant castings. This limited popularity is due largely to difficulties arising during casting. High casting temperatures involve excessive oxidation of the aluminium, the entrapped oxides of which cause unsoundness of the castings. Methods used to alleviate this are expensive.

11.2.6 Cupro-nickels   Since both copper and nickel are FCC metals with similar atomic dimensions and chemical properties it is not surprising that they form substitutional solid solutions in all proportions (Fig. 8.10). They are tough, ductile and reasonably strong when cold-worked. Because of their high cost they are used mainly where high ductility coupled with good corrosion resistance are required such as in the manufacture of our current 'silver' coinage.

Nickel-silvers are copper–nickel–zinc alloys used (in the silver-plated form) for cutlery manufacture, though now largely replaced by stainless steels for such purposes.

11.2.7 Beryllium Bronze   contains up to 2% Be and can be strengthened by precipitation treatment to give tensile strengths in the region of 1400 MPa. It is used as a tool material in the gas industries and explosives factories where sparks, generated from steel tools, might have disastrous results.

Details of some of the more important copper-base alloys are given in Table 11.1.

Aluminium and its Alloys

11.3   Because of its high affinity for oxygen, aluminium could not be extracted on a commercial scale until the electrolytic decomposition of aluminium oxide was introduced in 1886. Hence the technology of aluminium has been developed almost entirely in the twentieth century. Despite its great affinity for oxygen the corrosion-resistance of aluminium is high. This is because a thin but impervious film of oxide forms on the surface of the metal and protects it from further oxidation. The film can be further thickened by 'anodising' and since aluminium oxide is extremely hard, wear-resistance is also increased by the process. The high affinity of aluminium for oxygen makes it useful as a de-oxidant in steels.

Whilst the electrical conductivity of aluminium is less than a half that of copper, *weight for weight* it provides a better conductor, and, where space is available as in power-grid cables, it is widely used. However, pure aluminium is relatively soft and weak—it has a tensile strength of no more than 90 MPa in the annealed condition—and for most engineering purposes is used in the alloyed form. The strengths of many aluminium-base alloys can be further increased by *precipitation hardening* (7.5.2) to produce a strength/mass ratio – *or specific strength* – of the same order as for high-tensile steels. The greater relative volume of aluminium alloy involved for a specific force-bearing capacity means that greater flexibility in design is possible. Aluminium alloys used in both cast and wrought forms may be precipitation hardened if of suitable composition.

11.3.1 Aluminium Alloys Which are not Heat-treated   Although these alloys are not metallurgically exotic as are those which can be precipitation hardened, they none the less account for the greater proportion of the industrial output. The most widely used casting alloys are those containing 10–13% Si which imparts a considerable increase in casting fluidity. These alloys are of approximately eutectic composition (Fig. 11.12) so that their freezing range is very small making them particularly suitable for die casting where rapid ejection from the die is desirable in the interests of output. The rather coarse eutectic structure can be refined by a process

Table 11.1 Some Important Copper-base Alloys

	Composition (%) (Balance Cu)	Condition	Typical mechanical properties				Characteristics and uses
			0.1% P.S. (MPa)	Tensile strength (MPa)	Elongation (%)	Hardness (VPN)	
Brasses	30 Zn	Annealed Hard	77 510	325 695	70 5	65 185	*Cartridge brass*: deep-drawing brass, maximum ductility of the copper–zinc alloys
	37 Zn	Annealed Hard	95 540	340 725	55 4	65 185	*'Common brass'*: a general purpose brass suitable for limited forming by cold work
	40 Zn	Hot-rolled	110	370	40	75	Hot-rolled plate used for the plates of condensers. Also extruded rods and tubes. Limited capacity for cold work
	1.5 Mn 1.0 Al 1.0 Pb 0.7 Sn 0.7 Fe 37 Zn	Extruded	280	500	15	150	*High-tensile brass*: wrought sections for pump rods, etc. Cast alloys: marine propellors, rudders, water-turbine parts
Tin bronzes	3.75 Sn 0.10 P	Annealed Hard	110 620	340 740	65 5	60 210	*Low-tin bronze*: good elastic properties and corrosion resistance. Springs, instrument parts
	11 Sn 0.5 P	Sand-cast	125	280	15	90	*Cast phosphor bronze*: mainly bearings. Cast as sticks for machining of small bearing bushes
	10 Sn 2 Zn	Sand-cast	125	295	16	85	*'Admiralty gunmetal'*: pumps, valves, etc. particularly for marine purposes (good corrosion resistance). Also statuary
Aluminium bronzes	5 Al Ni and Mn up to 4% total	Annealed Hard	125 590	385 775	70 4	80 220	Imitation jewellery, etc. Excellent corrosion resistance and to oxidation on heating
	9.5 Al 2.5 Fe Ni, Mn up to 1.0% each	Cast	—	525	—	115	The best known aluminium-bronze for both sand- and die-casting. Corrosion-resistant castings
Cupro-nickels	25 Ni 0.25 Mn	Annealed Hard	— —	355 600	45 5	80 170	Mainly for coinage e.g. the current British 'silver' coinage
	68 Ni 1.25 Fe 1.25 Mn	Annealed Hard	215 570	540 725	45 20	120 220	*Monel metal*: good mechanical properties and good corrosion resistance. Mainly in chemical engineering plant

known as *modification*. This involves adding a small amount of sodium (about 0.01% of the mass of the charge) to the melt just before casting. This delays precipitation of silicon when the normal eutectic temperature is reached so that when nucleation begins a fine grain results from rapid crystallisation due to under-cooling. A shift of the eutectic composition to the right is also caused. Hence a 12% Si alloy when modified will contain an excess of primary α instead of embrittling primary silicon which appears in the structure of the unmodified 12% alloy (Fig. 11.12). Modification raises the tensile strength from 120 to 200 MPa and the % elongation from 5 to 15.

The cast aluminium–magnesium alloys are notable mainly for a good combination of corrosion resistance and rigidity, making them particularly suitable for moderately stressed parts working in marine atmospheres.

Those wrought alloys which are not heat-treated rely for their strength on the formation of solid solutions which are subsequently work-hardened

(i) 12% SILICON UN-MODIFIED   x75

since the composition is to the right of the eutectic point (11.6% Si) the structure consists of primary silicon and eutectic (which is coarse in structure).

(ii) 12% SILICON MODIFIED   x75

here 'modification' has displaced the eutectic point to 14.0% Si so that an alloy of similar silicon content to (i) now contains primary α and eutectic —which is fine-grained due to depression of the eutectic temperature.

Fig. 11.12 The 'modification' of aluminium–silicon alloys.

by cold-working processes. A mere 1% Mn is all that is necessary to strengthen aluminium sufficiently to make it suitable for the manufacture of milk-bottle caps and vehicle panelling, but where good corrosion resistance is also necessary then up to 5% Mg may be added.

11.3.2 Aluminium Alloys Which are Heat-treated   Although precipitation hardening can be applied to many alloys, both ferrous and non-ferrous, it is most commonly used to strengthen suitable aluminium-base alloys. 'Age-hardening', as it was then called, was in fact discovered in some aluminium-base alloys at the beginning of the twentieth century and subsequently developed for use in military aircraft during the First World War.

A number of wrought 'duralumin-type' alloys are in use but possibly the best known is that based on 4% Cu. At temperatures above that indicated by A (Fig. 11.13) the structure will consist entirely of α-solid solution. If

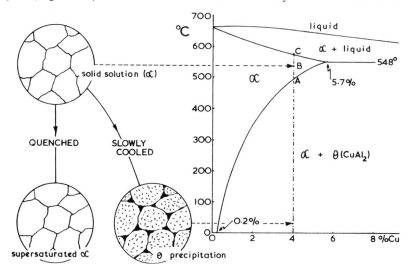

Fig. 11.13   The principle of 'solution treatment' of an aluminium–copper alloy.

the alloy is allowed to cool slowly to room temperature then the phase θ ($CuAl_2$) will be rejected as a non-coherent precipitate in accordance with the equilibrium diagram which indicates that the solid solubility of copper in aluminium falls from 5.7% at 548°C to 0.2% at 0°C. The resultant structure lacks strength because the α matrix contains so little copper in solid solution, whilst it will be brittle because of the presence of brittle particles of θ.

If the alloy is slowly reheated θ is gradually re-dissolved until at A the structure once more consists entirely of the uniform solid solution α. In industrial practice a solution-treatment temperature given by B will be used to ensure maximum solution of θ. Care must of course be taken that the temperature does not exceed that given by C or the alloy will begin to melt.

Fig. 11.14 A duralumin-type alloy (Cu 4.5; Si 0.75; Mg 0.7; Mn 0.6; Fe 0.4) in the extruded condition. Particles of various compounds (mainly CuAl$_2$ and an Al–Mn–Fe complex) elongated along the direction of extrusion. (Unetched.) ×250.

The alloy is quenched from B so that copper is retained in super-saturated solid solution at room temperature. This increases strength because the whole of the 4% Cu is now in solid solution and brittleness decreases because the brittle particles of θ (CuAl$_2$) are now absent. However, since the structure is not in equilibrium subsequent migration of copper atoms occurs to the extent that they form coherent precipitates (7.5.2.1) within the α lattice:

$$Cu \quad + \quad Al \quad \rightarrow \qquad \theta'$$

(Al lattice)   (Al lattice)   (Intermediate *coherent* precipitate)

Since the coherent precipitates impede the movement of dislocations the strength of the alloy increases.

11.3.3 The extent of the formation of coherent precipitates at ordinary temperatures is limited so that strength attains a fairly low maximum value in a few days and this process used to be called 'age-hardening'. At higher temperatures the formation of coherent precipitates proceeds further and so the strength continues to increase (Fig. 11.15). However, a point is reached where the thermal activation is such that tiny *non*-coherent particles of θ (CuAl$_2$) begin to form in accordance with phase equilibrium:

$$\theta' \quad \rightarrow \quad \theta\,(CuAl_2)$$

(coherent)   (non-coherent)

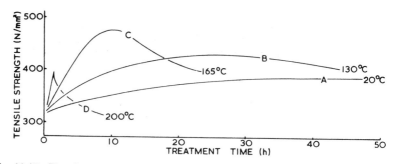

Fig. 11.15   The effects of time and temperature of precipitation treatment on the structure and tensile strength of a suitable alloy.

At this point strength and hardness begin to fall rapidly. Thus optimum strength, for the alloy relevant to Fig. 11.15, would be obtained by precipitation treatment at 165°C for 10 hours. Treatment at 200°C (curve D) would lead to the formation of non-coherent θ virtually 'overtaking' that of the formation of coherent θ'.

11.3.4   In addition to the 4% Cu duralumin-type of alloy, other wrought aluminium alloys which can be precipitation hardened are those containing small amounts of magnesium and silicon. These form the compound Mg_2Si the solubility of which, like that of $CuAl_2$, increases considerably with temperature.

After many years of development work precipitation-hardening alloys containing up to 2.5% lithium are being used in military aircraft. Since lithium has a relative density of only 0.534 alloys containing it have a relative density as low as 2.54, offering an increase in *specific modulus* of up to 25%. Precipitation hardening is due mainly to a phase δ' (Al_3Li), a precipitate coherent with the matrix.

A number of aluminium-based casting alloys can also be precipitation hardened with advantage. Possibly the best-known of these is the aircraft piston alloy known as 'Y alloy'.* This contains small amounts of copper, nickel, manganese and titanium and is notable for its tolerance of relatively high working temperatures.

Compositions and properties of some of the more important aluminium-base alloys are described in Table 11.2.

> | Q | An aluminium alloy (2% Cu; 0.3% Mg) in the form of rivets is solution treated at 495°C and used whilst it remains in this soft supersaturated solid-solution condition. It subsequently precipitation hardens at ambient temperature after the riveting operation is complete. During tests it was found that the solution-treated alloy began to harden in 4½ hours at ambient temperature (20°C) whilst if placed in boiling water (100°C) the precipitation hardening process began in only 3½ minutes.

*The reference code used during the development of this alloy at the NPL during the First World War.

It is required to refrigerate a batch of solution-treated rivets so that the onset of precipitation hardening is delayed in order that the rivets may be stored and used over a period of four days after solution treatment. What refrigeration temperature must be used so that the rivets will not fail during the riveting process?

A A modified form of the Arrhenius equation (5.8.6) may be used since an evaluation of the activation energy (Q) of the reaction is not required. Hence:

$$t = A.e^{X/T}$$

or $t = A.\exp(X/T)$ .... (1)

which can be written in logarithmic form as:

$$\ln t = \ln A + X/T$$

or $X = T(\ln t - \ln A)$ .... (2)

where t is the time required for precipitaton hardening to begin, T is the temperature (K) and A and K are constants for the system.
At 100°C (373K), $t = 3.5$ min. Therefore substituting in (2):

$$X = 373(\ln 3.5 - \ln A)$$ .... (3)

and at 20°C 293K, $t = 4.5$ hr. (270 min.). Therefore:

$$X = 293(\ln 270 - \ln A)$$ .... (4)

Combining (3) and (4) to eliminate X:

$$373(\ln 3.5 - \ln A) = 293(\ln 270 - \ln A)$$

and $373.\ln 3.5 = 293.\ln 270 + 80.\ln A$

Therefore

$$\ln A = \frac{(373. \ln 3.5) - (293. \ln 270)}{80}$$

$$= \frac{(373 \times 1.253) - (293 \times 5.598)}{80}$$

> $\log_e 3.5 = 1.253$
>
> $\log_e 270 = 5.598$

$$\ln A = -14.66$$

Substituting for $\ln A$ in (4)

$$X = 293(\ln 270 + 14.66)$$
$$= 293(5.598 - 14.66) = 5936$$

> $\log_e 270 = 5.598$

Let T_1 be the temperature corresponding to initiation of precipitaton in 4 days (5760 minutes) then, substituting for T_1, X and A in (2):

$$5936 = T_1(\ln 5760 + 14.66)$$
$$5936 = T_1(8.659 + 14.66)$$

> $\log_e 5760 = 8.659$

$$T_1 = \frac{5936}{23.32}$$

Table 11.2 Some Important Aluminium-base Alloys

	Composition (%) (balance Al)	Condition	Typical mechanical properties			Characteristics and uses
			0.1% P.S. (MPa)	Tensile strength (MPa)	Elongation (%)	
Wrought-alloys—not heat-treated	1.2 Mn	Soft	45	110	34	Domestic and commercial food containers; cooking utensils; panelling of land-transport vehicles; automobile radiators
		Hard	170	200	4	
	4.5 Mg 0.7 Mn	Soft	125	315	15	Unheated welded pressure vessels; marine, auto and aero parts; TV towers; drilling rigs; transportation equipment—uses where good corrosion resistance is needed
		½ hard	235	345	8	
Cast alloys—not heat-treated	5.0 Si 3.0 Cu 0.4 Mn	Sand cast	—	140	2	Sand castings and die castings. General purposes, where mechanical properties are of secondary importance
		Chill cast	—	160	3	
	11.5 Si	Sand cast	—	160	5	Sand castings; gravity and pressure die-castings. Excellent foundry properties. One of the most widely useful aluminium alloys ('modified'). Radiators, sumps, gear boxes and large castings
		Pressure die cast	—	190	7	
Wrought alloys—heat-treated	4.0 Cu 0.8 Mg 0.5 Si 0.7 Mn	Solution treated at 480°C, quenched and aged at room temperature for 4 days	280	400	10	General purposes—stressed parts in aircraft and other structures. The original 'duralumin'
	1.0 Mg 1.0 Si 0.7 Mn	Solution treated at 510°C; quenched and precipitation hardened at 175°C for 10 hours	150	250	20	Structural members for road, rail and sea transport vehicles; architectural work; ladders and scaffold tubes. High electrical conductivity—hence used in overhead lines
	1.6 Cu 2.5 Mg 5.8 Zn 0.15 Cr	Solution treated at 465°C; quenched and precipitation hardened at 120°C for 24 hours	590	650	11	Highly stressed aircraft parts such as booms. Other military equipment requiring a high strength/mass ratio. The strongest aluminium alloy produced commercially
Cast alloys—heat-treated	4.0 Cu 2.0 Ni 1.5 Mg 0.3 Si	Solution treated at 510°C; precipitation hardened in boiling water for 2 hours or aged at room temperature for 5 days	215	280	—	Pistons and cylinder heads for liquid and air-cooled engines. General purposes but now obsolescent. (The original Y alloy)
	11.5 Si 0.5 Mn 0.4 Mg	Solutions treated at 530°C for 3 hours; quenched in warm water; precipitation hardened at 160°C for 16 hours	—	290	2	Good corrosion resistance. High fluidity imparted by silicon—hence suitable for intricate castings

= 254.5K
= −18.5°C

In practice a temperature of −20°C would probably be used.

Magnesium-base Alloys

11.4 Despite its high affinity for oxygen magnesium, like aluminium, is a stable metal with a high resistance to corrosion at ordinary temperatures. Its chief attribute is its very low relative density (1.7) which makes it useful in the aircraft and aerospace industries. It has a low tensile strength but its alloys containing suitable amounts of aluminium, zinc or thorium can be strengthened by precipitation hardening. The solid solubility of each of these elements decreases as the temperature falls (Fig. 11.16) thus making

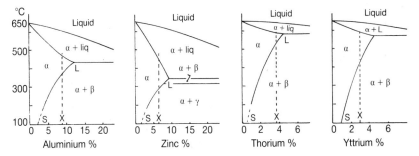

Fig. 11.16 The rapid changes in solid solubility with temperature of Al, Zn, Th and Y in magnesium (as shown by the sloping solvus line SL in each case) make possible the precipitation treatment of alloys of composition in the region of X.

the precipitation treatment possible. Both cast and wrought alloys are used, particularly as castings and forgings in the aircraft industries— landing wheels, crank cases, petrol tanks and many engine parts in both piston and jet engines. A typical general purpose casting alloy may contain 9 Al–0.5 Zn–0.2 Mn–bal. Mg, developing a tensile strength of up to 250 MPa in the heat-treated condition; whilst a wrought magnesium-base alloy containing 5 Zn–0.6 Zr–bal. Mg may reach a tensile strength of 315 MPa when precipitation hardened.

Nickel-base Alloys

11.5 Nickel is a valuable constituent both in corrosion-resistant alloys and, since it restrains grain-growth at high temperatures, in high-temperature alloys.

11.5.1 Corrosion-resistant Alloys Nickel is a principal constituent of many corrosion-resistant alloys used in the chemical industries. Many of these alloys are tough and ductile as well as being corrosion resistant. Thus, 'Corronel B' (66 Ni; 28 Mo; 6 Fe) is particularly resistant to mineral acids and acid chloride solutions and can be produced in the form of tubes and other sections for use in the chemical and petro-chemical industries. 'Hastelloy D' (85 Ni; 10 Si; 3 Cu; 1 Al) on the other hand is a tough, very hard casting alloy which is particularly resistant to hot concentrated sulphuric acid.

11.5.2 High-temperature Alloys Nickel–chromium alloys are particularly suitable for high-temperature working. Chromium imparts a low rate of oxidation (19.2) at all temperatures whilst nickel limits grain growth and the brittleness arising therefrom at high temperatures. Since these alloys also have a high specific resistance they are particularly suitable for resistance wire used in heater elements of electric fires, immersion heaters, toasters, etc. A grade of 'Brightray' containing 80 Ni–20 Cr has a maximum working temperature of 1150°C.

The best-known of the nickel–chromium base high-temperature alloys are those of the 'Nimonic' series which played a leading part in the development of the jet engine, but are now used for a large number of applications where a high creep strength at elevated temperatures is required. Most 'Nimonic' alloys contain approximately 75 Ni and 20 Cr to form a tough oxidation-resistant matrix at temperatures sometimes in excess of 1000°C. The alloy is stiffened for use at high temperatures by small additions of titanium, cobalt, zirconium, aluminium and carbon in suitable associations. These elements help to raise the limiting creep stress at high temperatures by the formation of dispersed particles of carbides or compounds such as $NiAl_3$.

Zinc-base Alloys

11.6 These are exclusively die-casting alloys which contain 4% Al and up to 2.7% Cu, the balance being zinc, and are used for the manufacture of a large number of domestic and engineering components. Door handles for motor cars, bodies for windscreen wiper motors, frames for electric fires, accurately scaled children's toys and parts for washing machines and refrigerators are typical examples.

Such alloys have quite good strength (320 MPa) and rigidity, but are sensitive to the presence of small amounts of impurities like cadmium, lead and tin which cause intercrystalline brittleness and some dimensional instability which manifests itself as swelling of the casting during use. These faults are largely due to intercrystalline corrosion for which reason good quality zinc-base die-casting alloys are generally made from 'four nines' zinc, i.e. zinc which is 99.99% pure.

Another series of zinc-base alloys contains 8, 11 and 27% Al respectively along with small amounts of Cu and Mg. The 8% alloy is used for pressure die-casting whilst the others are more suitable for gravity die casting. Like the 4% Al alloys they must be kept free of impurities if intercrystalline brittleness is to be avoided. After casting they age-harden naturally, rapid cooling during casting having suppressed the attainment of equilibrium.

Bearing Metals

11.7 The mechanical requirements of a bearing metal can only be met by the intelligent use of alloying. A bearing must be hard and wear-resistant with a low coefficient of friction but at the same time be tough, shock resistant and sufficiently ductile to allow for 'running in'. These properties of hardness, toughness and ductility cannot be found to the required degree in a single-phase alloy. Thus, intermetallic compounds are hard and have a low coefficient of friction but are extremely brittle, whilst pure metals and solid solutions though ductile are usually soft and with a relatively high coefficient of friction. A suitable combination of mechanical properties can, however, be obtained by using an alloy in which particles of a hard intermetallic compound are embedded in a matrix of ductile solid solution—or, in some cases, a eutectic of two solid solutions. During the 'running in' process the soft matrix tends to wear leaving the hard particles standing proud. This not only reduces the overall coefficient of friction of the bearing surface but also provides channels through which lubricant can flow.

11.7.1 'White'-bearing Metals These are either tin-base or lead-base. The former, the high-quality Babbitt metals, contain as much as 90% Sn and up to 10% Sb. These two metals form 'cuboid' crystals of the intermetallic compound SbSn which are easily identified in the microstructure and which constitute the hard, low-friction phase. In lead-free white metals the cuboids are held in a matrix of the solid solution α (Fig. 11.18) whilst in white metals to which lead has been added in amounts up to 80% in order to reduce cost, the SbSn cuboids will be in a matrix consisting of a eutectic of two solid solutions—one tin-rich and the other lead-rich.

Cuboids of SbSn separate from the liquid phase between X and 246°C and since the relative density of SbSn is less than that of the liquid, the cuboids float to the surface as they form. To prevent this type of segregation up to 3% Cu is added to white-bearing metals. This forms the intermetallic compound Cu_6Sn_5 which separates out as a network *before* the formation of SbSn cuboids. When these latter subsequently begin to form they become trapped by the network of Cu_6Sn_5 crystals and so prevented from segregating at the surface of the casting.

11.7.2. Copper-base-bearing Alloys These include the plain tin bronzes (10–15% Sn) and phosphor bronzes (10–13% Sn; 0.3–1.0% P). As the phase diagram (Fig. 11.2) indicates these alloys will be of $\alpha + \delta$ structure,

Fig. 11.17 A white-bearing metal containing 10% Sb; 3% Cu; 40% Sn; balance-Pb. A copper–tin compound (Cu_6Sn_5) crystallises first (light needles) forming a network which prevents the very light SbSn cuboids from segregating to the surface when they begin to form. The matrix is a eutectic of tin–antimony and tin–lead solid solutions. (Etched in 2% Nital.) ×70.

and though nominally the hard compound δ ($Cu_{31}Sn_8$) will be present as a eutectoid with α, in practice the layers of δ coagulate to form hard particles in a tough matrix of the solid solution α thus fulfilling the general requirements of a bearing metal.

11.7.3   Sintered bronze bearings produced by powder metallurgy (6.5.1)

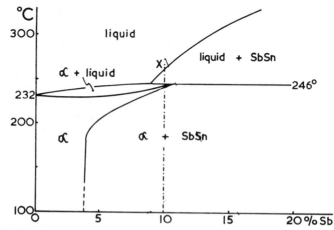

Fig. 11.18   Part of the tin–antimony phase diagram.

retain sufficient porosity to allow them to be impregnated with lubricating oil. These 'oil-less' bearings are widely used in vacuum cleaners, washing machines, gramophone motors and other applications where lubrication might be either uncertain or a lubrication system unjustifiably expensive.

Titanium Alloys

11.8   In terms of its occurrence in the Earth's crust titanium is a relatively abundant metal and, of the engineering metals, only aluminium, iron and magnesium are more plentiful. The amount of titanium ore within the range of mining operations is about one hundred times greater than that of copper. However, the very high affinity of titanium for both oxygen and nitrogen makes it difficult to extract whilst the molten metal itself reacts with all known refractories. Thus titanium is an expensive metal because of its cost of extraction and forming rather than any scarcity of its ores. Currently titanium billets cost about £5.00 per kg.

High-purity titanium has a relatively low tensile strength (216 MPa) and a high ductility (50%) but the strength can be raised considerably by alloying. It is a polymorphic element. The α phase (CPH) transforms on heating to 882.5°C to β (BBC) (Fig. 11.19(i)) and this change-point is

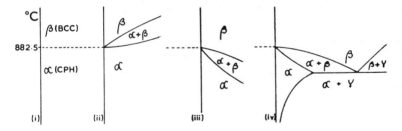

Fig. 11.19   Effects of alloying on the α ⇌ β transformation temperature in titanium.

affected by alloying as is the A_3 point in iron. Thus alloying elements which have a greater solubility in α than in β tend to stabilise α over a wider range of temperatures (Fig. 11.19(ii)). These elements include aluminium, which dissolves substitutionally, and oxygen, nitrogen and carbon all of which dissolve interstitially. Elements which dissolve more readily in β stabilise it in preference to α (Fig. 11.19(iii) and (iv)). They include the transition metals iron, chromium, molybdenum, etc. Alloys represented by a phase diagram of the type (iv) can be precipitation hardened as might be expected with a diagram of this form (7.5.2.1). A typical titanium-base alloy, IMI 700 (6 Al; 5 Zr; 3 Mo; 1 Cu) attains a tensile strength of 1540 MPa.

11.8.1   The relative density of titanium is only 4.5 and suitable alloys based on it have a high specific strength. Moreover creep properties up to

500°C are very satisfactory whilst the fatigue limit is also high. Titanium alloys therefore find application in the compressors of jet engines. In turbine engineering generally titanium alloys, being of low relative density, impose much lower centrifugal stresses on rotors and discs for a given blade size. Titanium alloys are also very resistant to erosion by wet steam and other media so that steam turbo-generators, gas turbines and condenser tubing all make use of these alloys.

Because of the high specific strength, high temperature resistance and good erosion resistance generally, many titanium alloys find use in supersonic aircraft—including 'Concorde'—as structural forgings. Such an alloy is IMI Ti680 (11 Sn; 2.25 Al; 4 Mo; 0.2 Si) which develops a strength of 1300 MPa.

An interesting development in the use of titanium is currently taking place in Japan—as a roofing material. Whilst titanium is of course expensive its high specific strength allows greater flexibility in design and its low relative density means that a much smaller load-bearing capacity is required of the roof members. Moreover it is of attractive appearance and requires minimum maintenance—important in a country like Japan where marine corrosion causes considerable problems with the more traditional metals used in architecture.

11.8.2   Titanium has a very high resistance to corrosion. Like aluminium it coats itself with an impervious closely adherent film of oxide which protects it from further corrosion so that, despite its relatively high cost, titanium is now used increasingly where high corrosion resistance in very hostile conditions is of primary importance.

When atmospheric pollution control is taken seriously and genuine attempts are made to reduce the vast amounts of sulphur dioxide reaching the atmosphere from power stations burning coal, some form of washing or 'scrubbing' of the flue gases is necessary. Wet lime or limestone is generally used for this purpose with the object of neutralising the sulphuric acid formed when SO_2 in the flue gas combines with water. Prior to neutralisation sulphuric acid will condense on some surfaces within the flue-gas desulphurisation (FGD) plant. FGD plants in the USA and more recently here at National Power's Drax Station have been fitted with titanium sheet linings. Although the initial cost is higher than the alternative 'glass' or borosilicate linings, annual maintenance costs are much lower and lining life much longer in the case of titanium, resulting in a lower overall annual cost when it is used.

Since titanium is superior to most other non-ferrous engineering alloys and stainless steels in terms of resistance to all forms of salt water corrosion it is potentially useful in equipment used on off-shore platforms.

12

Plastics Materials and Rubbers

12.1 The townships of mid-nineteenth-century Colorado attracted a motley assortment of adventurers, gamblers, saloon ladies and others bent on the acquisition of easy wealth. Amongst the more respectable establishments of these towns were billiard saloons which became increasingly popular. So much so that the quest for ivory led to the brutal slaughter of the African elephant to such an extent that ivory became scarce and hence very expensive. In an attempt to provide billiard balls John Hyatt sought a substitute. His early products consisted of a core of ivory dust bonded with shellac and an exterior coating of the somewhat unstable compound collodion. Unfortunately violent impact of the balls often caused the collodion to explode so that every gunman in the saloon instinctively 'pulled a gun'.*

However, in 1868 Hyatt successfully adopted a substance developed some fourteen years earlier by Alexander Parkes of Birmingham, England. It consisted of a mixture of cellulose nitrate and camphor and had been known as 'Parkesene'. This, the first of the man-made plastics, was renamed 'celluloid' and was used widely until a few years ago for the manufacture of many articles, in particular toys such as dolls. Its dangerous inflammability finally led to its obsolescence so that comparatively little of it is now used. When in 1907 Baekeland patented his process for the manufacture of 'bakelite' the plastics industry could be said to have 'arrived', for this material was quickly adopted by the rapidly expanding electrical trades as an insulation material. Now plastics materials of some thirty different compositions are used in our everyday artefacts—the modern motor car contains about 25% by weight of them.

*From *The First Century of Plastics—Celluloid and its Sequel* by M. Kaufman (The Plastics Institute, London).

12.1.1   The term 'plastics' is not one which is easy to define precisely. Many familiar 'plastic' articles like buttons, switchgear parts and egg-cups are in fact no longer plastic after being moulded to shape since they become hard and brittle and can no longer be deformed on heating.

There was little theoretical understanding of the nature of these materials until about 1920 when Herman Staudinger described them as 'macromolecules' and visualised their long thread-like form. He suggested that they could be seen using an ordinary optical microscope but for the fact that an individual molecule is so *thin* (if the average polythene molecule were magnified so that it appeared as a fibre 1 mm in diameter then it would be 5 m in length).

Plastics can be defined as organic materials containing molecules of high molecular weight (i.e. between 10^4 and 10^7) and which can be moulded to shape by the application of pressure at moderately high temperatures. Once moulded they may retain their plasticity in the manner of polythene or nylon or they may become permanently hard and brittle like bakelite. Some plastics, however, are compounded as liquids and proceed to 'set' without the application of either heat or pressure.

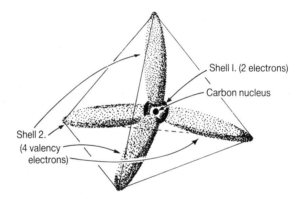

Shell I. (2 electrons)

Carbon nucleus

Shell 2.
(4 valency
electrons)

Fig. 12.1   The structure of the carbon atom.

12.1.2   Amongst the elements carbon and hydrogen, both with half-filled shells, are 'special cases' in that together they form probably 20 times as many compounds as there are compounds of all other chemical elements. Since carbon compounds are so numerous and their chemical properties so distinctive they are dealt with in a separate branch of chemistry—organic chemistry; so-called because many of these compounds were originally associated with living things.

By forming four covalencies the cabon atom has a fully-shared octet and therefore no 'lone-pair' of electrons. Because of this it cannot act as an electron donor (as do metallic atoms). At the same time it cannot be an electron acceptor (as do most non-metallic atoms) because there are no available orbitals. For these reasons carbon compounds tend to be very stable and carbon atoms can form very stable bonds with other carbon atoms. In fact carbon can form fairly stable *multiple* covalent bonds both with itself and with some other elements. The following three compounds with hydrogen exhibit single, double and triple bonds involving carbon atoms:

All three compounds are gases. Ethane is extremely stable chemically since a single bond covalently joins the carbon atoms to each other. Conversely the *triple bond* joining the two carbon atoms in the acetylene molecule is much less stable and is fairly easily ruptured. During my schooldays—rather long ago I'm afraid—the cyclist found his way at night by the dim light of an acetylene lamp inside which water dripped slowly on to calcium carbide to yield acetylene. This burned with an extremely sooty flame since with an inadequate air supply the weak triple bond broke to yield a large excess of unburnt carbon. (The released carbon particles were illuminated by that part of the gas which did burn, to give a very luminous flame despite the vast excess of soot.) In the gas ethylene the double bond, though stronger than the triple bond of acetylene, can nevertheless be made to break (by heating the gas at a pressure of some 1000 atmospheres) yielding free valencies which will instantaneously link up with free valencies from other like molecules to give a long-chain molecule up to 1200 carbon atoms in length. Applied heat is only required to initiate the reaction as, once started, it becomes strongly exothermic. The product is of course *poly*-ethylene or 'polythene' as it is commonly called.

12.1.3 It is thermodynamically more difficult to build up than to break down but in the present technological age the organic chemist has learned to reassemble some of the bits produced by these destructive processes. More important still he has built up complex compounds from these simple units to produce substances such as have never been known in Nature. Foremost amongst these substances are the *super-polymers*—or *plastics*—so much a part of life in modern times.

The synthesis of polythene from ethylene—a by-product of petroleum

Fig. 12.2 The polymerisation of ethylene. (i) Using orthodox chemical notation ('n' is approximately 600). (ii) Illustrated diagrammatically where ● = carbon and ○ = hydrogen.

'cracking'—has already been mentioned (2.3.3 and above). In this instance ethylene, C_2H_4 is known as the *monomer* whilst poly-ethylene $(C_2H_4)_n$ is the *poly-mer*. The single repeating unit, in this case $-C_2H_4-$, is often termed the *mer*. Such a process is known as *addition polymerisation* since a long succession of mers join together in this way forming a long chain-like molecule.

12.1.4　The number of mers which link up is usually referred to as the *degree of polymerisation*, n. As the degree of polymerisation increases and the size of the molecule increases in consequence so does the melting point of the polymer so produced. When polythene has a moleculer mass of only 200, n = approx. 7 (since the repeating unit $-C_2H_4-$ has a molecular mass of $[2 \times 12 + 4 \times 1] = 28$) and the polymer $(-C_2H_4-)_7$ is a viscous liquid at ambient temperature. As the molecular mass is increased to 400 (n = approx. 14) the polymer now has a melting point, T_m, of 62°C. T_m continues to rise with the degree of polymerisation until, when the molecular mass is in the region of 10 000 (n = approx. 715), T_m has reached 130°C.

　　Thus the degree of polymerisation in a plastics material is a measure of the *average* size of the molecule in terms of the number of monomer units—or mers—contained in the molecule. As indicated in Fig. 12.3 molecular size in a plastics material varies according to a Gaussian-type distribution.

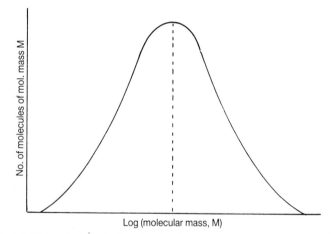

Fig. 12.3　A typical normal distribution of molecular masses in a thermoplastics material.

It forms a polymer by addition polymerisation with an average molecular mass of 4.5×10^4. What is the degree of polymerisation? (C = 12, H = 1 and O = 16)

| **A** | The molecular mass of the monomer is:

	Total relative mass
Hydrogen atoms—6 = (6 × 1)	= 6
Carbon atoms —4 = (4 × 12)	= 48
Oxygen atoms —2 = (2 × 16)	= 32
	86

Molecular mass of vinyl acetate   =   86

and   Degree of polymerisation $= \dfrac{4.5 \times 10^4}{86}$

= <u>523 (approx)</u>

That is, the average polymer chain in polyvinyl acetate will be about 1046 carbon atoms in length.

Thermoplastic Polymers

12.2 Compounds based on ethylene in which one or more of the hydrogen atoms have been replaced by a different atom or group of atoms are known collectively as *vinyl* compounds (Table 12.1). Some of these vinyl compounds will polymerise spontaneously at ordinary temperatures (Fig. 12.4) whilst others polymerise when heated in the presence of a catalyst. A typical vinyl compound may exist as a very mobile liquid (Fig. 12.4(i)). The liquid flows as easily as water because only weak van der Waals forces operate between the relatively small molecules. As polymerisation proceeds (Fig. 12.4(ii)) the molecules reach a size where the van der Waals forces become effective so that the liquid becomes relatively viscous. As polymerisation nears completion (Fig. 12.4(iii)) the molecules become very long so that the sum of van der Waals forces acting between adjacent molecules becomes considerable and the material so viscous as to be regarded as a solid. It no longer flows unless it is heated. This increases the energy of the molecules so that they are able to overcome the van der Waals forces acting between them. Such a substance is said to be *thermoplastic* since it can be softened repeatedly by the application of heat.

12.2.1 Polymers derived from a *single* monomer such as are listed in Table 12.1 are described as *homopolymers*. Sometimes two different monomers are used in the synthesis of the large molecule. The latter is then known as a *copolymer*. For example a mixture of vinyl chloride and vinyl acetate can be polymerised to produce the copolymer, polyvinyl chloride–acetate:

Table 12.1 *Common Vinyl Polymers*

VINYL CHLORIDE

POLY (VINYL CHLORIDE/ACETATE)

VINYL ACETATE

O = CARBON   O = HYDROGEN
⦻ = CHLORINE   Ac = ACETATE GROUP

Of course, as is indicated in Table 12.2 not all thermoplastics are vinyl compounds.

NATURE OF MOLECULE   CHANGE IN VISCOSITY

(i)

(ii)

(iii)

Fig. 12.4 Stages in the polymerisation of a vinyl compound and the changes in viscosity which ensue.

> **Q** The average molecular mass of a sample of commercial grade polypropylene (which is known to contain some polythene) is 33 400 and its degree of polymerisation is 850.
>
> Calculate the relative proportion of ethylene monomers in the material.
> (C = 12.01, H = 1.01, O = 16.00)

| **A** | Ethylene monomer—C_2H_4 = 2C + 4H |

Molecular mass of C_2H_4 = $(12.01 \times 2) + (1.01 \times 4)$
= 28.06

Propylene monomer—$C_2H_3.CH_3$ = 3C + 6H

Molecular mass of $C_2H_3.CH_3$ = $(12.01 \times 3) + (1.01 \times 6)$
= 42.09

Mean molecular mass of monomers in the copolymer = $\dfrac{33\,400}{850}$
= 39.29

Let the proportion of the ethylene monomer in the copolymer = P.

$(28.06P) + (42.09(1 - P)) = 39.29$
$28.06P + 42.09 - 42.09P = 39.29$
$\therefore P = 0.199$

About one fifth of the material consists of ethylene monomers.

Thermosetting Polymers

12.3 The polymers mentioned so far have all been of the *thermoplastic* type. That is they can be softened and remoulded repeatedly by the application of heat. A number of polymeric substances, however, are of the *thermosetting* variety. These are plastic in the primary stages of manufacture but once moulded to shape they 'set' and cannot subsequently be softened by re-heating. Thermosetting is due to the formation of covalent bonds between chain molecules.

12.3.1 In 1907 a Belgian chemist Dr Leo Baekeland developed the first synthetic plastics material. Recognising the importance of his discovery he patented it and named the product 'bakelite'. This is a thermosetting material made from the two organic chemicals phenol, $C_6H_5.OH$, and formaldehyde, $H.CHO$. Under suitable conditions these two substances can be made to react with each other (Fig. 12.5). As a result of this reaction a chain-type of molecule is produced. However, this method of polymerisation differs from that involved in addition polymerisation as

Fig. 12.5 Formation of the 'novolak'. The phenol molecules could equally well be linked at positions ② and ③ .

typified by the poly-vinyls. Here parts of the reacting molecules are eliminated (in this particular instance as molecules of water, H_2O) and this type of reaction is termed *condensation polymerisation*.

In the above reaction only sufficient formaldehyde is used to link the phenol molecules in a simple linear manner (Fig. 12.6). Consequently at

Fig. 12.6 Chain molecule at the 'novolak' stage. (i) The phenol groups covalently linked by $-CH_2-$ groups. (ii) Only van der Waals forces will operate between the resultant chain molecules.

this stage the material, known in the trade as a 'first-stage resin' or 'novolak', is still thermoplastic. It is allowed to cool and is then powdered and mixed with hexamethylene tetramine. During the subsequent mould-ing processes the hexamethylene tetramine decomposes, releasing more formaldehyde which forms cross links between the chain-molecules of the novolak (Fig. 12.7). The formation of these cross links means that the chain-molecules are now joined by covalent bonds which are very strong when compared with the relatively weak van der Waals forces which hold together the chain molecules in a thermoplastic material. For this reason thermosetting materials are generally stronger at elevated temperatures than are thermoplastic materials in which van der Waals forces are readily overcome by thermal vibrations within the molecules. Thermoplastic

Fig. 12.7 The type of cross-linking which results in thermosetting. (i) Covalent linking between chains by means of other phenol and $-CH_2-$ groups. (ii) Covalent linking now holds chain molecules rigidly together.

Table 12.2 Some Important Thermoplastic Polymers

Type	Structure and/or composition	Relative density	Typical mechanical properties			Max. service temperature (°C)	T_g (°C)
			Tensile strength (MPa)	Elongation (%)	Impact (J)		
Polyethylene ('polythene')	$\left[\begin{array}{cc} H & H \\ -C-C- \\ H & H \end{array} \right]_n$ Homopolymer—with various degrees of chain branching, which governs both flexibility and density	Low dens. -0.93 / High dens. -0.96	11.0 / 31.0	90–650 / 50–800	No fracture / 2.0–18.0	85 / 125	-120
Polypropylene	$\left[\begin{array}{cc} H & H \\ -C-C- \\ H & CH_3 \end{array} \right]_n$ Homopolymer	0.91	30.0–35.0	50–600	1.0–10.0	150	-27
Polystyrene	$\left[\begin{array}{cc} H & H \\ -C-C- \\ H & C_6H_5 \end{array} \right]_n$ $C_6H_5 =$ (benzene ring) Properties depend largely upon structure	1.07	28.0–53.0	1.0–35	0.25–2.5	65–85	100
Polyvinyl alcohol	$\left[\begin{array}{cc} H & H \\ -C-C- \\ H & OH \end{array} \right]_n$ Some acetate groups may be present, replacing the hydroxyl groups	1.26	120.0	400	—	100	85
Polyvinyl chloride (PVC)	$\left[\begin{array}{cc} H & H \\ -C-C- \\ H & Cl \end{array} \right]_n$ The pure polymer is rigid but is often plasticised (12.6) to produce a more flexible material	Rigid form— 1.37 / Flexible form— 1.35	Rigid form— 49.0 / Flexible form— 7.0–25.0	10–130 / 240–380	1.5–18.0 / —	70 / 60–105	87

Polymer	Description / monomer structure						
Polyvinyl acetate (PVA)	$\begin{bmatrix} H & H \\ \mid & \mid \\ -C & -C- \\ \mid & \mid \\ H & COO.CH_3 \end{bmatrix}_n$	1.19	Too soft (showing excessive cold flow) for use in moulded plastics generally				29
Polyvinyl acetate/chloride	A co-polymer of the above two monomers	1.35	30.0–35.0	200–400	0.7–1.5	—	—
Polytetrafluoroethylene (PTFE or 'Teflon')	$\begin{bmatrix} F & F \\ \mid & \mid \\ -C & -C- \\ \mid & \mid \\ F & F \end{bmatrix}_n$	2.17	17.0–25.0	200–600	3.0–5.0	260	126
Polyvinylidene chloride	Homopolymer, or, generally, a copolymer with about 20% vinyl chloride $\begin{bmatrix} H & Cl \\ \mid & \mid \\ -C & -C- \\ \mid & \mid \\ H & Cl \end{bmatrix}_n$	1.68	25.0–35.0	up to 200	0.4–1.3	60	−17
(PMM or 'Perspex')	$\begin{bmatrix} H & CH_3 \\ \mid & \mid \\ -C & -C- \\ \mid & \mid \\ H & COO.CH_3 \end{bmatrix}_n$	1.18	50.0–70.0	3.0–8.0	0.5–0.7	95	0
Acrylonitrile-butadiene-styrene (ABS)	Copolymers containing varying proportions of the monomers: (acrylonitrile) $\begin{matrix} H & H \\ \mid & \mid \\ -C & -C- \\ \mid & \mid \\ H & CN \end{matrix}$ (butadiene) $\begin{matrix} H & H & H & H \\ \mid & \mid & \mid & \mid \\ -C & -C=C & -C- \\ \mid & & & \mid \\ H & & & H \end{matrix}$ (styrene) $\begin{matrix} H & H \\ \mid & \mid \\ -C & -C- \\ \mid & \mid \\ H & C_6H_5 \end{matrix}$	1.05	30.0–35.0	10.0–140	7.0–12.0	100	−55

Table 12.2 continued

Polyamides (nylons)	Polymers in which various aliphatic groups separate recurring amide groups	'Nylon 66' 1.14	50.0–87.5	60.0–300.0	1.5–15.0	120	50
Polyimides		1.42	90	6–8	—	350	235
Polyesters (thermoplastic types)	poly(ethylene terephthalate)	1.4	over 175	60–110	1.0	150	69
Acetals (polyformaldehyde or polyoxymethylene)	Homopolymer derived from formaldehyde. Copolymer-extra CH_2 mers increase stability at high temperatures	1.41	50.0–70.0	15.0–75.0	0.5–2.0	105	–73
Polycarbonates		1.2 (unfilled)	60.0–70.0	60–100	10.0–20.0	130	150

Polymer	Structure / Notes	Density					
Polysulphones	[chemical structure: aryl ring–SO₂–aryl ring–O–]	1.24	70	—	—	175	230
Polyether ether ketone (PEEK)	[chemical structure: aryl ring–C(=O)–aryl ring–O–]	1.26 amorp. 1.32 cryst.	92	—	8	150	144
Polyurethanes (thermoplastic types)	Polymers prepared from di-isicyanates and polyethers or polyesters $\left[-O-C-N-R-N-C-O- \right]_n$ (with O and H on the carbamate groups)	1.21	up to 55.0	up to 700	—	—	—
Cellulose nitrate	[chemical structure of cellulose monomer with CH₂OH, OH groups]ₙ Some, or all, of the hydroxyl groups (OH) of the cellulose monomer (above) are nitrated: $>\!C-OH \xrightarrow{\text{HNO}_3} >\!C-O.NO_2$	1.40	35.0–70.0	10–40	3.0–11.0	—	53
Cellulose acetate	Various degrees of acetylation of the cellulose monomer (above) are possible—about 80% of the available hydroxyl groups are generally acetylated	1.28	24.0–65.0	5–55	0.7–7.0	70	120
Cellulose acetate/butyrate	Esters with usually 20–40% butyrate groups	1.20	17.0–50.0	8–80	0.7–6.5	70	122
Ethyl celluloses	About 85% of the hydroxyl groups of cellulose (above) are etherified	1.15	40.0–60.0	10–40	4.5–8.0	60	43

Table 12.2 *continued*

	Properties & Characteristics	Uses
Polyethylene ('polythene')	One of the most versatile thermoplastics. Tough and flexible over a wide temperature range. Good dimensional stability. Easily moulded. Good resistance to most common solvents. Good weathering properties but some deterioration on long exposure to light	Used largely in packaging as bags and squeeze-bottles. Also a wide range of household goods such as buckets, bowls and other containers. Piping, chemical equipment, coating for cables and wires
Polypropylene	Similar properties to polythene but has better heat resistance. Good mechanical properties. Excellent resistance to chemical attack by acids, alkalis and salts even at high temperatures. Tough, rigid and light in weight	Mouldings for hospital and laboratory equipment; chemical plant and domestic hardware. Sheets used in packaging. Monofilaments used as ropes, fishing nets and filter cloths
Polystyrene	A tough, dense plastic—hard and rigid. Good dimensional stability. Moulds with high surface gloss. Good strength but inclined to be brittle due to high T_g. Retains properties at low temperatures. Attacked by petrol and other organic solvents. Can be foamed to produce rigid but extremely light cellular material	Refrigerator trays, boxes and many articles of household holloware. Toys, display figures, etc. etc. Rigid foams used in building and refrigeration as insulation against heat and sound, e.g. ceiling tiles, and in packaging for insulation against shock. Buoyancy equipment.
Polyvinyl alcohol	Tough and strong. Flexible. Resistant to many common organic solvents but dissolves slowly in boiling water. Softens above 100°C. Processed by moulding, extrusion and solution coating	In water solution can be used for sizing papers and fabrics. Also in adhesives. High solubility in water limits its use
Polyvinyl chloride (PVC)	In the unplasticised form is tough and hard. If plasticised it becomes soft, flexible and rubbery. Good dimensional stability. Good resistance to water, acids, alkalis and most common solvents. Plasticised material stiffens with age	The rigid form is used in mouldings of all kinds. The flexible form in imitation leather cloth, table cloths, raincoats; etc. Also for piping, guttering and electrical cable covering. Safety helmets; chemical tank linings
Polyvinyl acetate (PVA)	Soft, with poor dimensional stability. Good adhesive properties. Resistant to water but dissolves in many common organic solvents	Since it is soluble in many organic solvents it is used in lacquers, paints, coatings and adhesives
Polyvinyl acetate/chloride	Tough, more flexible than PVC. Good dimensional stability. Good resistance to acids, alkalis and most common solvents. Good electrical insulation	The unplasticised sheet is used for packaging. Other uses include hand luggage, upholstery, covering for wires and cables, long-playing gramophone records, chemical equipment
Polytetrafluoroethylene (PTFE or 'Teflon')	Tough and flexible. Excellent heat-resistance. Does not burn. *Not attacked by any known reagent and no known solvent.* Excellent electrical insulation. Waxy surface and low coefficient of friction. Relatively expensive	Bearings, fuel hoses, gaskets and tapes. Non-stick coatings for fry-pans and other cooking utensils. Also in chemical industry because of its resistance to attack
Polyvinylidene chloride	Very strong and tough. Flexible. Good resistance to attack by acids and alkalis and most organic solvents. Excellent weather-resisting properties	Food wrapping, outdoor furniture fabrics, public transport seat coverings, chemical equipment, screens, filters, wire coating, filaments for doll's hair

Polymer	Properties	Applications
Polymethyl Methacrylate (PMM or 'Perspex')	Excellent transmission of light. Strong and rigid but easily scratched. Softens in boiling water but unattacked by most household chemicals. Attacked by petrol and many organic solvents. Good electrical insulation	Lenses, dentures, telephones, knobs and handles. Aircraft windows, sinks, building panels, lighting fittings, baths, protective shields, dials, corrugated roof lights. Fancy goods, advertising displays, ornamental lighting. Adhesives, surface coatings
Acrylonitrile-butadiene-styrene (ABS)	High impact value along with high strength (resists impact by flying stones). High dimensional stability over widely differing conditions. Remains tough at sub-zero temperatures. Resists attack by acids, alkalis and many petroleum-type liquids	Pipes, protective helmets, valves, refrigerator parts, battery cases, water pumps, tool handles, textile bobbins, radio cabinets. Now widely used in the automobile industry for radiator grills and other fittings because of its high impact value and the fact that it can be chromium plated
Polyamides (nylons)	Very strong and tough. Flexible. Good resistance to abrasion. Good dimensional stability. Absorbs water but has good resistance to most common solvents. Deteriorates with long exposure to outdoors. Good electrical insulation	A number of different varieties are in use. Gears, valves, electrical equipment, handles, knobs, bearings, cams, shock absorbers, combs. Surgical and pharmaceutical packing. Raincoats. Filaments for bristles, climbing ropes, fishing lines and textiles
Polyimides	Excellent electrical insulation, solvent resistance and flame resistance. Very good abrasion resistance and heat resistance. Also resistant to degradation by oxidation and to most chemicals (except strong bases)	Compressor seals in jet engines, data processing equipment (pressure discs, bearings). Shafts in shut-off valves. Soldering and welding equipment (good heat stability and deformation resistant, though inferior to PEEK). Laminates (with glass and carbon fibre) in aircraft industry—supersonic air-liners. Laminates can be used up to 250°C and, intermittently, up to 400°C
Polyesters (thermoplastic types)	Highly crystalline but if quenched from a melt to below T_g an amorphous structure results. This crystallises if annealed at above T_g. Used as fibre or film— strong and stable. Good dimensional stability and resistance to many organic solvents but attacked by strong acids and alkalis. Excellent electrical insulation	Widely used as a textile fibre ('Terylene', 'Dacron'). In the electrical industries as insulation. Drums and loudspeaker cones, recording tapes, drafting materials, photographic films for special purposes, book covering, packaging. Gaskets, mould release foils, insulating tapes. In moulded form: bottles (beer, soft drinks, fruit concentrates) wide-neck jars (e.g. coffee).
Acetals (polyformaldehyde or polyoxymethylene)	Very strong and stiff—creep resistant. Good fatigue endurance. High crystallinity and high melting point make it suitable for replacing some metals. Good electrical properties. Resistant to alkalis and many solvents but attacked by mineral acids. Slight water absorption. Adversely affected by prolonged exposure to ultra-violet light—not recommended for outdoor use in strong sunlight	Impellors and other parts for water pumps, housings, washing machines and extractor fans. Instrument panels and housings for automobiles. Plumbing fittings. Pipe clips (replacing the well-known Jubilee clips). Bearings, cams and gears. Hinges and window catches. Parts for aerosol containers
Polycarbonates	Good impact strength and heat resistance. Good dimensional stability. Resistant to oils and most solvents. Surface takes a high gloss	Electronic equipment, aircraft parts, fittings for automobiles
Polysulphones	Excellent creep resistance. Good temperature resistance, rigidity, transparency and self-extinguishing characteristics. Superior to polycarbonates in creep resistance	Printed circuit boards—heat resistant to soldering. Hair driers, oven, iron and fan-heaters. Microwave oven parts, chemical plant, pumps, coffee machines. Lamp bulb bases and TV components

Table 12.2 continued

	Properties & Characteristics	Uses
Polyether ether ketone (PEEK)	High oxidation stability, low flammability and very low smoke emission. High softening point. Superior toughness and good fatigue resistance. Low moisture absorption	Mouldings for aggressive environments—nuclear plant, oil wells, high-pressure steam valves, aero- and automobile engines, chemical plant. Wire coverings—resistant to cuts and rupture at sharp corners. Carbon fibre—PEEK composite up to a tensile strength of 2700 MPa
Polyurethanes (thermoplastic types)	Physical properties of the thermoplastic varieties are similar to nylon. Good dimensional stability. Strong and tough. Low water absorption and resistant to most common solvents	Packaging film, filaments for bristles, etc. Textile coatings, tank linings, surgical gloves, gears, hosing. Elastic thread for foundation garments
Cellulose nitrate	Tough and flexible. Easily worked. Softens in boiling water. Good resistance to water but dissolves in some organic solvents such as alcohol, ketones and esters. Not suitable for exposure outdoors. Its extreme inflammability limits its use now that many more suitable compounds are available	Solutions are used for lacquers and leather cloth. Combs, toys (though inadvisable due to high inflammability), handles, spectacle frames, brushes, windows, piano keys, drawing instruments, pens and pencils, handle grips
Cellulose acetate	Tough and durable. Good impact strength. High absorption of water which affects dimensions. Not suitable outdoors. Softens in boiling water. Resistant to most household chemicals but is attacked by acids, alkalis and alcohol	Toys, containers, lampshades, machine guards, brush handles, spectacle frames. Photographic film base, book covers, packaging. Beading, insulation, handle covers. Combs, brushes, steering wheels, oil pumps, buttons, door handles, radio cabinets
Cellulose acetate/butyrate	Tough and flexible. Better resistance to water than the acetate. Better dimensional stability. Softens in boiling water. Resistant to most household chemicals but not organic solvents	Machine guards, advertising displays. Beading special purpose hoses. Hands for brushes and cutlery. Pens and pencils. Furniture
Ethyl celluloses	Tough and flexible. Good impact strength at low temperatures. Good dimensional stability. Low absorption of water and good resistance to acids and alkalis, but dissolves in many organic solvents. Not suitable for outdoor use. Good electrical insulation	Packaging, watch 'glasses'. Handle and grip covers, special hoses. Telephones, furniture, automobile fittings, tool handles, vacuum cleaner parts, strippable protective coatings

polymers are therefore often described as *linear polymers* whilst the more rigid thermosetting materials are known as *network polymers*. The 'network' is of course three dimensional and not two dimensional as diagrams drawn on flat paper may suggest.

12.3.2 Cold-setting Plastics Similar in principle to the thermosetting materials are the cold-setting plastics. These are formed by reagents which will undergo polymerisation when they are mixed at ordinary temperatures. In addition to moulding materials some varnishes and adhesives are of this type.

The Physical Nature of Solid Polymers

12.4 So far we have been considering the form of single polymer molecules. With the exception of 'three-dimensional' condensation polymers of the thermosetting type these molecules can be visualised as thread-like entities bonded to their neighbours by van der Waals forces. Some molecules like those of polyethylene, are 'symmetrical' in section whilst others may carry relatively bulky side groups, non-symmetrically placed. Thus, in polystyrene these appendages consist of phenyl groups ⬡, and in polypropylene the group, $-CH_3$. Moreover, any sample of a pure polymer will contain molecules of varying length. Thus, whilst a polyethylene molecule consists *on average* of a chain some 1200 carbon atoms in length, individual molecules may contain many more—or less—than the average number.

Likewise varying degrees of 'side-branching' can occur in polyethylene molecules (Fig. 12.8), producing branches of varying lengths. The presence

Fig. 12.8 Side branching in polyethylene (polythene) molecules.

of these side branches has considerable effect on the mechanical properties, generally increasing the strength and hardness.

The forces acting between individual atoms within a polymer molecule are of the primary valency type—strong covalent bonds (2.2)—whilst forces acting between adjacent molecules are of a secondary nature known collectively as van der Waals forces (2.4). Naturally the magnitude of these relatively weak secondary bonding forces govern the ease with which molecules may be separated and so affect such properties as solubility, viscosity and frictional properties. Whilst strength depends largely upon the covalent bonding within the molecule chains it is also affected by van der Waals forces operating between adjacent chains.

Table 12.3 *Some Important Thermosetting Plastics*

Type	Structure and/or composition	Relative density	Typical mechanical properties			
			Tensile strength (MPa)	Elongation (%)	Impact (J)	Max. service temperature (°C)
Casein plastics	Natural protein (skimmed milk) cross-linked with formaldehyde (thermoplastic when not cross-linked in this way)	1.34	55–70	2.5–4	1.5–2	150
Phenol formaldehyde (bakelite)	See Fig. 12.5 A condensation product of either phenol or cresol with formaldehyde	1.35 ('filled' with wood flour)	35–55	1	0.3–1.5	75
Urea formaldehyde	A condensation product of urea and formaldehyde	1.5 (cellulose filled)	50–75	1.0	0.3–0.5	75
Melamine formaldehyde	A condensation product of melamine and formaldehyde	1.5 (cellulose filled)	56–80	0.7	0.2–0.45	100

Type	Description					
Epoxides	Polymers generally cross-linked by amines	1.15 (rigid) and unfilled)	35–80	5–10	0.5–1.5	200
Polyesters (i) unsaturated polyesters	Unsaturated linear polymers cross-linked by various vinyl monomers	1.12	55	2	0.7	220
(ii) alkyd resins	Condensation products of, for example, glycol and phthalic anhydride	2.0	25	—	0.25	150
Polyurethanes (thermosetting variety)	They contain the group H O / ...-N-C-O-... / Foams are produced due to the release of carbon dioxide simultaneously with the cross-linking process	1.2 (solid polymer)	up to 55	up to 500	—	Rigid foams up to 150
Silicones	(structure)	1.88	40	—	0.4	450

Type	Properties and characteristics	Uses
Casein plastics	Tough and rigid. Takes good polish. Absorbs water but is not affected by most common solvents. Brittle below 0°C. Not recommended for exposure to weather. A poor electrical insulator	Imitation tortoiseshell and 'mother o'pearl-buckles, brush backs, buttons, ornament pens and pencils, knitting pins, umbrella handles
Phenol formaldehyde (bakelite)	In the 'raw' state phenolics are brittle—hence 'filled' with fibrous materials to increase strength. Wide variation of properties upon	Electrical equipment, handles, buttons, radio cabinets, furniture, vacuum cleaner parts, cameras, ash trays, engine ignition

Table 12.3 continued

Type	Properties and characteristics	Uses
Phenol formaldehyde (bakelite) (cont.)	filler used. Brittle in thin sections. Absorbs water but resistant to alcohol, oils and most common solvents. Does not soften—decomposes above 200°C	equipment. Cheap jewellery, ornaments. Laminated materials—electrical equipment, gears, bearings, aircraft parts, chemical equipment, clutch and brake linings
Urea formaldehyde	Properties vary considerably with fillers or laminating materials. Hard and rigid. Scratch resistant. Tough if a suitable filler is used. Absorbs water with some loss in dimensional stability. Resistant to most solvents and common reagents. Unaffected by detergents. Good electrical resistance	As a syrup or resin: a binder for foundry core production, finishes for textiles and papers, adhesives, laminates. Mouldings: electrical equipment (plugs and switches), buttons, buckles, bottle tops, knobs, cups, saucers and plates, toys, clock cases, radio cabinets, kitchen ware. Laminates: building panels
Melamine formaldehyde	Properties are generally similar to those of urea formaldehyde but with improved resistance to heat. Also harder than UF and absorbs less water. Excellent electrical properties	As syrups and resins: surface coating of papers and textiles, laminates. Mouldings: electrical equipment, table ware (particularly cups and saucers), knobs and handles. Laminates: electrical equipment, panels
Epoxides	Commonly used in conjunction with glass fibre (available to D.I.Y. experts) for increased strength. Good adhesion to metals. Good resistance to water and most solvents and common reagents. Excellent heat resistance. Good electrical insulation	Sold as resins and syrups. Used as adhesives for gluing metal, low-pressure laminations, surface coatings, casting and 'potting', as a putty for repairing castings. Laminates (with glass fibre): boat hulls, table tops, laboratory furniture
Polyesters (i) unsaturated polyesters	Mechanical properties variable, depending upon polymer used. Generally good impact resistance and surface hardness. Good resistance to water and most common solvents but attacked by conc. acids and alkalis. Resistant to weather	Adhesives, surface coatings. Lampshades, radio grilles, refrigerator parts, laminates for boat hulls, car bodies and wheel barrows. Helmets, fishing rods and archery bows
(ii) alkyd resins	Excellent heat resistance, retaining good dimensional stability. Good resistance to water and many organic solvents. Excellent electrical properties	Liquid resins are used as paints and enamels for cars, refrigerators, washing machines, etc. Mouldings: for electrical equipment of cars, fuses, switches, TV components, etc.
Polyurethanes (thermosetting variety)	Strong, shock-resistant and similar in many ways to nylons. Good adhesive properties for bonding synthetic rubbers and fibres in tyres. Good dimensional stability. Good resistance to wear. Resistant to most common solvents, oils and chemicals	Gears and bearings. Electrical equipment. Knobs, handles, cable covering, packaging film. Footwear, paints and lacquers. Foams: upholstery, toilet sponges, insulation. Rigid foams: reinforcement for aircraft wings
Silicones	Available as oils, greases, rubbers, thermoplastic and thermosetting polymers, depending upon the monomeric material and polymerising conditions. Water-repellent and very resistant to weather. Not attacked by mineral acids. Excellent dielectric properties even when exposed to moisture	Water-proof coatings for fabrics. Anti-foaming agents. Hydraulic fluids. Mouldings: electrical equipment, e.g. coil formers, switch parts, induction heating equipment, insulation for motors and generator coils. Gaskets, seals in chemical plant

12.4.1 Many solid substances exist in purely crystalline form. This is usually so when the particles of which they are composed, whether ions or molecules, are small and are able to fall easily into a predetermined regular pattern. The long polymer chains in plastics materials, however, are likely to become entangled, particularly if large side appendages are present. Consequently only a limited degree of crystalline arrangement is generally possible in the case of these substances and limited regions occur in which the linear chains are arranged in an ordered pattern (Fig. 12.9). These

Fig. 12.9 The existence of 'crystallites' in a solid polymer.

ordered regions are known as *crystallites*. They can be detected by the use of X-ray analysis (3.4) which produces appropriate diffraction patterns. Those polymers which are completely amorphous of course produce no such clear patterns.

In the solid state, therefore, polymers can be regarded as consisting of both crystalline and amorphous regions in relative amounts which vary from one polymer to another and which depend upon the molecular structure as well as upon the previous thermal history. Those polymers in which the molecular structure is regular and the section highly symmetrical, show a high proportion of crystallites. Such polymers include

$$\text{polythene} - \begin{bmatrix} H \\ | \\ -C- \\ | \\ H \end{bmatrix} \text{ and polytetrafluoroethylene } \begin{bmatrix} F \\ | \\ -C- \\ | \\ F \end{bmatrix} \text{ where short re-}$$

petitive units of high symmetry allow chains to align themselves. Bulky side chains, such as are present in polyvinyl acetate, on the other hand, impede the formation of crystalline regions and such substances are largely amorphous. Crystalline regions vary in length from as much as 50 nm for strongly crystalline materials like polythene to as little as 2 nm for poorly crystalline polymers like polyvinyl chloride. Highly crystalline polymers have a crystallinity of 80–90%.

When a polymer crystallises freely in the absence of any external forces the crystallites so formed are randomly oriented, but if an external stress is applied in a single direction the crystallites orientate themselves so that their longitudinal axes are parallel to the direction of the applied stress. During the process of cold stretching the degree of crystallinity of a partially crystalline polymer does not increase appreciably but in the case of amorphous polymers or those of very low crystallinity, some increase in crystallinity may occur. As the degree of crystallinity and the extent of orientation increase so do the tensile strength and stiffness.

12.4.2 Melting Points of Crystalline Polymers A pure and completely crystalline solid, for example a metal or an inorganic salt, melts at a single definite temperature called the melting point. An amorphous solid, however, merely becomes progressively less rigid as it is heated but shows no sharp transition from solid to liquid. This is roughly what one would expect since, as with a liquid (which is amorphous), viscosity decreases progressively as van der Waals forces decrease with rise in temperature. For those polymers which are particularly crystalline, however, a melting curve similar to that shown in Fig. 12.10 is obtained. Here the change in

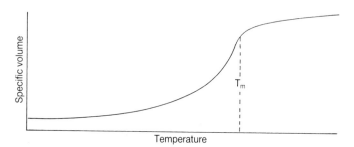

Fig. 12.10 The melting curve for a partially crystalline polymer.

specific volume is used as an indication of the change in structure. Melting begins as the smaller, less perfect, crystallites change to an amorphous structure and this is indicated by the gradual change in the shape of the curve. At a high temperature there is a fairly sudden change in the specific volume with temperature, and this is regarded as the melting point. The melting point, T_m, of a crystalline polymer is defined as the temperature at which the crystallinity disappears and the structure becomes completely amorphous. Melting point is, to some extent, a measure of the van der Waals forces acting between molecules. Thus, increase in crystallinity of polythene leads to an increase in melting point presumably because inter-molecular forces are greater in the crystalline regions where chains are more closely aligned.

12.4.3 Glass Transition Temperature Under normal conditions polymers may be either soft, flexible substances or hard brittle glassy materials. If a soft, flexible polymer is cooled sufficiently a temperature is reached where it becomes hard and glassy. Thus, a soft rubber ball when cooled in liquid air becomes hard and brittle and will shatter to fragments if an attempt is made to bounce it whilst it is still at a low temperature. The temperature at which this change in properties takes place is known as the *glass transition temperature*. For some polymers the glass transition temperature is above room temperature so that these materials are normally hard but will become soft and pliable when heated. Polystyrene (Table 12.2) is a polymer of this type.

The glass transition temperature, T_g, can be assessed by measuring the change in some property such as specific volume or specific heat with change in temperature and is indicated by a change in slope of the curve

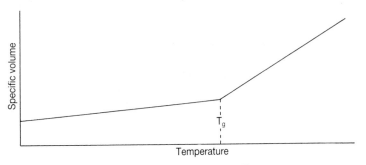

Fig. 12.11   Derivation of the glass transition temperature, T_g.

produced (Fig. 12.11). The glass transition temperature is less well defined than is the crystalline melting point, T_m.

12.4.4   The glass transition temperature is a property associated with the amorphous regions in a polymer and is usually described as the temperature at which, on heating, the polymer chains acquire sufficient thermal energy to vibrate in a co-ordinated manner. Below T_g the atoms, apart from restricted vibrations about equilibrium positions, are in a rigid frozen state and the polymer is hard and glassy as a result. Above T_g the

Table 12.4   *Melting Point (T_m) and Glass Transition Temperature (T_g) for Some Polymers*

Polymer	Melting point (T_m) (°C)	Glass transition temperature (T_g) (°C)
Natural rubber	28	−73
Polyethylene (50% crystalline)	120	−90
Polyethylene (80% crystalline)	135	−90
Polypropylene	175	−27
Polyvinylidene chloride	198	−17
Polyvinyl chloride	212	87
Polytetrafluoroethylene	327	126

co-ordinated vibrations referred to above impart to the polymer its rubber-like properties. The presence of plasticisers and other additives which decrease the magnitude of the van der Waals forces give rise to a reduction in T_g. Conversely the presence of bulky radical groups attached to the chain, making the molecule non-symmetrical, increase T_g significantly. Thus PVC (87°C) and, more particularly, polystyrene (100°C) have higher values of T_g than polyethylene (−90°C).

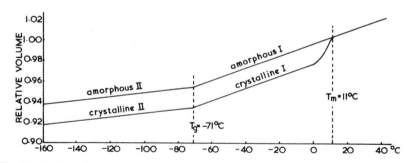

Fig. 12.12 The volume–temperature relationship of a high polymer. The upper curve is for a completely amorphous polymer while the lower is for a partly crystalline one.

The Mechanical Properties of Polymers

12.5 Engineers are familiar with the elastic properties of metals and with the extent to which they are reasonably constant under normal conditions. Even with metals, of course, large deviations occur when high temperatures or large deformations are involved, whilst the effects of creep over long periods of time are also well known. With plastics, however, such deviations occur when relatively normal conditions prevail making it impossible to relate deformation in plastics in terms of simple elastic constants. As one might expect from a study of the structure of a polymer, deformation is very dependent upon *time*. Molecular response is slow to reach equilibrium with the external forces and so the material continues to deform, or creep, almost indefinitely. When the applied forces are removed some recovery may occur but a degree of permanent deformation may also persist.

12.5.1 Deformation in plastics is also very dependent upon temperature and this must be very seriously considered in design to a far greater extent than is necessary with metals. As temperature increases the weak van der Waals forces are further diminished and slip occurs more easily between adjacent chains. When a crystalline material, such as a metal, is stressed atoms are progressively displaced from their equilibrium positions until new internal forces set up balance the externally applied forces. Assuming that the distortion produced is not too great, the atoms will return to their original positions when the stress is removed. Larger stresses will produce permanent deformation but provided that the stress does not reach these values the change in shape of the material is instantaneous and elastic.

This mode of deformation may also occur in a plastic upon the application of stress. When two atoms within a chain are held in fixed positions relative to each other by strong covalent bonds, some displacement of the atoms relative to each other occurs instantaneously on the application of stress (Fig. 12.13). On the removal of stress the atoms immediately return to their original positions. In highly cross-linked

UNSTRESSED   STRESSED

Fig. 12.13   Purely elastic deformation in chain molecules.

polymers and in thermoplastics below their glass-transition point, atoms can be regarded as being fixed in this manner so that this is the only type of deformation possible.

In thermoplastics above T_g this type of elastic deformation will be due not only to bonds between individual atoms in chains but also, if crystallites are present, to inter-molecular forces as well. In the amorphous regions of such thermoplastics, however, the molecules are only attracted to each other by relatively weak van der Waals forces, so that they are able to 'slip'

UNSTRESSED   STRESSED

Fig. 12.14   The slip of chain molecules into new positions.

into new permanent positions relative to each other (Fig. 12.14). This type of movement is not instantaneous but is dependent upon the viscosity of the material. *Viscoelastic* deformation thus occurs.

12.5.2   In addition to elastic deformation which is dependent upon small movements of atoms relative to each other, and plastic deformation caused by wholesale slipping of chain molecules past one another into permanent new positions, further deformation occurs due to straightening of coiled and folded chain molecules themselves in the direction of the applied stress (Fig. 12.15). Essentially this type of distortion is elastic since it is due to

UNSTRESSED   STRESSED

Fig. 12.15   Deformation in chain polymers due to chain straightening.

chain-straightening which will be a reversible process when stress is relaxed. It is also a time-dependent change so that the resultant deformation is viscoelastic. Straightening of molecules tends to take place more quickly than slipping so that much of the initial deformation is elastic.

12.5.3   'Spring and dashpot' models are widely used to represent the viscoelastic modes of deformation of a completely amorphous thermoplastic material above T_g (Fig. 12.16). Here it is assumed that the springs obey Hooke's Law, i.e. strain (ε) is proportional to stress (σ) or:

$$\varepsilon = \frac{\sigma}{E} \tag{1}$$

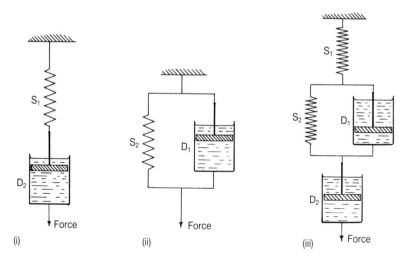

Fig. 12.16 'Spring and dashpot' models: (i) a Maxwell element; (ii) a Voigt-Kelvin element; (iii) a combination model to illustrate the method of deformation in a stressed thermoplastic polymer.

where E is Young's Modulus; whilst the dashpots contain a Newtonian fluid in which the deformation/time relationship, $d\varepsilon/dt$, is linear for a given stress, i.e.

$$d\varepsilon/dt = \sigma/\eta \qquad (2)$$

where η is the viscosity of the fluid.

The element of the model in which the spring and dashpot are in series (Fig. 12.16(i)) is usually called a Maxwell element whilst the element in which the spring and dashpot are in parallel (Fig. 12.16(ii)) is termed a Voigt-Kelvin element.

The spring S_1 represents the instantaneous elastic deformation of the atomic bonds between individual atoms, whilst the spring S_2 represents the elastic deformation arising from chain straightening, the dashpot D_1 introducing the time-dependence of that deformation. Dashpot D_2 simulates the non-elastic, non-reversible time-dependent deformation due to molecules slipping relative to each other into new positions.

Under the action of a force the various components of the system shown in Fig. 12.16 will give strain/time diagrams as indicated in Fig. 12.17. If we now consider the sudden applications of a force, F, to the system as a whole we obtain a strain/time relationship like that indicated in Fig. 12.18. The sudden application of force results in an elastic deformation S_1 (ii) which is followed by a viscoelastic response S_2 in the S_2/D_1 unit coupled with viscous flow in the dashpot D_2 (iii) in time, t_1. When the displacing force is removed the elastic element (S_1) relaxes immediately (iv) and the viscoelastic element slowly (v) during time $(t_2 - t_1)$, but the viscous flow D_2 is never recovered (v).

It should be appreciated that these models only exhibit the *general* characteristics of the viscoelastic behaviour of polymers and are necessarily

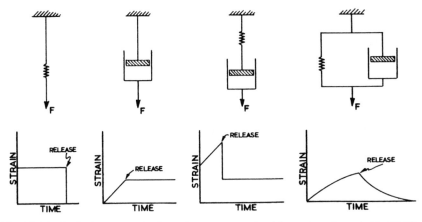

Fig. 12.17 Strain-time curves for the various components of the system shown in Fig. 12.16.

oversimplified. There is considerable deviation both from the Hookean elastic response and from Newtonian flow.

12.5.4 Stress-strain Measurements The force-extension type of test already well established with metals was soon applied as a matter of course to polymer materials. Unfortunately tests carried out at a single temperature and a single rate of force application are relatively meaningless when applied to polymers and in order to obtain reasonably useful data a series of tensile tests carried out over a range of temperatures and application rates are necessary. The most successful method of testing is to stretch the

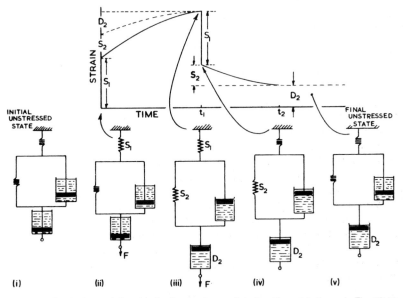

Fig. 12.18 Strain-time relationship for the 'spring and dashpot' model shown in Fig. 12.16.

test piece at a constant rate and measure the tensile force produced in it by means of strain gauges or by some electronic device.

Stress-strain curves for different types of polymer were classified some years ago by Carswell and Nason into five main groups (Fig. 12.19). Weak, soft polymers such as would be of little use practically have not only a low tensile strength and modulus of elasticity but also a relatively low elongation (Fig. 12.19(i)). For many of the very soft thermoplastics the stress/strain diagram does not show any significant linear portion and it is customary to measure E by reference to the *secant modulus* (4.1.6.1) at a strain of, typically, 0.2%. Hard and brittle materials, such as polystyrene, polymethyl methacrylate and some of the phenolic resins at room temperature, have fairly high tensile strengths and quite high moduli of elasticity, but they often break with an elongation of no more than 1% (Fig. 12.19(ii)). Hard and strong polymers (Fig. 12.19(iii))—some rigid poly-

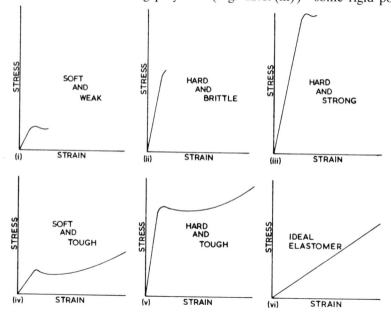

Fig. 12.19   Type of stress-strain diagram for different polymer groups (after Carswell and Nason).

vinyl chlorides and modified polystyrene—have a relatively high tensile strength and modulus of elasticity coupled with a moderate elongation of up to 10%; whilst many soft, tough materials such as plasticised polyvinyl chlorides and some rubbers have low yield values and elastic moduli but high elongations of up to 500% or more, coupled with moderately high tensile strengths (Fig. 12.19(iv)). Tough, hard polymers including nylons and cellulosics are characterised by high yields, high tensile strengths and high elastic moduli. They also elongate considerably, generally by 'necking' (Fig. 12.19(v)).

12.5.5   Tensile properties are highly dependent upon temperature

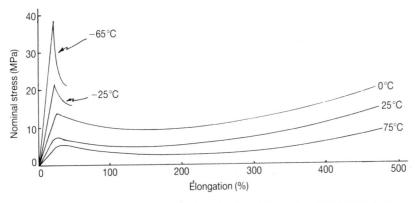

Fig. 12.20 Tensile stress-strain curves for low-density polythene at various temperatures.

(Fig. 12.20) and a series of tests is necessary in order to give worthwhile information about the material. For rigid polymers elongation usually increases as temperature increases though for rubbers elongation may increase as the temperature falls within certain limits.

The rate at which deformation is produced also has a considerable effect on the tensile properties of a polymer as indicated in Fig. 12.21. As the rate

Fig. 12.21 The effect of rate of deformation on the stress–strain characteristics of polymethyl methacrylate.

of elongation is increased so the tensile strength increases. Thus an increase in the rate of deformation has a similar effect to a decrease in temperature.

12.5.6 *Creep*, or gradual extension under a constant force, is a phenomenon which must be considered in the case of metals, particularly if they are destined to work at high temperatures (16.4). Polymers, however, are much more prone to this type of deformation as has been suggested in the foregoing section. In the form of an engineering component a rigid

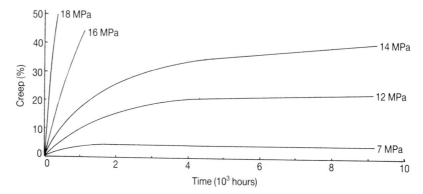

Fig. 12.22 Creep curves for cellulose acetate at 25°C.

polymer must be able to withstand reasonable tensile or compressive forces over long periods of time without alteration in shape. Even a soft plastic raincoat must be capable of supporting its own weight so that it can hang in the wardrobe for long periods without gradually flowing on to the floor.

The series of creep curves shown in Fig. 12.22 are typical of most thermoplastic polymers. The test piece undergoes an initial elongation as soon as the specified force is applied and this is followed by a period during which there is a rapid rate of creep. This creep rate then decreases to a constant value, which approaches zero for small values of stress. As would be expected the greater the applied force the greater the creep rate and elongation.

Creep properties of a polymer are very dependent upon temperature. At temperatures well below the glass-transition temperature, T_g, a polymer tends to be rigid and have a low creep rate for a given stress. As the temperature is increased the creep rate and the amount of elongation increase. At T_g a given load gives a much greater elongation than the same load at lower temperatures. Even more significant is the great increase in the rate of elongation for, whilst at temperatures well above T_g greater *elongations* will be obtained, the actual *creep rate* decreases.

Under these circumstances the creep rate of a polymer can be estimated using the Voigt-Kelvin model (Fig. 12.16(iii)). From (1) and (2) (12.5.3):

$$\sigma_s = E\varepsilon \quad \text{(spring)}$$

$$\sigma_d = \eta d\varepsilon/dt \quad \text{(dashpot)}$$

and in the Voigt-Kelvin model:

$$\sigma = \sigma_s + \sigma_d$$

$$= E\varepsilon + \eta d\varepsilon/dt$$

$$\therefore d\varepsilon/dt = \frac{(\sigma - E\varepsilon)}{\eta}$$

from which it can be shown that:

$$\varepsilon = \frac{\sigma}{E}[1 - \exp(-E.t/\eta)]$$

This is an Arrhenius type relationship (5.8.6) which, in addition to creep, governs many time- and temperature-related processes such as diffusion (7.4.2), oxidation (19.2.1) and corrosion. It states that the rate of creep (or diffusion, etc.) *increases* exponentially with temperature, or, that the time for a given amount of creep (or diffusion, etc.) *decreases* exponentially with temperature.

Plasticisation

12.6 Plasticisers are substances which are added to plastics materials in order to improve their flow properties and in some cases to reduce the brittleness of the product. Two principal groups of plasticisers are in use. Some substances—called 'primary plasticisers'—contain polar groups which neutralise the fields of force of the polymer groups, thus reducing the van der Waals forces between adjacent polymer chains. In many cases, however, the plasticiser is a compatible and inert material without polar groups but which exists dispersed throughout the polymer providing mechanical 'spacers' which separate the polymer chains and so reduce the van der Waals forces of attraction between them. These are termed 'secondary plasticisers' (Fig. 12.23(ii)).

(i) STRONG INTER-MOLECULAR FORCES   (ii) MOLECULES OF PLASTICIZER

Fig. 12.23 The function of a secondary plasticiser in separating polymer chains (ii) and so reducing intermolecular attraction (van der Waals forces) (i).

12.6.1 Internal Plasticisation This involves the polymer chain by introducing small amounts of another monomer unit during polymerisation. This second monomer requires long and flexible side groups. Thus PVC can be plasticised by adding 15% of vinyl acetate during polymerisation. The presence of the occasional bulky acetate groups in the polymer so formed act as spacers and so reduce the inter-molecular forces between adjacent chains. This in turn increases the amount of viscous movement possible. In this case increase in plasticity depends upon the formation of a new copolymer rather than upon the use of a plasticiser and is less often employed.

12.6.2 External Plasticisation This process is more commonly used. Here the plasticiser is added to the polymer *after* polymerisation is complete. It is often a compatible non-volatile solvent liquid which will disperse in molecular form throughout the polymer virtually forming a very viscous solution with the latter. Only the amorphous regions are so affected and a highly crystalline polymer cannot successfully be plasticised in this way since insufficient solvent will be absorbed.

An alternative form of external plasticisation involves the use of a non-solvent oil. This disperses as tiny droplets throughout the plastics material. These droplets act as spacers between the polymer chains decreasing the van der Waals forces in the usual way.

Fillers and other Additives

12.7   To a large extent fillers are used in plastics materials in the interests of economy. Often up to 80% of the additive is used. In general fillers are finely powdered naturally occurring non-polymic materials which are readily available at low cost. Phenolic (bakelite) type resins are nearly always compounded, or filled with substances such as wood flour, short cellulose fibres from wood pulp, cotton flock, mica and asbestos. In addition to the economic advantages obtained from the increase in bulk so produced, these materials greatly improve the strength, toughness and dimensional stability of the product. Mica improves electrical resistance whilst graphite reduces friction. Asbestos, of course, improves temperature resistance.

Some soft thermoplastics are blended with up to 80% mineral solids such as crushed quartz, clay or limestone. In materials of this type the plastics material functions as an inter-particle adhesive in much the same way as cement in concrete. On the whole, however, fillers are more often used with the thermosetting moulding compounds of the phenol-formaldehyde type.

In some cases the polymer is the vehicle for a filling material with special properties. Thus, some magnetic materials can be bonded in particle form with a suitable polymer so that better control of eddy currents is possible and more complex shapes more easily produced.

In addition to such materials as colourants, other additions may also be made to plastics materials. Thus *antioxidants* are sometimes used to prevent or inhibit the oxidation of polymers. The antioxidants are themselves substances which readily oxidise and in this way protect the polymer itself. *Stabilisers* are also sometimes used, for example, to prevent damage to a polymer by long exposure to ultra-violet light which often causes photo-decomposition as evidenced by a yellowish tinge and a reduction in transparency as well as a serious deterioration in mechanical properties. The addition of carbon black to the polymer renders it opaque to light and in this way protects it.

Foamed or 'Expanded' Plastics Materials

12.8   Foamed rubber has been in use for many years but more recently a number of polymers have become available in this form. Of the thermoplastic polymers 'expanded' polystyrene is well-known as an extremely lightweight packing material, also as a thermal insulator and for the manufacture of ceiling tiles for the enthusiastic home decorator. Up to 97% of its volume consists of air bubbles making it an excellent flotation material, since it has only about one-tenth the relative density of cork and does not become waterlogged.

Chief of the thermosetting polymers available as foams are the polyurethanes. These are available in the flexible form for the manufacture of upholstery and sponges whilst rigid polyurethane foams are used for the internal reinforcement of aircraft wings.

Two varieties of foamed polymers are manufactured—*open* (interconnecting) or *closed* cellular structures—though in practice foams are likely to contain some of each type of structure with one predominating. As suggested above, thermoplastics, thermosetting polymers and elastomers can all be foam expanded and a number of methods are available for introducing the gas pockets into the polymer:

(i)   By mechanically mixing air with the polymer whilst it is still in the molten state or dissolved in some solvent which will evaporate.

(ii) By dissolving a gas such as carbon dioxide in the mix by the use of pressure. During the hardening process pressure is reduced so that the gas is released from solution—rather like the bubbles of carbon dioxide in mineral waters. Since hardening is in progress the gas bubbles are trapped producing a foam.

(iii) Similar in principle is the addition of a volatile component to the mix. During the hardening process this vaporises forming bubbles.

(iv) Alternatively a gas may be evolved during a cross-linking process, e.g. as a result of condensation polymerisation (12.3.1). This gas or vapour can be utilised to produce the foaming process during hardening.

(v)   A 'blowing agent' may be added which will release gas in the mix during heating. This is useful for foaming thermoplastics. A common reagent is sodium bicarbonate which reacts in the same way as it does in baking powder, that is, by releasing bubbles of carbon dioxide on heating thus causing the cake to 'rise':

$$2\,NaHCO_3 \xrightarrow{\text{Heat}} Na_2CO_3 + H_2O + CO_2$$

Rubbers

12.9   Rubber is a very important material used in a wide variety of products ranging from tyres for heavy goods vehicles and motor cars to condoms and chewing gum. It is an *elastomer*. This is a substance in which the arrangement of the polymer molecules allows considerable reversible extension to take place at normal temperatures. Elastomers exist as long chain molecules which are irregularly coiled, bent and generally entangled when in the unstressed state. Since the coils are not arranged in a regular pattern the structure is amorphous. When a tensile force sufficient to stretch the material is applied, the molecules partially disentangle and straighten out so that they become orientated in the general direction in which stretching occurs. Since the molecules become more closely aligned along the direction of stretching, the forces of attraction between adjacent molecules also increase so that the material becomes stronger and stiffer.

As temperature is reduced polymers become less elastic and pass into a glassy state. This occurs at the glass transition temperature, T_g, and this varies from one polymer to another. T_g for most useful elastomers is well below room temperature. This can be demonstrated by cooling a rubber ball to a very low temperature in liquid air. When an attempt is made to bounce the ball it shatters in a manner similar to glass. Under normal conditions the molecules of an elastomer are in constant motion with adjacent sections vibrating relative to each other. As T_g is reached the molecular structure becomes rigid so that movement is restricted. In crystalline polymers the presence of multiple and hence fairly strong van der Waals forces acting between adjacent molecules will raise T_g to very near the softening temperature of the polymer. Consequently such materials are not elastomeric.

12.9.1 Natural rubber is a polymer derived from the sap of the rubber tree, *Hevea brasiliensis*. Since this tree is native to South America it is believed that Christopher Columbus was the first European to handle a piece of rubber. It seems that some tribes of South American Indians played ball games from very early times.

The monomer from which rubber polymerises is isoprene, C_5H_8. During the natural polymerisation process, double bond positions in the molecule are rearranged (Fig. 12.24(i)) so that isoprene units link up in the usual

(i) isoprene

| O carbon |
| • hydrogen |
| ⊘ -CH₃ |

(ii) 'cis' CONFIGURATION   (iii) 'trans' CONFIGURATION

Fig. 12.24

manner of polymers to form a chain molecule some 20 000 carbon atoms—or approximately 0.0025 mm—in length. This type of chain molecule can display the property of *stereoisomerism*. For our purposes this can be explained in a simplified manner as indicated in Fig. 12.24(ii) and (iii). Here it will be noticed that the CH_3 groups can be arranged on different 'sides' of the chain. The arrangement indicated in Fig. 12.24(ii) is termed the 'cis'-configuration and is the structure present in rubber, whilst that shown in Fig. 12.24(iii) is known as the 'trans'-configuration and is the

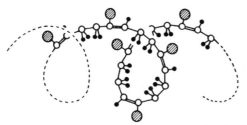

Fig. 12.25 The ability of *cis*-polyisoprene (rubber) to form a coiled molecule.

molecular pattern of *gutta-percha*. Rubber is an elastomer whilst gutta-percha is inflexible and hard by comparison. This could well be due to the fact that a chain molecule of the 'cis' type (rubber) is able to bend and coil easily in one direction, i.e. 'away' from the CH_3 groups (Fig. 12.25) whilst the relatively even distribution of CH_3 groups on both 'sides' of the molecule in the 'trans' configuration (gutta-percha) limits flexibility of the molecule in either direction. Since the double bond cannot rotate, the CH_3 groups are in fixed positions. Hence both the cis- and trans-configurations are stable and one form cannot convert to the other.

12.9.2 In its natural state rubber is both elastomeric and thermoplastic. Being thermoplastic it becomes *permanently* deformed when it is stretched since van der Waals forces operate between the chain molecules only at a limited number of points. In this state rubber is useless as an engineering material, but when rubber is 'vulcanised' these van der Waals forces are replaced by strong permanent covalent bonds and the material becomes thermosetting (Fig. 12.26).

Fig. 12.26 Long-chain molecules in rubber. Because of their 'folded' form they become extended in tension, but return to their original shapes when the stress is removed. In 'raw' rubber (i) a steady tensile force will cause individual molecules to slide slowly past each other into new positions, so that when the force is removed some plastic deformation remains, although the elastic deformation has disappeared. By 'vulcanising' the raw rubber (ii) the chain molecules are chemically bonded so that no permanent plastic deformation can occur and only elastic deformation is possible.

12.9.3 The process of vulcanisation was introduced by Goodyear in 1839 and involves the use of sulphur to provide cross links between the polyisoprene molecules. The raw rubber is first 'masticated' or kneaded between rolls so that a degree of depolymerisation occurs and the material becomes more plastic as a result. It can then be mixed with sufficient sulphur—or other vulcanising agents—before being heated to 200°C to cause cross-linking between the polyisoprene molecules (Fig. 12.27).

Adjacent cis-polyisoprene chains

● = H; O = C; ◍ = CH₃; ◉ = S

Fig. 12.27 Cross-linking by sulphur atoms in the vulcanisation of natural rubber. (The diagram indicates all available cross-linking positions as being used up, but this would not be the case in *soft* rubber.)

As the relatively unstable double bonds in the polyisoprene chain are broken, sulphur atoms form covalent bonds with adjacent chains. The greater the amount of sulphur used the higher the degree of cross-linking between adjacent chains and the harder and more rigid the vulcanised rubber becomes since movement of one chain relative to its neighbours becomes more difficult. When an excess of sulphur is used so that all available bond sites are used up a hard rigid solid is produced and elastomeric properties disappear. Such a material was produced in the early years of the twentieth century under such trade names as 'Vulcanite' and 'Ebonite' and used in the manufacture of 'fountain pens' and coil formers used by amateur radio constructors. Such purposes have long been replaced by modern rigid plastics materials but 'Ebonite' sheet is still used as a lining for tanks carrying corrosive liquids, particularly those containing the Cl^- ion, since it has a higher resistance to Cl^- because of its fully cross-linked structure and the consequent absence of available reactive sites.

Rubber used in the manufacture of automobile tyres is only vulcanised to the extent of about 5% of the available bond positions. This is sufficient to reduce the plasticity of the rubber by introducing enough sulphur cross links but at the same time retaining most of the elastomeric properties.

12.9.4 In addition to the vulcanising agent raw rubber is generally compounded with other ingredients or 'fillers'. Carbon is added to render rubber tyres opaque and so restrict damage by UV light. Some increase in both strength and abrasion resistance is afforded by the addition of carbon. The unsaturated double bonds still remaining in rubber after vulcanisation are susceptible to oxidation, particularly by ozone. This effectively reduces elasticity by forming cross links in those remaining positions not already linked by sulphur atoms. As a result the rubber becomes brittle or 'perished'. To reduce this tendency various anti-oxidants are incorporated in the mixture.

The Mechanical Properties of Rubber

12.10 The resistance of rubber to deformation in tension at first decreases (as with metals) but then increases as indicated in the stress/strain diagram (Fig. 12.28). Rubber does not obey Hooke's Law when in tension.

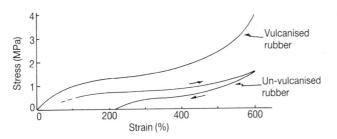

Fig. 12.28 The stress–strain curves for raw rubber and vulcanised rubber. For raw rubber the stress was relaxed when a strain of 600% had been reached. A *permanent* extension of more than 200% remained.

This is due to the fact that in materials with high elasticity the considerable extension is accompanied by a big reduction in cross-section whereas in metals, which obey Hooke's Law, this reduction is negligible. The ratio of lateral contraction to longitudinal extension (Poisson's Ratio) is therefore important in considering the tensile extension of rubbers.

Unlike metals generally, vulcanised rubbers do not yield under stress, that is extension is entirely elastic. The tensile strength is about 27 MPa with an elongation of up to 700%. As might be expected Young's Modulus is relatively low since a large amount of strain is produced by a small stress and for natural rubber (NR) the value of E is about 17 MPa. Comparable mechanical properties are obtained for most synthetic rubbers so that the choice between NR and one of the synthetic rubbers is more likely to depend upon properties such as wear resistance, heat dissipation, ageing and chemical resistance to particular acids, alkalis and solvents. The term *drift* as applied to rubbers is analogous to creep in metals. It refers to permanent distortion of rubber under long-time loading. As with creep in metals the rate of drift increases with temperature and live loads produce greater drift than do static loads.

The effects of temperature on rubbers must therefore be considered carefully. High temperatures accelerate oxidation degradation and whilst antioxidants can help against this, an upper limit of 100°C for general purpose rubbers under long service conditions and 250°C for special elastomers is usually applied. All rubbers become brittle when the temperature is sufficiently low (12.9) so that a minimum working temperature is about −40°C for most synthetic rubbers and as low as −65°C for NR.

Synthetic Rubbers

12.11   The Japanese invasion of Britain's Far Eastern rubber plantations in 1941 led to a serious crisis in the war potential of the Allies. The situation was saved by the development of the American synthetic rubber known as GR–S, a copolymer of styrene and butadiene. Currently its production exceeds that of NR.

Most synthetic rubbers are polymerised from organic compounds whose names end in –*diene* (hence they are often called 'dienes'). These are of the general formula:

$$\overset{|}{C}=\overset{|}{\underset{|}{C}}-\overset{|}{C}=\overset{|}{\underset{|}{C}}$$

Two double C=C bonds separated in this way by a single bond are known as *conjugated double bonds*. During the polymerisation process one of these double bonds disappears to produce the valency which joins this monomer to another in forming the polymer:

$$\overset{|}{C}=\overset{|}{\underset{|}{C}}-\overset{|}{C}=\overset{|}{\underset{|}{C}} \;\rightarrow\; \cdots\; \overset{|}{C}-\overset{|}{\underset{|}{C}}=\overset{|}{C}-\overset{|}{\underset{|}{C}}\; \cdots$$

However, one double bond remains in the 'rearranged' unit and it is this double bond which is available for cross-linking during the vulcanisation process. Table 12.5 indicates the relationships between monomers and polymers in the more important rubbers.

The chemistry of some synthetic rubbers is different. Neither IIR or EPR (Table 12.5) are derived from 'dienes' and when polymerisation is completed no double bonds remain for vulcanisation. Some cross-linking does occur in EPR, thought to be due to small crystallites forming in groups of polypropylene units in adjacent chains, thus 'locking' these chains together. Nevertheless a suitable diene (e.g. 2% isoprene) is often used so that EPR becomes EPDR. Butyl rubber is similarly treated for vulcanisation.

12.11.1   *Styrene-butadiene rubber* (SBR) (Table 12.5) is produced by copolymerising about 78% butadiene with styrene. It represents the bulk of synthetic rubbers produced, its cost being relatively low making it useful for car tyres. It is a good general purpose rubber with excellent abrasion resistance but its fatigue resistance is inferior to that of NR. It differs from NR in its ageing behaviour for, whereas oxidation causes *scission* (19.8.2)

Table 12.5 *Some synthetic rubbers*

in NR leading to softening, in SBR cross links tend to form making it harder and more brittle.

In addition to car tyres SBR is used for hose, belting, flooring, shoe soles and electrical insulation. It is less suitable than NR for the tyres of HGVs where heat tends to build up more quickly with SBR. Operational temperature range is between −40 and 80°C.

12.11.2 *Polybutadiene rubber* (BR) is made from the single monomer butadiene. Its strength and tear resistance are generally inferior to NR but it is highly resilient. Hence it is rarely used alone but blended in amounts up to 40% with either NR or SBR to improve the abrasion resistance of tyres in which it is most often used. Temperature range: −40 to 100°C.

12.11.3 *Butyl rubber* (IIR) is produced by co-polymerising isobutylene with small amounts of isoprene. Since the product has a low degree of 'unsaturation' (few double bonds) it is chemically very stable and has a good resistance to heat and to many chemicals but not to oils. It is vulcanised in the same way as natural rubber. Butyl rubber is cheaper than natural rubber and since it is resistant to ozone because of its low unsaturation, it is less liable to 'perish'. It is also impermeable to air and other gases making it useful for inner tubes and tubeless tyres, air bags, sporting equipment and moulded diaphragms. Being impervious to many chemicals it is also used for hoses, tank linings and conveyor belts. Generally its physical properties are inferior to NR and SBR. Temperature range: −50 to 125°C.

12.11.4 *Polychloroprene ('neoprene') rubber* (CR) is a polymer of chloroprene (chlorobutadiene) and is generally inferior to nitrile rubber in its resistance to oils but better than either NR, SBR or IIR. Moreover its mechanical properties are superior to those of most synthetic rubbers and only slightly inferior to those of NR. It is a relatively expensive material used for gaskets, protective gloves, adhesives, hose, conveyor belts and electrical insulation. A major disadvantage is its poor flexibility at low temperatures. Temperature range: −40 to 100°C.

12.11.5 *Isoprene (polyisoprene) rubber* (IR) is a synthetic rubber with virtually the same composition and structure as NR. Its properties are slightly more uniform than NR because of the greater 'cleanliness' of its raw material, isoprene, but it is less easy to process and is of lower strength.

12.11.6 *Nitrile rubber* (NBR) is a copolymer of acrylonitrile and butadiene. It has excellent resistance to grease, oil and solvents but is attacked more easily by alkalis than is NR. It is used extensively as an adhesive and to impregnate, or as a coating for, paper products. Other uses include heels for shoes and as a lining for hoses and fuel tanks carrying petroleum products. Temperature range: −40 to 100°C.

12.11.7 *Ethylene-propylene rubber* (EPM) is a copolymer of the two unsaturated hydrocarbons (alkenes) ethylene and iso-propylene. It is a non-crystalline product with rubbery properties. Because of the chemical saturation of its structure following polymerisation it is chemically inert and has good resistance to ageing (as does polythene). It is a lightweight rubber with up to 65% of either monomer. A further *diene* monomer is often added to provide sites for cross-linking (then designated as EPDM). It has good resistance to weathering and ozone attack, but a rather high creep rate and poor resistance to solvents. Temperature range: −50 to 125°C.

12.11.8 Of some dozen other synthetic rubbers available *Silicone rubber* (Q) is very resistant to weathering and is good for both high and low temperature use. Temperature range: −60 to 200°C; whilst *Fluorocarbon rubber* (FPM) has, like PTFE, a high resistance to many aggressive chemicals (fluorine being the most strongly electronegative of all elements and therefore forming very stable compounds). It is also useful at high temperatures. Temperature range: −30 to 250°C.

13 ────────────────────────

Ceramics and Glasses

13.1 The atoms or molecules of any crystalline solid are permanently located with respect to their neighbours. Atoms vibrate continuously about equilibrium positions because of the energy they have received in the form of heat. The total internal energy of a solid increases as its temperature rises and for this reason both the frequency and amplitude of atomic vibrations increase so that atoms or molecules are forced further away from their equilibrium positions. This movement away from original positions manifests itself as *thermal expansion* and also explains the storing of energy by a body in terms of its heat capacity (*specific heat*).

Atomic vibrations also furnish one method by which heat can travel through a substance, that is, by *conduction*. Metals are particularly good conductors of heat since their valency electrons, constituting a common 'cloud', are able to move freely under the influence of a temperature gradient just as they will under the action of an electrical potential difference.

Materials such as glasses, rubbers, cellulose and many of the plastics materials are usually considered as solids in the sense that they are of definite and permanent shape. On the other hand they are not crystalline since their molecules are not arranged in a regular pattern. For this reason, as well as for others such as the lack of a definite melting point, they are often regarded as super-cooled liquids. When heated to a sufficiently high temperature range they become increasingly plastic as the temperature rises. In this respect they resemble viscous liquids which become more mobile at higher temperatures. Pure crystalline solids on the other hand pass from a strictly geometrical lattice structure to a liquid state at a single fixed temperature.

13.1.1 Most ceramics are hard and brittle and, with the exception of glasses, crystalline in nature. In metals the crystalline structure is relatively simple and operates through the metallic bond which renders them ductile, but in ceramics chemical bonding is either covalent (diamond, silicon carbide, silicon nitride) or ionic (magnesium oxide) or in some cases a combination of the two (aluminium oxide).

As the Periodic Classification (Table 1.3) indicates carbon and silicon are chemically similar elements and each forms important ceramic materials. Each has four valency electrons in its outer shell and this results in their forming four equivalent electron orbitals which of course repel each other strongly so that they are directed towards the corners of a regular tetrahedron (Fig. 12.1). A stable 'octet' is then achieved by the formation of four covalencies. This tetrahedral form is reflected in the crystalline structure of diamond (Fig. 3.28) and of many silicon compounds (Figs 13.3, 13.4, 13.5 and 13.6). The great strength of covalent bonds of this type results in the extreme hardness of many of the compounds of carbon and silicon such as diamond, silicon carbide, silica, silicon nitride, tungsten carbide, glass, etc.

13.1.2 The term 'ceramics' is applied to a range of inorganic materials of widely varying uses. Generally these materials are non-metallic and in most cases have been treated at a high temperature at some stage during manufacture. The word 'ceramic' is derived from the Greek, 'keramos'— or 'potter's clay', though the group of materials now so described includes glass products; cements and plasters; some abrasives and cutting tool materials; building materials such as bricks, tiles and drain pipes; various electrical insulation materials; refractory linings for furnaces; porcelain and other refractory coatings for metals; as well as the more traditional uses in pottery, crockery and sanitary ware.

From the point of view of composition and structure ceramics can be classified conveniently into four main groups:

(i)   *Amorphous ceramics* these are substances referred to generally as 'glasses'. They include those such as 'obsidian'* which occur naturally, and man-made glasses used for the manufacture of bottles, windows and lenses;

(ii) *Crystalline ceramics* these may be single-phase materials like magnesium oxide or aluminium oxide, or various mixtures of materials such as these. In addition some carbides and nitrides belong to this group;

(iii) *Bonded ceramics* materials in which individual crystals are bonded together by a glassy matrix as in a large number of products derived from clay;

(iv) *Cements* a number of these are crystalline but some may contain both crystalline and amorphous phases.

---

*Used in Victorian times as a black semi-precious stone for ladies' jewellery. I once literally stumbled across an outcrop ridge of it a mile or so long in some remote mountains in Iceland. Small bits of it were quite expensive in the souvenir shops of Reykjavik!

Structures of Ceramics

13.2   Ceramics are far less ductile than metals and tend to fracture immediately any attempt is made to deform them by mechanical work. They are often of complex chemical composition and their structures may also be relatively complex. These structures fall into two main groups:

13.2.1 Simple Crystal Structures   These include structures in which the bonding is either ionic or covalent. Thus magnesium oxide, MgO, is an ionic compound of cubic structure (Fig. 13.1) similar to that of sodium

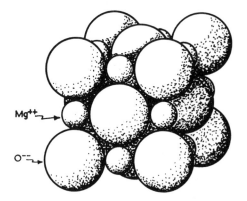

Fig. 13.1   The crystal structure of magnesium oxide in which each O^{--} ion is surrounded by six Mg^{++} ions and, similarly, each Mg^{++} ion is surrounded by six O^{--} ions.

chloride. Magnesium oxide is a refractory material used in the linings of furnaces for the manufacture of steel by basic processes and can withstand temperatures in excess of 2000°C.

Silicon carbide is also a ceramic material of simple composition and structure. Here the bonding is covalent and the structure tetrahedral such that each silicon atom is surrounded by four carbon atoms and, conversely, each carbon atom by four silicon atoms. As a result the structure is very like that of diamond. As mentioned earlier (13.1.1) both carbon and silicon are quadrivalent members of Group IV. Due to the stable covalent bonds which operate and also the symmetry of the structure, silicon carbide is a very hard material. Though less hard than diamond which it resembles in structure it is nevertheless a well-known abrasive material.

Some relatively simple crystalline ceramics contain a mixture of covalent and ionic bonds. Such a material is alumina (aluminium oxide, Al_2O_3) (Fig. 13.2). Although the structure is basically close-packed hexagonal, one third of the aluminium ion/atom sites remain vacant in order to satisfy the valency requirements of both elements.

13.2.2 Complex Silicate Structures   The majority of ceramic materials, and particular those derived from clay, sand or cement, contain the

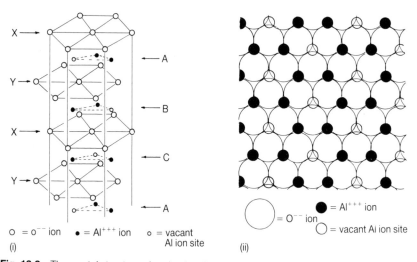

Fig. 13.2 The crystal structure of α-alumina. (i) A spatial representation in which the aluminium ion/atom sites (A, B, C) are situated between hexagonal layers (X, Y) of the oxygen ions; (ii) a 'plan view' showing that only two-thirds of the aluminium ion/atom sites are occupied; thus satisfying the valencies, i.e. conforming to the formula—Al_2O_3.

element silicon in the form of silicates. The similarity between silicon and carbon in respect of many chemical properties has already been mentioned. Just as carbon polymer molecules are built up from a succession of repeating 'mers' (12.1.3) so are many silicates formed as chain, sheet or framework structures. The most simple silicon–oxygen 'mer' is the 'SiO_4' group in which a small silicon atom is surrounded tetrahedrally by four oxygen atoms (Fig. 13.3). In this way the quadrivalency of the silicon is

Fig. 13.3 The tetrahedral arrangement of the $(SiO_4)^{4-}$ ion. (iii) Indicates the unsatisfied valencies of the oxygen atoms.

satisfied but the bivalency of the oxygen atoms is not. Hence to satisfy the valency of the oxygen atoms, SiO_4 groups are frequently ionically bonded to metallic atoms thus producing negatively charged silicate ions $(SiO_4)^{4-}$ and positively charged metallic ions, e.g. Ca^{2+}; Na^+; Mg^{2+} and Al^{3+}. The positive and negative ions then form a crystal structure.

Alternatively the electron deficiency of the oxygen can be made up by combination with another silicon atom thus making it part of an adjacent tetrahedron (Fig. 13.4). The resultant unit $(Si_2O_7)^{6-}$ is present in only a few minerals.

An example of a mineral based on the $(SiO_4)^{4-}$ ion is magnesium silicate

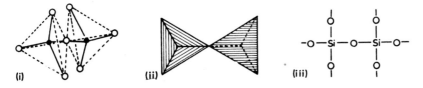

Fig. 13.4 The structure of the $(Si_2O_7)^{6-}$ group. (i) 'Picture view'. (ii) 'Plan view'. (iii) Valency diagram.

or *forsterite*. The empirical formula, that is, that formula which indicates the ratio of atoms present in a compound, is Mg_2SiO_4, but it must be remembered that, as in the case of sodium chloride (3.2.2), the unit involved here is the crystal and not the molecule. *Forsterite* is useful as a refractory material since its melting point is in the region of 1900°C.

13.2.3 The structure depicted in Fig. 13.5 is fundamentally that of a chain based on silicate ions $(SiO_3)^{2-}$, in which adjacent chains are ionically bonded by bridging metallic ions such as Ca^{2+}. Herein lies the difference in mechanical properties between an organic polymer and a silicate ceramic. In the case of the organic polymer adjacent chain molecules are held together by weak van der Waals forces but in the ceramic these forces are replaced by stronger ionic bonds making ceramics hard and relatively strong. Nevertheless the ionic bonds linking the silicate chains to the metallic 'bridges' are weaker than the covalent bonds operating within the silicate group. Consequently fracture usually occurs through these ionic bonds, that is, parallel to the direction of the silicate chains.

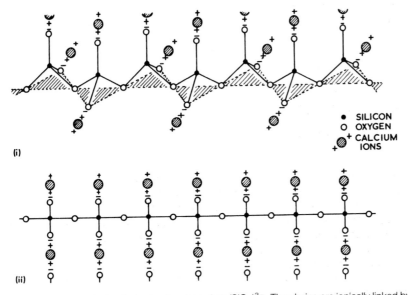

Fig. 13.5 A silicate single-chain structure based on $(SiO_3)^{3-}$. The chains are ionically linked by Ca^{++} ions.

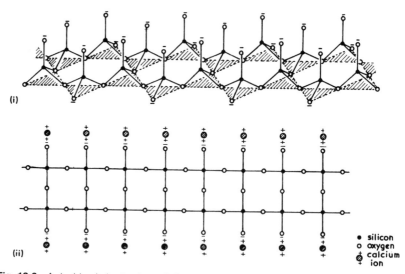

(i)

(ii)

● silicon
○ oxygen
⊘ calcium
┼ ion

Fig. 13.6 A double-chain structure of silicate tetrahedra.

Double chains of silicate tetrahedra are also formed, i.e. $(Si_4O_{11})^{6-}$ and these are in general more important than the single-chain compounds. Here the oxygen atoms forming the bases of tetrahedra are shared with adjacent units (Fig. 13.6). The double-chain units are linked by metallic ions forming characteristically rigid crystalline structures. In some cases the chains are held together by van der Waals forces as well as by ionic bonds thus allowing the material to be separated into fibres. Asbestos is an example of this class of mineral which have empirical formulae of the type $3MgO.2SiO_2.H_2O$.

13.2.4 In many minerals of the group which include clays, talc and micas the silicate tetrahedra are linked in sheet form instead of in chain form. Figure 13.7 represents a 'plan' view of this system in which the positions of the oxygen atoms at the apices of the tetrahedral units are shown. The

'apex'
oxygen
atom

silicon
atom

'basal' oxygen atoms

Fig. 13.7 A sheet arrangement of silicate tetrahedra.

empirical formula of the silicate unit is $(Si_2O_5)^{2-}$. On the basal plane of the sheet all oxygen valencies are satisfied, i.e. each oxygen atom has 8 available electrons. The mineral *kaolinite*, the simplest type of clay, has aluminium ions (Al^{3+}) which satisfy the valencies of the upper oxygen atoms; the remaining two valencies of the Al^{3+} ions then in turn being satisfied by $(OH)^-$ ions which are also present. Since all valencies within the sheet structure are now satisfied adjacent silicate sheets can attract each other only by weak polarisation forces of the van der Waals type. This explains the plasticity of clay and the fact that mica is readily separable into extremely thin sheets.

When water is added to clay the H_2O molecules are attracted to the surface layers of the sheets by polarisation forces (2.4) thus providing a boundary lubricant between adjacent sheets (Fig. 13.8). When water is

$\bullet$ = OH$^-$

O = Al^{+++}

◯ = O^{--}

$\bullet$ = Si

water molecule

silicate sheet in kaolinite

water molecules

polarisation charges

Fig. 13.8   The absorption of water at the surfaces of the silicate sheets in clay (kaolinite). This absorption is due to the attraction between polarisation forces on the surfaces of the kaolinite sheets and the hydrogen bond of the water molecules (2.4).

dried out the plasticity of the clay is correspondingly reduced and when the clay is baked cross-linking between adjacent sheets produces a strong brittle structure. This is a non-reversible reaction comparable with the thermosetting of bakelite and the clay has now permanently lost its plasticity.

13.2.5   The development of the silicate tetrahedral structure into three dimensions produces a framework type of arrangement which is in effect a giant molecule (3.8.1) of empirical formula $(SiO_2)_n$. Each silicon atom is surrounded by four oxygen atoms and each oxygen atom is jointly shared by an adjacent unit. In this way all valency bonds are used up and a hard, rigid framework structure is produced. The simplest of these structures is that of *cristobalite* whilst another is that of *quartz*. Both of these materials are hard and fracture in an irregular manner since there is no preferred direction of cleavage, bonding forces being equal in all directions. Other framework structures include various aluminosilicates.

Shaping of Ceramics

13.3 Various types of clay are the most widely used raw materials for ceramics manufacture. They can generally be used in the 'as mined' condition when mixed with water which acts as a plasticiser (13.2.4). Although clays differ in composition the main constituent is an alumino-silicate mineral such as *kaolinite* ($Al_2O_3.2SiO_2.2H_2O$). The kaolinite content varies according to the type of clay. English china clay as mined in Cornwall may contain up to 95% kaolinite whilst clays used in brick manufacture may contain as little as 30%. Most other ceramic materials have to be manufactured or at least purified by some chemical process. The powders so produced are then compacted and sintered to achieve a coherent solid mass. Glasses however are shaped whilst in a liquid state (13.6.1).

13.3.1 Shaping Clay Products Mention of the potter's wheel may conjure up visions of 'bearded weirdies' wearing sandals, living on a diet of stewed nettles and producing potters' knick-knacks for the local tourist trade. Nevertheless the potter's wheel still has commercial use though industry generally employs mechanical operations with a standard product in view.

13.3.1.1 Hydroplastic forming of wet clay is most commonly carried out by extrusion using a 'screw pump' similar in design to that in Fig. 6.27. After being extruded the clay is strong enough to be handled prior to firing. Most bricks are now made from extruded stock, i.e. 'parted off' and embossed.

13.3.1.2 Slip casting is used to produce a variety of household and decorative pottery from china clay, particularly when complex or irregular shapes are involved. Here the clay is held as a suspension, or 'slip', of particles in water. This is then poured into a mould made from some *absorbent* material such as Plaster of Paris which takes up water from the slip leaving a uniform layer of clay particles on the mould walls. When the layer reaches the required thickness excess slip is poured away. After the necessary drying period which allows the article to become strong enough it is removed from the mould (Fig. 13.9) prior to firing.

13.3.2 Forming Ceramics from Powders A number of variants of a basic pressure-application process are used, principally in making engineering ceramics.

13.3.2.1 Cold pressing/sintering—many engineering ceramics are made by cold pressing the dry or slightly damp powder to produce a compact which is strong enough to be handled. The compact is then sintered at a temperature high enough to cause fusion at the particle boundaries as is the case with some powder-metallurgy products (6.5.1).

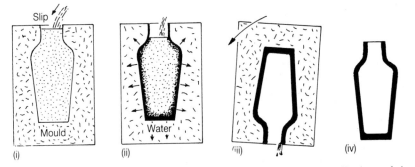

Fig. 13.9 Slip casting of pottery (i) 'slip' is poured into an absorbent mould; (ii) a layer of clay forms as the mould surface absorbs water; (iii) when the shell is of suitbable thickness excess slip is poured away; (iv) the resultant casting.

13.3.2.2 Isostatic pressing (Fig. 13.10) is used to obtain a more uniform density particularly in components of intricate shape. The ceramic powder is contained in a rubber mould which is then immersed in a liquid bath to which a high pressure is transmitted.

Fig. 13.10 The isostatic pressing of a ceramic component.

13.3.2.3 Hot pressing is used to produce a denser product in those materials which tend to suffer from high porosity, but because of the high temperature and pressures involved, it is limited to the manufacture of simple shapes. Silicon nitride, Si_3N_4, is shaped by *hot isostatic pressing* at temperatures in the region of 1800°C, using graphite dies and pressures of 100 MPa.

13.3.3 The Heat-treatment of Ceramics   Most clay-based ceramics need to be carefully dried. As the film of water which separates the ceramic particles is removed shrinkage inevitably occurs but warping and cracking can be prevented by controlling the temperature and humidity of the environment to constant values.

13.3.3.1   Drying of the ceramic is followed by firing which is carried out at a temperature which will lead to sintering followed by vitrification. Sintering involves the compacting of the material by prolonged heat-treatment at a high temperature which is nevertheless below the melting point of the main phase of the ceramic. Diffusion across the particle boundaries occurs, accompanied by some recrystallisation which completes the bonding of the particles.

Clay-based ceramics depend upon vitrification occurring during firing. Here considerable amounts of liquid—up to 35%—are formed and, on cooling from the firing temperature, this liquid phase forms a glass which is evident as a glazed finish on the surface.

Properties of Ceramics

13.4   Until recent years most ceramic materials consisted of crystalline particles cemented together by glass, which, as we shall see later, is an amorphous substance formed when acidic silica combines with a basic metallic oxide at high temperatures. These materials, mainly alumino-silicates of varying compositions, were used as furnace linings; for the manufacture of pottery and electrical insulators and for many building materials such as bricks and tiles. However, modern technologies of sintering have produced structures which are wholly crystalline. Some of these materials maintain high strength at elevated temperatures since low melting point glasses are absent.

13.4.1 Strength   The high *compressive* strength of ceramics has long been known to man and was well established by the time Pheidias constructed his Parthenon in Athens in about 450 B.C. In more recent times Messrs Josiah Wedgwood and Sons Ltd published a photograph showing an eleven-tonne double-decker bus supported by six of their fine bone china coffee cups.

Although ceramics are hard and reasonably strong they lack ductility and this, in turn, prevents any stress concentrations which may be present from being relieved by the onset of plastic flow. Whereas metals are malleable and ductile most ceramics, particularly where covalent bonds are present, are brittle. This is due largely to the overall *width* of dislocations present in the structure and thus in turn is dependent on the nature of interatomic forces. In a metal the ions are held in position by their mutual attraction for the 'flexible' electron cloud, whilst in ceramics such as diamond the strong covalent bonds are strictly directional. These factors in turn affect the equilibrium width of any dislocations present (Fig. 13.11). The width of a dislocation is defined as the distance over which the displacement of atoms or ions is greater than $b/4$ where b is the Burgers vector of the material (Fig. 5.10).

13.4.1.1   In Fig. 13.11 it will be seen that the width, W, of the dislocation in the case of a metal (i) is much greater than in a covalently bonded material (ii). In fact it is estimated that with malleable metals such as gold, copper and aluminium dislocation widths of about ten ion spacings are present (Fig. 13.11(i)). When the width is great the force necessary to

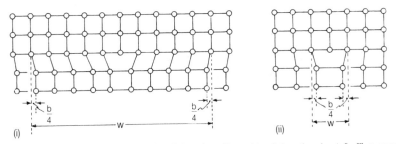

Fig. 13.11 Illustrating the width of a dislocation: (i) a wide dislocation (metal); (ii) a narrow dislocation (covalently-bonded material).

move such a dislocation is small because the highly distorted region at the centre of the dislocation is not critically localised on a single particular crystal site. In (ii), however, a large stress will be necessary to move the dislocation away from its strongly localised site. At the same time there is a comparative lack of slip systems in a ceramic material and these influences will combine to cause the interatomic forces of attraction in the region of the dislocation to be exceeded so that fracture occurs. This difference in the width of dislocations as between metals and covalently- or ionically-bonded materials is responsible for the high malleability and toughness of metals and, consequently, their strength and reliability as engineering materials.

13.4.1.2 Since ceramics tend to suffer from the presence of microcracks which act as stress raisers it follows that *tensile* stresses must generally be kept low if sudden failure is to be avoided. Moreover the incidence of such faults in a ceramic is much more variable than in a metal where the quality of the structure is relatively easy to control. Consequently a bigger 'scatter' of measured strengths is to be expected in ceramics. Their *compressive* strengths, however, are high compared with their tensile strengths. Retention of strength at high temperatures is one of the useful features of such materials. For example, titanium diboride will maintain a compressive strength of 250 MPa at 2000°C making it one of the strongest materials known at that temperature.

13.4.1.3 In most ceramics ductility is, for all practical considerations, zero. This again is because many ceramics are hexagonal in crystal structure (Fig. 13.2) in which only two slip planes are available. Moreover bonding, being covalent, is strongly directional between specific atoms making slip virtually impossible. In practice of course such internal imperfections as porosity further reduce strength and ductility. Such faults are more difficult to control during manufacture than they are in metals since most engineering ceramics are compacted from powders which suggest that some porosity is inevitable. A few ceramics such as MgO in which bonding is primarily ionic exhibit limited ductility when in the form of a single crystal but in general ceramics can only be classed as 'brittle'.

13.4.2 Creep   This takes place in ceramic materials at relatively high temperatures since ceramics are stable *crystalline* solids with high melting points. Glasses on the other hand are non-crystalline and have fairly low softening temperatures so that considerable creep will occur at moderate temperatures. With these materials steady-state creep occurs according to an Arrhenius-type expression (16.4):

$$\frac{d\varepsilon}{dt} = A\sigma^n \exp(-Q/RT)$$

where A is a constant, Q is the activation energy (also a constant for the condition), R is the universal gas constant, T the absolute temperature, σ the stress and n the time exponent which for many ceramics and glasses is in the region of 10.

13.4.3 Hardness   Ceramic materials generally are hard when compared with most other engineering materials. This hardness makes certain ceramics useful as abrasives or, suitably bonded, as cutting tool materials. Thus, cubic boron nitride is almost as hard as diamond whilst materials like silicon carbide, alumina (aluminium oxide) and beryllium oxide are already well known. Some hard ceramics are listed in Table 13.1, hardened tool steel being included for comparison.

Table 13.1   *Hardness of Some Ceramics*

Material	Knoop hardness number
Diamond	7000
Cubic boron nitride	6900
Titanium diboride	3300
Boron carbide	2900
Silicon carbide	2600
Silicon nitride	2600
Alumina	2000
Beryllium oxide	1220
Hardened steel	700

The hardest of these substances depend for their properties on the very strong stable nature of their directional covalent bonds.

13.4.4 Refractoriness   generally defined as the ability of a material to withstand high temperatures without appreciable deformation under service conditions. Some refractories like fireclay and high-alumina brick soften gradually over a range of temperature and may ultimately collapse under load at a temperature well below the normal fusion temperature. Materials most widely used as refractories in furnaces are clays containing various proportions of silica (SiO_2) and alumina (Al_2O_3). As the thermal equilibrium diagram (Fig. 13.12) indicates mixtures containing in the region of 4–8% alumina are best avoided for use at high temperatures since they will melt completely at a little above the eutectic temperature of 1595°C. As the amount of alumina increases above 5.5% the quantity of liquid (produced at 1595°C) will decrease and so the bricks will be stronger

Fig. 13.12   The silica–alumina phase diagram. (Crystobalite and tridymite are allotropic forms of silica.)

at that temperature. For severe conditions where temperatures reach 1800°C the alumina content must exceed 71.8%.

Certain ceramic materials have very high melting points indeed and for special purposes these may be used, assuming that their relative high cost is justified. A few of these high temperature ceramics are included in Table 13.2.

Table 13.2   *Melting Points of Some High-temperature Refractory Materials*

Material	Melting point (°C)
Hafnium carbide	3900
Tantalum carbide	3890
Thorium oxide	3315
Magnesium oxide	2800
Zirconium oxide	2600
Beryllium oxide	2550
Aluminium oxide	2050
Silicon nitride	1900
Fused silica glass	1680 (approx.)
Soda glass	340–430 (approx.)

Again, as with hardness, this property can be thought of as a function of chemical bond strength. Strongly bonded materials require a large input of thermal energy to disrupt the atomic bonds.

13.4.5 Electrical Properties   Until a short while ago the use of ceramics in the electrical industries was restricted to insulation. Today ceramic materials include many substances varying in properties from low-loss high-frequency insulators to semi-conductors, conductors and ferro-magnets. Consequently they are used as thermistors, thermoelectrics, transistors and storage cells in memory systems. Indeed the current race to develop a really high-temperature superconductor (18.4.2) is based at present on sophisticated ceramics.

Some Engineering Ceramics

13.5 Modern ceramics used in engineering are either:

(i) *hard* materials, usually oxides, carbides, nitrides or borides of metals or metalloids; or
(ii) *strong* materials such as carbon or certain glasses in fibre form.

The hard materials are used as cutting tools or in the form of wear-resistant or penetration-resistant surfacing substances both for military purposes and in the auto-engineering industries.

13.5.1 Alumina (Al_2O_3) produced from the ore bauxite which is also the source of metallic aluminium. As mined, bauxite is a reddish-brown mineral because of the presence of impurities such as iron oxide and silica. The crushed bauxite is digested with hot caustic soda under pressure when the aluminium oxide dissolves as sodium aluminate, $NaAlO_2$:

$$Al_2O_3 + 2NaOH \rightarrow 2NaAlO_2 + H_2O$$

The solution is filtered so that solid impurities are eliminated. Sodium aluminate is unstable so that when CO_2 is passed through the solution aluminium hydroxide, $Al(OH)_3$, is precipitated. This is filtered off, the remainder of impurities passing away in the filtrate. The pure $Al(OH)_3$ is then dried and calcined at 1100°C when the hydroxide decomposes to give pure alumina:

$$2Al(OH)_3 \rightarrow Al_2O_3 + 3H_2O$$

In the manufacture of armour plating alumina is compacted as a dry or slightly damp powder at high pressure so that a dense plate, strong enough to be handled, is formed. This is then densified by sintering which involves the consolidification of the powder by prolonged heating at a very high temperature but below the melting point.

Pure alumina has a very low thermal conductivity which makes it susceptible to over-heating and hence rapid wear when used as a cutting tool at high machining speeds. Thermal conductivity is increased by adding some titanium carbide which in turn reduces wear by over-heating. Such tools have to be hot-pressed from powder with consequent limitations on shape. An alumina–titanium nitride mixture, however, can be shaped by the orthodox cold pressing/sintering method. It has a still greater thermal shock resistance than pure alumina and is also harder, making it more suitable for machining harder steels and cast irons under conditions of high cutting forces and high temperatures.

Much alumina is used as military armour plating. It has the disadvantage of a high relative density of 3.8 but is relatively inexpensive. Such armour in the form of moulded breast plates is used by specialist units in anti-terrorist, counter-revolutionary operations where snipers using high-powered rifles may be encountered. Its hardness, wear resistance and *chemical inertness* have led to the increasing use of alumina in hip-joint replacements. Since alumina has a low electrical conductivity it is used for the ceramic bodies of sparking plugs.

13.5.2 Silicon Nitride and 'Sialons'   Silicon nitride, Si_3N_4, is a stable covalent compound with low thermal expansion which makes it useful for metal cutting since in addition to great hardness it possesses resistance to thermal shock. Silicon nitride itself cannot be sintered successfully since a high degree of porosity is retained. Consequently required shapes are *preformed* in silicon to produce porous compacts which are then heated in a nitrogen atmosphere just below the melting point of silicon (1412°C) to produce the nitride.

13.5.2.1   Since Si_3N_4 cannot be sintered additions of basic oxides such as MgO, Al_2O_3 and Y_2O_3 are made so that, on heating, a liquid phase is formed by reaction with acidic SiO_2 which is present on the suface of the Si_3N_4 grains. On solidification the resultant microstructure is a two-phase system consisting of crystalline β–Si_3N_4 embedded in an intercrystalline bonding phase in the form of a glass. This glass can be more or less devitrified by slow cooling or by further heat-treatment. Silicon nitride tool properties are largely dependent on the amount and composition of the intergranular phase, particularly at high temperatures.

13.5.2.2   During the 1970s it was discovered that O^{2-} could be substituted for N^{3-} in the structure of the β–Si_3N_4 crystal whilst Al^{3+} could be substituted for Si^{4+} in suitable proportions which would maintain a balance of valencies. Depending upon the relative proportions of the reactants used the resultant solid compound, virtually a solid solution, can have a structure of either β–Si_3N_4 or α–Si_3N_4. These compounds are known collectively as 'Sialons' (an acronym on the chemical symbols of the principal ingredient elements: Si–Al–O–N).
    The β–sialons are of the general 'formula':

$$Si_{6-Z}Al_ZO_ZN_{8-Z}$$

where Z represents the degree of substitution of silicon and nitrogen by aluminium and oxygen respectively, and: $0 < Z \leqslant 4.2$. The α–sialons are of composition:

$$M_X(Si, Al)_{12}(O, N)_{16}$$

where $X \leqslant 2$ and M is a metallic ion and can be Li^{1+}, Be^{2+}, Mg^{2+}, Sc^{3+}, Ca^{2+}, or a lanthanide, but Y^{3+} is generally used.
    So far the β–sialons based on β–Si_3N_4 are those used in any quantity—as cutting tool materials for machining metals; tube- and wire-drawing dies; bearings; welding shields; nozzles; rock-cutting and coal-cutting plant; containers for conveying liquid metals and in turbine blades. A rupture modulus in the region of 850 MPa is obtained.
    The microstructure of a sialon consists of crystals of the sialon phase held in a glassy or partly crystalline matrix. The crystals can be either β–sialon

Fig. 13.13 The formation of sialon ceramics.

or a mixture of α– and β–sialons. Generally the hardness of the ceramic increases as the proportion of the α–sialon increases but the outstanding property of sialons is that they retain their hardness at much higher temperatures than does alumina. Toughness is also an important property though this is less than in a cemented carbide (14.2) of similar hardness. In the foundry where a combination of high-temperature and thermal-shock resistance is necessary sialons are used in the manufacture of small crucibles, impellors, thermocouple sheaths, radiant heater tubes and the like for use at temperatures up to 1250°C.

13.5.3 Zirconia (ZrO_2)   a hard refractory material which has been used for many years as a high temperature furnace and crucible refractory at temperatures up to 2500°C. It is a polymorphic compound existing in three different crystalline phases. The cubic form, stable above 2380°C changes below that temperature to a tetragonal form and on further cooling to below 1170°C a monoclinic form becomes stable:

$$\text{Cubic } ZrO_2 \underset{}{\overset{2380°C}{\rightleftarrows}} \text{ tetragonal } ZrO_2 \underset{}{\overset{1170°C}{\rightleftarrows}} \text{ monoclinic } ZrO_2$$

However, the final transformation (below 1170°C) is martensitic in nature and in fact shares some of the features of the austenite→martensite transformation (9.6.1) in steels. In particular the polymorphic change is accompanied by a sudden volume change of some 3%. Since zirconia is a brittle substance this sudden expansion causes severe cracking, making it useless to produce components from *pure* ZrO_2. However, additions of other metallic oxides will suppress the martensitic transformation to below ambient temperatures[*] and so stabilise the tetragonal or even cubic phase. Varying amounts of lime and magnesia were added for this purpose when zirconia was used as a furnace refractory but yttria (yttrium oxide, Y_2O_3) is now used as a stabiliser in modern zirconia ceramics.

13.5.3.1 Reference to the ZrO_2–Y_2O_3 phase diagram (Fig. 13.14) indicates that an addition of more than approximately 18% wt Y_2O_3 stabilises the cubic (c) structure at ambient temperatures producing a fully-stabilised zirconia (FSZ). Whilst the cubic phase is rather brittle such a structure would be free from cracking associated with volume changes (brought

[*] This is comparable with the effect of nickel in stabilising austenite at ambient temperature in steels.

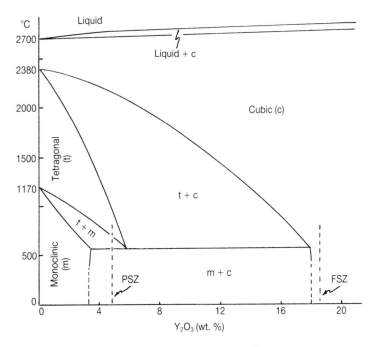

°C
Liquid

2700

Liquid + c

2380

2000   Cubic (c)

Tetragonal (t)

1500

1170   t + c

t + m

Monoclinic (m)

500   m + c

PSZ   FSZ

0

0   4   8   12   16   20

Y$_2$O$_3$ (wt. %)

Fig. 13.14   The zirconia-rich part of the ZrO$_2$–Y$_2$O$_3$ phase diagram.

about by temperature variations) such as would occur in unalloyed zirconia. Whilst such an 'alloy' has some electrical uses, e.g. in fuel cells, a partially stabilised zirconia (PSZ) is more useful as an engineering ceramic. Thus the addition of 5% wt Y$_2$O$_3$ will stabilise the tetragonal (t) phase at ambient temperatures as is indicated in Fig. 13.14. The t phase so stabilised is tough and reasonably shock resistant.

13.5.3.2   A PSZ is produced by sintering an *ultrafine* ZrO$_2$ powder with 5% wt Y$_2$O$_3$ at 1400–1500°C and, as Fig. 13.14 indicates, this will produce a tetragonal structure (fine-grained because of the ultrafine powder used). On cooling a proportion of these t grains transform martensitically to the monoclinic (m) form but the remainder persist in the *metastable* t form. Those grains remaining in the t state do so because elastic strains internally balanced in the matrix oppose further volume increase associated with the t→m (martensite) transformation.

Unlike the martensite transformation in steel which increases brittleness, the formation of the martensitic phase here has a toughening effect. The volume change associated with the t→m transformation results in the production of a pre-stressed material. Thus any crack being propagated through the structure is able to release the stresses in neighbouring grains—which immediately undergo the t→m transformation. Because the volume of m is greater than t from which it is derived the resultant volume increase generates compressive stresses near the tip of the crack and extra energy is required before the crack can travel further. It is believed, therefore, that the martensitic-type t→m transformation occurring at the crack 'front' is the principal energy-absorbing mechanism, which in turn increases the toughness of the material.

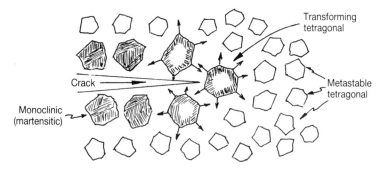

Fig. 13.15   The mechanism by which the tetragonal→monoclinic (martensitic) transformation toughens zirconia/yttria ceramics.

13.5.3.3   PSZs have a low thermal conductivity at all temperatures but a high coefficient of thermal expansion. Both of these properties make PSZs suitable as protective coatings on metals working at high temperatures such as superalloy rotor blades in jet turbines. The high coefficient of thermal expansion reduces the thermal expansion mismatch between PSZ coating and the metal substrate. A typical PSZ is 'Cerafine' containing 5% wt Y_2O_3.

The 'environmentally friendly' properties of zirconia when used *inside* the human body have led to its use in implantology. This biocompatibility, combined with a bending strength and toughness which are double that of the currently used alumina, have led to its adoption for hip joint replacements in both North America and France. An yttria-stabilised zirconia with the trade name *Prozyr* is used.

13.5.4 Some Very Hard Engineering Ceramics   Among cutting tool materials one must not forget the hardest and most effective of all—*diamond*. The structure of this allotrope of carbon has been mentioned (3.8.1). It is strong and hard because of the very stable directional covalent bonds which bind each atom to four near neighbours. Naturally occurring diamonds come mainly from Australia, Zaire, Botswana, CIS, and South Africa. Whilst these naturally occurring stones are superior as a tool material for some applications, synthetic diamonds now play a major rôle as an abrasive grit. Diamonds can only crystallise from the liquid phase, a situation only possible at extremely high temperatures and pressures. Several claims to have synthesised diamonds were made during the nineteenth century but it is doubtful if any of these were authentic. The development of high-temperature presses enabled the Swedish company

ASEA to succeed in 1953. Graphite capsules in the presence of metallic solvents were subjected to a temperature of 3000°C and a pressure of over 90 000 atmospheres in a hydraulic press. Technology has improved since then and the manufacture of this synthetic diamond material appears as a rough black grit—which could scarcely qualify as 'a girl's best friend'.

13.5.4.1 '*Cubic' boron nitride*, BN, is next only to diamond in terms of hardness. It is manufactured in a similar way to synthetic diamond and is consequently as expensive. Though less reactive with ferrous metals at temperatures above 1000°C it will transform to its softer hexagonal crystal form. Hence it is not used for high-speed machining but rather for dealing with hardened alloy steels and chill-cast irons.

13.5.4.2   Boron carbide, B_4C, is extremely useful as a ceramic armour material where its low relative density makes it valuable in aeronautical applications, e.g. battlefield helicopters and personal armour for aircrew.

13.5.4.3   Titanium diboride, TiB_2, has been developed for similar purposes in ground warfare since it has a higher relative density. It provides protection against heavy threats such as anti-tank projectiles. It is usually hot-pressed and has both high hardness and a high elastic modulus.

Glasses

13.6   The word 'glass' will immediately suggest to the reader the hard brittle stuff of which window panes and beer tankards are made but in scientific terms it has a wider significance. Glass comprises a range of substances from boiled sweets and toffee apples to some sophisticated industrial materials. One property all glasses have in common is that during manufacture they must be cooled quickly enough to prevent crystallisation from taking place so that the glassy (amorphous) state is retained at ambient temperatures. With orthodox glasses this is easy to achieve. For example the 'blank' from which a large optical telescope mirror is ground may be cooled at a rate of no more than 1°C per day (to avoid subsequent distortion arising from internal stress) yet it still retains its amorphous state. In an attempt to achieve a glassy state in metals, however, we must cool them at a rate in the region of 10^6°C per second. Between these extremes are many materials such as plastics which retain a glassy state during manageable rates of cooling but in many of which a mixture of both glassy and crystalline regions occurs (12.4.1).

   This situation exists because in both glasses (based on silica) and polymers the chemical 'unit' is very large. In polymers we have immense chain molecules in which atoms are held together by powerful covalent bonds whilst the molecules themselves are attracted to each other by weak van der Waals forces, the *total* effect of which is nevertheless considerable. In glasses large silicate networks, again covalently bonded internally, are held together by ionic bonds provided by metallic ions within the structure. In either case movement of these huge 'chemical units' with respect to their neighbours is difficult even when they are thermally activated and so it is easy to cool the material past its normal melting point (T_m) without crystallisation taking place. In a metal, however, the 'chemical unit' is very small viz. the atom—or more properly, the ion, which is strongly attracted to the surrounding very mobile 'cloud' of electrons. It therefore moves quickly and easily into an equilibrium position on reaching T_m so that cooling rates of $10^6\,^\circ$C/s or more must be used to prevent this from taking place.

If silica is melted and then cooled extremely slowly it will crystallise at a particular temperature, T_m, the normal melting point of silica and this process is accompanied by a reduction in volume as atoms take up regular positions in a crystalline lattice (broken line in Fig. 13.16.(i)). This regular

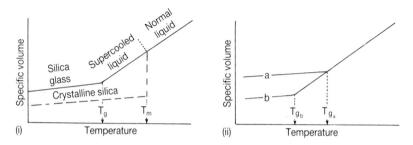

Fig. 13.16 (i) The glass transition (or fictive) temperature for silica glass; (ii) the effect of cooling rate on the glass transition temperature (cooling rate 'a' is greater than cooling rate 'b').

crystalline structure is depicted as a two-dimensional representation in Fig. 13.19(i). If, however, molten silica is cooled more rapidly then it is unable to attain the long-range order of a fully-crystalline structure since its atoms are covalently bonded to neighbours and are therefore unable to move as freely as those in the amorphous state of a molten metal or a molten salt. Nevertheless thermal contraction does take place due to limited rearrangement of the atoms, leading to their more efficient packing. This type of contraction is common to all liquid glasses (Fig. 13.16). With further reduction in temperature a point is reached where a sudden change in the coefficient of expansion occurs. Below this temperature, called the *glass transition temperature* (or *fictive temperature*), T_g (12.4.3), there is no further rearrangement of atom packing and such small amounts of contraction as take place are due solely to a reduction in the thermal vibrations of the atoms. T_g is not a well defined temperature and depends upon the prevailing cooling rate (Fig. 13.16(ii)). The more rapid the cooling rate, the higher is T_g.

Since the slope of the curve between T_g and T_m is the same as that above T_m this indicates that, during rapid cooling, there is no change in the structure at T_m, i.e. between T_m and T_g the structure is that of a super-cooled liquid. Moreover unlike the slowly-cooled silica at T_m there is no evolution of latent heat at T_g. Thus the *glassy state* prevailing below T_g is very like the liquid state.

This lower coefficient of expansion is roughly the same as that for *crystals* of similar composition to the glass (Fig. 13.16). In rigid crystalline materials contraction can only result from decreased thermal vibrations. Hence the term *glass* could well be defined as a substance having expansion characteristics similar to those indicated in Fig. 13.16. Other physical properties which are affected in a similar manner to the coefficient of expansion include density, electrical resistivity, optical properties and viscosity.

A glass will sometimes crystallise if given sufficient time—perhaps several hundred years. This process, known as *devitrification*, sometimes occurs with very old stained-glass windows and is generally manifested as a considerable increase in brittleness as crystallisation takes place. Glass does not always devitrify, of course, and some glass at least 3000 years old has retained its vitreous—or 'glassy'—properties. However, if molecular mobility is increased by raising the temperature the tendency for devitrification is encouraged. Devitrification may also be precipitated by the presence of impurity particles in the glass. Such nucleation of crystallisation occurs in a similar way in molten metals due to the presence of particules of slag or dross (3.7.2).

13.6.1 The Manufacture of Glass   Glass was one of the earliest man-made materials, though prior to the present technological age, its composition fell within a limited range. Ordinary commercial glass is produced from various inorganic oxides of which silica in the form of sand is generally the most important constituent. The principal ingredients of common soda glass are silica sand, lime (from limestone) and soda ash (crude sodium carbonate). Since glass is one of the easiest products to recycle a large amount of scrap glass, known as 'cullet', constitutes the charge to the 'tank' furnace where both raw materials and cullet are heated to 1590°C. The acidic silica then reacts with basic lime and soda to form the mixed silicates we know as glass. The gas-fired furnaces hold up to 250 tonnes of molten glass.

The furnace charge can take up to 90% cullet which is derived mainly from bottle banks. Unfortunately of the six billion glass bottles produced in Britain annually only one billion reaches the bottle bank—the rest end up 'in the bin'. Few industries are able to use scrap so efficiently and from the 'green' aspect it is a great pity so much potential cullet is wasted.

13.6.1.1 Sheet and Plate Glass   Early sheet glass was manufactured by blowing a cylinder which was then cut and, whilst still hot, opened out to form a reasonably flat sheet. In the modern Float Process (Fig. 13.17) the required mixture of silica and metallic oxides along with cullet is melted in an open-hearth furnace much like that used until recently for steelmaking. In fact the original Siemens brothers glass-making furnace of 1856 was adapted shortly after that date for steelmaking.

In the Float Process a continuous ribbon of molten glass up to 4 m wide

Fig. 13.17   The principle of the float-bath for the manufacture of flat glass.

flows from the melting furnace and floats along the surface of a bath of molten tin. The speed of the ribbon is adjusted so that it is held at a high temperature for long enough for any irregularities to be removed so that the surface becomes flat, smooth and parallel. To prevent oxidation of the tin and consequent contamination of the glass the 'float bath' operates under a nitrogen atmosphere. The glass cools sufficiently before it leaves the float 'tank' so that it is hard enough to be drawn into the annealing *lehr* without the use of rollers which might mark the surface. The main purpose of the annealing lehr is to ensure slow, uniform cooling, thus limiting internal stresses which would arise from rapid cooling. When allowed to flow under normal conditions imposed by such factors as density and surface tension the thickness of the glass produced is about 6 mm but working conditions can be adjusted to give thicknesses between 2.5 mm and 25 mm.

An important feature of the Float Process is that modifications of the suface properties of the glass can be carried out during the process. Metallic ions such as those of copper can be made to replace existing metallic ions by passing an electric current between the molten tin bath and an electrode above the surface of the glass. Typical products of this *electro-float process* are anti-glare glass and glass which reduces heat transmission.

13.6.1.2 Glass Blowing   The development of modern glass blowing began in Venice in the twelfth century and by 1292 it had become necessary for the industry to seek seclusion on the nearby Island of Murano to avoid industrial espionage. By the middle of the nineteenth century Flemish glass blowers were settling here in the Midlands of England. At that time my great-grandfather, as a glass-blower, was earning a wage some ten times the average for those days. The need for skill and powerful lungs is now no more and the blowing of bottles, jars and other holloware is an automated process using metal moulds. Usually a mixture of pressing and blowing operations is employed (Fig. 13.18) to shape a container from the initial 'gob' of molten glass. Mouth blowing is now restricted to the manufacture of artistic glass and high quality tableware.

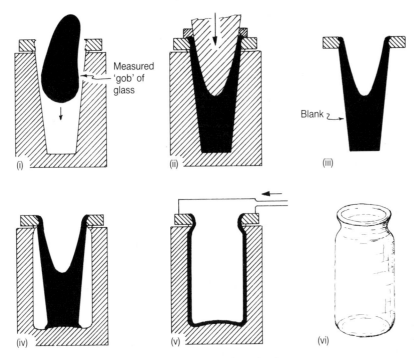

Fig. 13.18 A 'press-and-blow' method for producing a jam jar.

The Structure of Glass

13.6.2 As we have seen glasses differ from solid crystalline substances in respect of a number of properties. They are plastic at high temperatures and rigid at low temperatures but under normal manufacturing conditions they do not crystallise. Thus glass is in a condition which is continuous with, and analogous to, the liquid state of the substance but which, as a result of having been cooled from a fused state, has acquired a high degree of viscosity so as to be, for most practical purposes, rigid.

13.6.3 The structure of *crystalline* silica, quartz, is fundamentally tetrahedral (Fig. 13.3) as determined by X-ray crystallographic methods. In this three-dimensional structure each silicon atom is chemically bonded to four oxygen atoms placed at the corners of the tetrahedron, each oxygen atom being shared with an adjacent silicon atom from another tetrahedron. However, for the sake of clarity we will consider a two-dimensional adaptation of this structure as shown in Fig. 13.19(i). Here the structure is one of long-range order, that is, a regular lattice structure is apparent. If, however, fused silica is cooled at such a rate as to make normal crystallisation impossible, a structure such as is suggested in Fig. 13.19(ii) is obtained

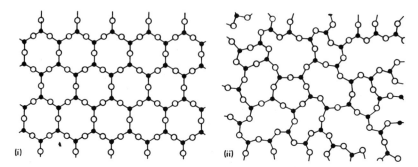

Fig. 13.19 A two-dimensional representation of the structure (i) crystalline silica (quartz) and (ii) silica glass. Since the diagram is two-dimensional the silicon atoms are shown bonded to only three oxygen atoms instead of four as in Fig. 13.3. In (i) there is long-range order whilst in (ii) only short-range order prevails, i.e. individual silica tetrahedra themselves.

for silica glass. Here the long-range order is replaced by a short-range one in which individual tetrahedra still persist as they do in crystalline quartz, but the random orientation of individual tetrahedra, resulting from local distortions, destroys the periodicity of the lattice network and the structure is no longer crystalline.

When metallic oxides such as soda (Na_2O) and lime (CaO) are added to the silica the ratio of oxygen ions to silicon ions is increased because one Si^{4+} ion must be removed for every *four* Na^{1+} ions added if valency bonds are to be satisfied. The extra oxygen ions cannot be incorporated into each of two tetrahedra and as a consequence the network loses its continuity in regions where these ions are present. The structure is therefore less rigid at high temperatures and soda glass is more easily worked than pure silica glass. Na^{1+} and Ca^{2+} ions take up suitable positions available within the structure (Fig. 13.20). Since Na^{1+} ions are relatively small they are able to migrate through the comparatively open network with relative ease even at room temperatures. This influences some of the properties of soda glass.

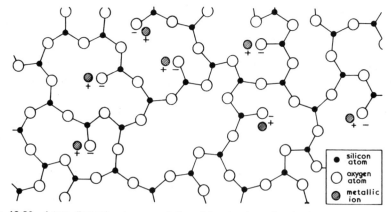

	silicon atom
●	silicon atom
○	oxygen atom
◍	metallic ion

Fig. 13.20 A two-dimensional representation of the structure of soda glass. Again the silicon atoms are shown bonded to only three oxygen atoms.

13.6.4 Special Glasses   Ordinary soda glass is of little use in coping with sudden temperature changes. If boiling water is poured into a cold soda-glass jar the latter usually shatters because soda glass has a relatively high coefficient of expansion but a low thermal conductivity. Differential expansion between the inner and outer layers of the jar sets up high stresses with which soda glass is unable to cope and fracture occurs. Unlike soda glass, temperature resistant glasses such as 'Pyrex' have only small amounts of 'free ions' (Na^+). Instead up to 20% of 'glass formers' such as boron oxide, B_2O_3, are added to the silica. These glass formers increase the network formation and continuity within the structure. In turn this leads to higher viscosity, resistance to chemical attack and, most important, a very *low* coefficient of expansion because the network is not disrupted. The low coefficient of expansion makes 'Pyrex' suitable for kitchen, laboratory and industrial uses as the tendency to shatter with sudden temperature change is much reduced. In modern glass-ceramics, controlled devitrification is employed to initiate large numbers of nuclei for crystallisation in glasses of suitable composition to produce polycrystalline materials. After shaping, the material is given a two-stage heat-treatment. First, a low-temperature treatment is given to promote the formation of large numbers of crystal nuclei within the glass and this is followed by a higher temperature treatment at which crystal growth rate is at a maximum. This continues until the crystalline content may reach 90% and only a little glass remains. The mechanical properties of such a glass-ceramic are intermediate between those of a glass and of a true ceramic such as Al_2O_3 or MgO. They are used in cooking ware (e.g. 'Pyroceram' and 'Pyrosil'), heat exchangers and telescope mirror blanks. Other glass-ceramics based on magnesia-alumino-silicates and nucleated to give crystals of cordierite and crystobalite (Fig. 13.12) are used in missile nose cones and heat exchangers. Ceramics of this type are not only resistant to thermal shock but are transparent to radar signals.

13.6.5 The Mechanical Properties of Glass   Metals can be deformed because slip, i.e. the movement of dislocations, can occur readily along such crystallographic planes as are available. However, glass cannot deform in this way because it has no crystal lattice. Only viscous flow can take place in glass and this only by the thermally assisted movement of molecules past each other. But since a glass at ambient temperature is, by definition, below T_g, very little thermally assisted movement of molecules is possible. Thus if stress is applied, molecules are unable to move significant distances relative to one another making glass brittle at ambient temperatures.

The rate of viscous flow is dependent mainly upon the prevailing temperature but is also dependent upon the composition and structure of the glass. Small applied stresses cause the more highly strained bonds within the structure to be ruptured. The bond rearrangement which follows this process leads to a small degree of permanent strain. As this process continues more bonds become strained in turn to the extent where they will break and so allow viscous flow to continue. As would be expected the

Table 13.3 *Compositions and Uses of Some Glasses*

Type of glass	Approx. composition (%)	Properties and uses
Soda glass	72 SiO_2 15 Na_2O 9 CaO 4 MgO	Windows, plate glass, bottles and other containers
Lead glasses	60 SiO_2 15 $Na_2O + K_2O$ 25 PbO	Has a high electrical resistance—used for lamps, valves, etc.
	40 SiO_2 7.5 $Na_2O + K_2O$ 47.5 PbO 5 Al_2O_3	High refractive index and dispersive power—used for lenses and other optics. Also for 'crystal' glass tableware
Borosilicate glass	70 SiO_2 7 $Na_2O + K_2O$ 3 Al_2O_3 20 B_2O_3	Low expansion and good resistance to chemicals—used for heat-resistant ware ('Pyrex') and laboratory apparatus. Can 'seal' to a number of metals
Aluminosilicate glasses	35 SiO_2 30 CaO 25 Al_2O_3 5 B_2O_3 5 MgO	Has a high softening range with T_g as high as 800°C
Soft solder glass	5 SiO_2 15 B_2O_3 64 PbO 16 ZnO	Melting point below 400°C
Silicon-free glass	36 B_2O_3 10 MgO 27 Al_2O_3 27 BaO	Sodium vapour discharge lamps
'Lindemann glass'	83 B_2O_3 15 Li_2O 2 BeO	Low absorption of X-rays since all cations (Be^{3+}; Li^{1+}) and B have low atomic masses
High-cadmium glass	26 SiO_2 64 CdO 2 Al_2O_3 8 CaF_2	Absorbing neutrons
'Vycor'	96 SiO_2 3 B_2O_3 1 Na_2O	Very low coefficient of expansion—missile nose cones and windows of space vehicles

stresses necessary to cause a prescribed degree of viscous flow are less at elevated temperatures since bond-strength becomes smaller at high temperatures.

The amount of viscous flow in glass at ambient temperatures is extremely small since it requires thermal assistance. However, glass will flow very slowly under its own weight and a number of cases have been reported where large plate-glass windows have become measurably thicker at the lower edge over a period of years.

There is little in common between the mechanical properties of glasses and metals when under the action of applied forces. Such differences can be classified as follows:

(i)   under short-time testing methods, glasses are brittle at ambient temperatures. They are elastic right up to the point of fracture and fail without any previous yield or plastic deformation. Even the most brittle of metals show some plastic flow;

(ii)   although an external load may be applied in compression, failure in glass always results from a tensile component of the stress. The strength of such materials can therefore best be described in terms of tensile strength;

(iii)   the time for which a static load is applied has a great influence on the strength of glass in the long term. Thus the extrapolated infinite-time modulus of rupture for glasses is usually between a third and a half of that for short-time loading.

13.6.6 The Heat-treatment of Glass   Since glass is a poor conductor of heat it follows that during shaping a considerable temperature gradient may be set up across the section of the work piece. Consequently its surface, since it cools more rapidly, will cool to below T_g and become relatively stiff while the interior, still at a higher temperature, remains in a more mobile liquid state above T_g. The slower cooling rate is also related to a lower T_g value (Fig. 13.16(ii)) accompanied by large volume changes during the liquid→glass transition, so that during cooling the tendency is for the contraction of the core to be greater than that for the surface. This will inevitably lead to the formation of internal stresses which may result in failure either during final stages of manufacture or in subsequent service. Therefore an annealing process is necessary at a temperature high enough to allow sufficient movement of molecules into positions approaching equilibrium. Failure to anneal glass leaves it weak and brittle.

13.6.6.1 Annealing   This is usually carried out immediately after the shaping process. The glass—whether plate glass, bottles or jam jars—is passed slowly through a conveyor-hearth type of furnace known as a *lehr*, where the entry temperature is near to T_g and the exit temperature little above ambient. The small amount of viscous flow which takes place during annealing leads to stress relaxation within the glass which is thus rendered relatively stress-free. This treatment bears some similarity to the low-temperature stress-relief annealing given to some cold-worked metals and alloys such as 70–30 brass in that atoms move short distances into positions nearer to equilibrium. At the same time it contrasts sharply with those annealing processes which lead to complete recrystallisation in metals.

13.6.6.2 Tempering   Glass is sometimes toughened by a tempering process which seeks to reduce the formation of surface cracks by putting the surface in a state of compression. The principle involved is similar to that applied in the pre-stressing of concrete in order to prevent cracks from developing in its surface (15.10.2). Tempering of glass is achieved by heating it to much the same temperature as that used for annealing it, and then cooling the surface rapidly by means of air jets. In common with other substances glass contracts as it cools. Initially the outside surface cools, contracts and hardens more quickly than the inside. The latter, being still soft, will yield in the early stages of cooling. During the final stages of cooling, however, both inside and outside behave in an elastic manner but their contractions are already out of step. Hence, when the glass has finished cooling the outer surface is in a state of compression whilst the inside is in tension. Since the surface layers are in compression these compressive forces will balance any moderate tensile forces to which the glass may be subjected during service. Glass invariably fails as a result of tensile components of forces even when these forces are, overall, compressive. Glass thus treated has been used for a long time in motor-car windscreens and is likely to be developed in the future for many industrial uses.

Strain patterns in transparent materials such as 'perspex' and glass become visible when viewed by polarised light and to a lesser extent by white light which becomes partly polarised when reflected from non-metallic surfaces. For this reason it is often possible to see the pattern of air-jet impingements used for cooling the glass when a car windscreen is inspected at an oblique angle or, preferably, by using polarising spectacles.

13.6.7 Optical Fibres   In recent years the well-known electronic systems of communications have been unable to deal with the ever-increasing volume of information to be transmitted. Hence much higher transmission frequencies were required so that communications engineering moved into the visible spectrum where the movement of photons (unit of light energy) is used instead of the movement of electrons. The actual fibres consist of a pure silica tube containing a core of silica which is 'doped' with another oxide such as germanium oxide, GeO_2, to give a slightly higher refractive index. (Germanium is a 'metalloid' member of the carbon-silicon Group IV so it can be expected to 'fit' into the structures of glasses based on silica—Ge of course has a valency of 4.) The light pulses are guided along the core by internal reflection at the interface between the core material and the pure silica coating. For light to be transmitted over a significant distance the core material must have a very high transparency and consequently be of a very high purity so that light pulses could be transmitted over kilometres rather than metres; but this was not the only problem encountered in the early stages of development of optical fibres.

As Fig. 13.21 indicates, in a single fibre system containing a core of uniform refractive index, light can travel by a variety of different paths.

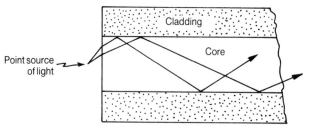

Fig. 13.21   Since the angle of incidence can vary there are many paths by which light can be transmitted.

This will lead to distortion of the signal since the light pulses reach their destination at different intervals. This problem is overcome by using a core material in which the refractive index varies from a maximum at the centre of the core to that of the cladding at the interface in a near-parabolic manner. Then, those light rays taking a longer path travel faster so that all rays reach their destination simultaneously. Cores vary in diameter between 5 and 50 μm and fibres are used in the form of cable systems. Research is directed at reducing the loss of optical signal which, since 1970, has already received a twenty-fold reduction.

It is more than 100 years since Tyndall demonstrated to the Royal Society that light could be guided by a jet of water using internal reflections. Now optical fibres using this principle enable us to see round corners and consequently have many uses. Medical endoscopes are used to make visual examinations of hollow organs such as stomach and bowel, whilst in natural history studies similar instruments are available to peep into—and photograph—the nests of ants and termites.

13.6.8 Metallic Glasses   The difficulty involved in producing cooling rates high enough to achieve the glassy state in metals was touched on at the beginning of this section (13.6). Nevertheless various ingenious methods have been developed to attain the high cooling rates of $10^6\,°C/s$ or so. Perhaps the most successful has been some form of the 'melt spinner' (Fig. 13.22). This consists essentially of a copper wheel rotating at a peripheral speed of up to 50 m/s and a jet of molten metal projected under pressure at the rim. This forms a 'pool' which freezes very quickly and is separated from the wheel by centrifugal force, forming a continuous ribbon which is wound onto a spool (Fig. 3.23).

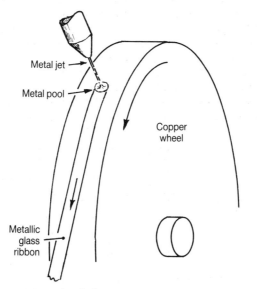

Fig. 13.22   Principles of the melt-spinning process.

Fig. 13.23 The production of thin ribbon of 80Fe–20B metallic glass by a melt-spinning process (Allied Signal Inc., USA). The amorphous ribbon is produced at a speed of 90 km/h.

The metallic glasses formed by this process are usually alloys which fall into two categories:

(i) metal-metal alloys in which one or both elements is a transition metal, e.g. Cu(50)–Zr(50); Ni(60)–Nb(40);

(ii) metal-metalloid (or non-metal) alloys where the metal is usually a transition metal, e.g. Pd(80)–Si(20); Fe(80)–B(20); Fe(40)–Ni(40)–P(14)–B(6).

The type of structure envisaged in a metallic glass is of a *dense random packing* of spheres such as would be obtained by tipping steel balls into a flexible bag. Since 80 metal–20 metalloid mixtures seem to be most effective in producing metallic glasses, it is supposed that in this dense packing the smaller metalloid atoms are surrounded by larger metallic atoms.

Some of these metallic glasses are strongly ferromagnetic being extremely 'soft' magnetically with very high permeability. Since they are also very strong mechanically, materials such as Fe(40)–Ni(40)–P(14)–B(6) are used in the form of fabric (woven from ribbon) as a magnetic shielding material. Minature transformers and magnetic cores are also based on some of these glasses.

Though metallic glasses are very strong and stiff they are capable of only very limited plastic deformation. In tension this is usually less than 1% at fracture. Probably the best known is Fe(80)–B(20) (18.6.8) which is stronger than the best carbon fibre and almost as stiff. The mechanical properties of these fibres have yet to be fully explored.

14

Composite Materials

14.1 New jargon appears frequently in the field of technology. The vocabulary of the materials scientist was enriched by the term 'composite materials' a few decades ago. Nevertheless the principles of these materials are not new. The structure of a massive tree, one of the largest and oldest living things on Earth and which has evolved over a period of millions of years, is an example of what we now call a 'fibre-reinforced composite'. As recently as the 1940s plastics material reinforced with glass fibre was based on the natural structure of bamboo.

Composite materials—or 'composites'—can be divided into two main groups.

14.1.1 Particle Composites in which some matrix material is 'filled' with another material in particle form. Such a composite is designed to fulfil one or more of the following functions:

(i) to provide dispersion of a very hard material in a tough, shock-resistant matrix as in cemented carbides for cutting tools;
(ii) to furnish a material of increased tensile strength, e.g. the use of finely-dispersed Al_2O_3 in metallic aluminium;
(iii) to reduce overall costs of some material by 'bulking up' with a cheap, low-grade 'filler'. Thus bakelite mouldings have long been 'filled' with sawdust, wood flour or minerals such as fine sand or ground limestone.

14.1.2 Fibre-reinforced Composites in which the object is to produce a material of high specific strength or modulus. This is the more important type of composite from an engineering point of view and will be discussed more fully in the next chapter.

14.1.3 If a material contains particles these suffer elastic strains when the material is stressed. In this way the particles contribute to the load-carrying capacity of the material as well as providing obstacles to the movement of dislocations, assuming that the particles themselves are strong. Thus if the volume of such strong particles in the composite is proportionally large they will provide a high strength by producing a high load-carrying capacity. The most satisfactory method of increasing tensile strength is to use the particles in the form of long fibres. When stressed the matrix may begin to flow but in doing so will cause a force to be set up at the surface of the fibre. If the fibre is long enough this transmitted force will ultimately lead to its fracture and the fibre will therefore have contributed fully to the strength of the composite material. Obviously strength will be directional, i.e. at a maximum value parallel to the direction of the fibres.

14.1.4   The nature of the interface between the particles or fibres and the matrix has a bearing on the extent to which the load will be transferred from the matrix to the strengthening material. Cohesion at the interface may be achieved by one or more of the following methods:

(i)   *mechanical bonding*   this involves a high enough coefficient of friction acting between the surfaces. This applies mainly to some fibre-reinforced composites e.g. the fibres in a simple hemp rope;
(ii)   *physical bonding*   depending upon van der Waals forces acting between surface molecules;
(iii)   *chemical bonding*   at the interface—this can, however, give rise to weak brittle compounds in some instances;
(iv)   multiple bonds formed by solid solution effects.

14.1.5   Generally speaking sophisticated engineering composites fall into three main groups:

(i)   those in which the matrix is relatively weak and brittle but is reinforced by fibres or platelets of high tensile strength. Such materials are useful when high tensile strength and stiffness are required as in a fishing rod or the blades of a turbine rotor;
(ii)   materials in which a tough and ductile matrix carries the second component in the form of hard spherical particles. These composites are characterised by great hardness and high compressive strength and are represented by tool and die materials;
(iii)   dispersion-hardened materials in which the strength of the matrix is increased by the action of microscopical particles in impeding the movement of a 'dislocation front'. Since such dislocations cannot pass *through* such particles—assuming that they are harder and stronger than the matrix—'loops' are formed (Fig. 14.1). These dislocation loops then act as further barriers to the progress of any dislocation fronts which may follow.

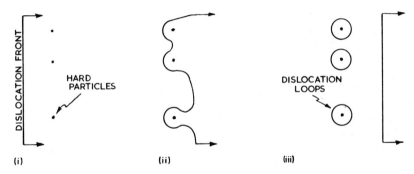

Fig. 14.1 The formation of dislocation loops (dispersion hardening).

Particle-hardened Composites

14.2 These are commonly materials in which the ductility and toughness of a metallic matrix is combined with the hardness and strength of ceramic particles. Usually known as *cermets* these materials are widely used as cutting-tool tips, drilling bits, friction materials for brakes and in the compounding of fuel elements for nuclear power production. The principles of particle hardening were widely investigated in the production of materials for use in gas turbines. Such materials require a high creep-resistance at high temperatures. Unfortunately cermets proved to be generally inferior to 'Nimonic' alloys (11.5.2) in respect of relative brittleness.

Cemented carbides are possibly the best known materials of this type and have been used for many years. They consist of hard particles of tungsten or titanium carbide in a matrix of tough, ductile cobalt. The product is manufactured by compressing a mixture of carbide powder and cobalt powder to the extent where cold-welding occurs at points between the cobalt particles. This process is followed by 'sintering'—that is, heating the component at some temperature above that of recrystallisation for the cobalt so that a continuous tough matrix of cobalt is formed (6.5.1).

14.2.1 Cermets are also made by causing a liquid metal to infiltrate around particles of a solid ceramic. Ideally a strong positive bond should be formed at the metal/ceramic interface. This may be achieved either by a chemical reaction between metal and ceramic, resulting in the formation of a film of intermetallic compound bridging the two phases, or by a simple solid solution of the metal and ceramic forming at the interface. Since intermetallic compounds are generally weak and brittle solid solution formation would seem to be the more effective.

In order that any form of bonding may occur it is necessary for the molten metal to wet the surface of the ceramic. This wetting ability is related to the surface tension of the liquid and to the surface energy of the ceramic, and is a function of the angle of contact, θ, between liquid and

Fig. 14.2   Wetting of a ceramic surface. (i) Good wettability. (ii) Poor wettability.

solid (Fig. 14.2). The relation between the surface energies is given by the equation:

$$\gamma_{SV} - \gamma_{SL} = \gamma_{LV} \cos \theta$$

Wettability is at a maximum in the ideal case, viz. when $\theta \to 0$, but this state of affairs is never approached in practice. The quality and nature of the solid surface is generally the limiting factor and films of oxide or other matter cause a progressive increase of θ and a consequent reduction of adhesion energy between the surfaces.

Typical cermets are based on hard carbides, oxides, borides or nitrides bonded with a suitable tough strong metal. The main groups of cermets are listed in Table 14.1.

Cermets may contain up to 80% of the cermic phase with the remainder bonding metal.

Dispersion-hardened Materials

14.3   Some composite materials containing particles which are roughly spherical in form are developed for tensile strength rather than for hardness. Of primary importance in dispersion strengthening is alumina, Al_2O_3, in the form of very small particles dispersed throughout a suitable matrix. Possibly the easiest material in which to achieve this state of affairs is aluminium itself. If aluminium is produced in a powdered form and then ground, in the presence of oxygen under pressure, much of the surface film of oxide so formed disintegrates and forms very fine powder intermingled with the aluminium particles. The mixture is then sintered by powder-metallurgy techniques to give a more or less homogeneous aluminium matrix containing about 6% of alumina particles. This product is known as 'sintered aluminium powder'—or SAP—and shows a definite advantage over ordinary aluminium in terms of tensile strength particularly at high temperatures. The alumina has a dispersion-hardening effect (14.1.5).

The high strength attained in some aluminium alloys as a result of precipitation hardening (11.3.2) is unfortunately lost at high temperatures because of the formation of large particles of non-coherent precipitates. In this respect SAP shows an advantage in retaining strength at relatively high temperatures (Fig. 14.3). Alumina particles are utilised for the dispersion strengthening of other materials, in particular silver and nickel.

14.3.1   Conventional melting/casting techniques are unsuitable for producing dispersion-hardened materials. Since the particle mineral—or 'dis-

Table 14.1 *Principal Cermet Groups*

Group	Ceramic	Bonding metal or alloy	Uses
Carbides	Tungsten carbide WC Titanium carbide TiC Molybdenum carbide Mo_2C Silicon carbide SiC Chromium carbide Cr_3C_2	Cobalt Molybdenum, cobalt, tungsten or nickel–chromium Cobalt or nickel Cobalt or chromium Nickel	Cutting tool bits fabricated by powder-metallurgy Gauge blocks, slip gauges, wire-drawing dies
Oxides	Aluminium oxide Al_2O_3 Magnesium oxide MgO Chromium oxide Cr_2O_3 Uranium oxide UO_2	Cobalt, iron or chromium Magnesium, aluminium, cobalt, iron or nickel Chromium Stainless steel	Notable for their high-temperature properties—rocket motor and jet engine parts. Also 'throw-away' tool bits. Nuclear fuel elements
Borides	Titanium boride TiB_2 Chromium boride Cr_3B_2 Molybdenum boride Mo_2B	Cobalt or nickel Nickel Nickel or nickel–chromium	Mainly as cutting tool tips

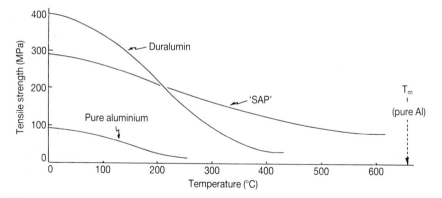

Fig. 14.3 The relationship between tensile strength and temperature for pure aluminium, duralumin and sintered aluminium powder (SAP).

persoid'—is usually of low relative density it will float to the surface of the molten matrix alloy giving a non-uniform composite structure. A process known as 'mechanical alloying' is under development in which both dispersoid and matrix, in powder form, are treated in a high-speed ball mill. Here the high-energy impact of hard steel balls upon the powder particles causes a degree of 'mechanical alloying' between the particles. Final consolidation and shaping is by orthodox powder metallurgy or extrusion.

14.3.2 Early dispersion-strengthened high-temperature alloys contained thorium oxide, ThO_2, as the dispersoid but this mildly radioactive substance has been largely replaced by yttrium oxide, Y_2O_3, in modern superalloys used in aerospace. A feature of such materials is that high strength is retained at high temperatures approaching T_m.

Both Al_2O_3 and Al_4C_3 are used as dispersoids in some aluminium alloys as strength derived from 'mechanical alloying' is often superior to that obtained by heat treatment (precipitation hardening). Moreover strength is still maintained at temperatures at which precipitation-hardened materials would soften. Precipitation-hardened aluminium–lithium alloys (11.3.4) have been developed to reduce the relative density of strong alloys for the aerospace industry but improved properties are obtained by using Li_2O as a dispersion hardener instead since there is always the danger of introducing brittleness during precipitation treatment due to the formation of brittle films of intermetallic compound. The dispersion hardened alloy Inco MAP-Al 9052 (Table 14.3) offers 8% lower relative density and 10% greater stiffness than the equivalent conventional precipitation-hardened alloy. Both alloys in Table 14.3 are basically simple non-heat-treatable corrosion-resistant alloys (11.3.1) which are dispersion hardened by the presence of Al_2O_3, Al_4C_3 and, in the second alloy, Li_2O.

Table 14.2 Some Dispersion-strengthened Alloys for Aerospace Applications

Alloy	Composition (%)								Uses
	Ni	Fe	Al	Cr	C	Y_2O_3	Ti	Other	
Incoloy MA 956	—	Bal.	4.5	2.0	—	0.5	0.5	—	Combustion chamber, afterburner, turbine casing sections (where resistance to creep, oxidation and attack by sulphur gases is necessary)
Incoloy MA 754	Bal.	1.0	0.3	20	0.05	0.6	0.5	—	Brazed nozzle guide vane and band assemblies (US military aero-engines)
Incoloy MA 6000	Bal.	—	4.5	15	0.05	1.1	2.5	W –4.0 Mo –2.0 Ta –2.0 B –0.01 Zr –0.15	First and second stage turbine vanes and blades (machined from solid bar); forged aerofoil components

Table 14.3 Dispersion Hardened Aluminium Alloys

Alloy	Composition (%)					Typical mechanical properties					Uses
	Al	Mg	Li	C	O	Relative density	0.2% yield strength (MPa)	Tensile strength (MPa)	Elongation (%)	Young's Modulus (GPa)	
Inco MAP Al 9052	Bal.	4	—	1.1	0.6	2.66	380	450	13	71	Military aerospace uses where a combination of light weight, high strength and corrosion-resistance is required
Inco MAP Al 9052 XL	Bal.	4	1.3	1.1	0.6	2.58	448	517	9	78	Developed for air-frame structures

Composite Bearing Materials

14.4 A bearing material requires good wear-resistance and a low coefficient of friction combined with compressive strength, toughness and rigidity. As far as metallic bearings are concerned these properties do not occur together in a single-phase alloy. A solid solution, though tough and strong, has a poor wear-resistance and often a high coefficient of friction whilst an intermetallic compound, though hard and rigid and having a low coefficient of friction, is brittle and weak. Thus a common bearing alloy—such as a Babbitt metal (11.7.1)—has a duplex structure consisting of hard particles of an intermediate phase embedded in a tough solid solution. The latter wears, leaving the particles of compound standing in relief and providing the low-friction bearing surface, whilst the wearing of the solid solution matrix provides channels which assist the flow of liquid lubricant traditionally used in orthodox engineering practice. Friction between surfaces is due largely to adhesion and this is at a maximum with metals which form a solid solution in each other. Often wear-resistant surfaces exhibit high coefficients of friction and only a few solid materials give both low friction and resistance to wear. Chief of these are graphite and some other lamellar solids and some polymers.

14.4.1 A large number of modern mechanisms particularly in aerospace engineering, require *solid* lubrication in some form. This may be due to the fact that liquid lubricants are unstable at high temperatures or that their transmission to the working surfaces may be erratic or, alternatively, expensive to promote. Some materials such as graphite are intrinsically low-friction substances but the most successful method of using the solid lubricant is to incorporate it in some form of composite so that the lubricant is in effect 'built in'. Metallic or polymeric matrices containing either lamellar solids (graphite or molybdenum disulphide) or polymers (such as PTFE) are the most widely suggested.

  Sintered bronze bearings containing small amounts of graphite have been used for many years as have various mixtures of graphite and metals employed as brush materials in electrical machinery. The function of graphite as a lubricant depends upon its lamellar structure (3.8.2). Since only weak van der Waals forces operate between the layered structure of carbon atoms the latter slide over each other with ease and provide the lubrication effect. Other solids of lamellar structure include the sulphides, selenides and tellurides of molybdenum, tungsten, niobium and tantalum. Of these the best-known is molybdenum disulphide which has been used as a high-pressure lubricant for some time. A number of MoS_2/metal composites have been used but possibly the most successful so far are those containing MoS_2 bonded with up to 16% iron and 4% platinum. MoS_2/nickel composites give possibly the best results in terms of low friction ($\mu = 0.03 - 0.2$) though nickel is rather a poor binder. Composites containing calcium fluoride, CaF_2, as the solid lubricant in a nickel/chromium matrix are also promising, having a coefficient of friction of about 0.2 and a working temperature up to 800°C.

The coefficient of friction of PTFE (Table 12.2) is lower than that of any other solid material ($\mu = 0.03 - 0.1$) but its mechanical properties as a single-bearing material are poor. For this reason it is best used in a composite. Bronzes impregnated with PRFE are in use though the maximum working temperature of such bearings is limited by PTFE to about 300°C. Nylon is also widely used as a bearing material but its main disadvantage is its dimensional instability resulting from absorption of water.

Cement, Mortar and Concrete

14.5   The Gallerus Oratory in the south-west of Ireland is an example of dry stone walling, being built entirely without the use of mortar. After some 1500 years it is still perfectly preserved since there is no mortar to deteriorate. Mortar had, however, been used by the Egyptians some 4000 years earlier in building the pyramids. This mortar was based on lime containing a little gypsum. Then, as now, these mortars contained a cement which when mixed with water to form a paste, set and progressively hardened with the passage of time.

Limestone, $CaCO_3$, is a cheap plentiful material which, when 'calcined' at a high temperature forms *quicklime*:

$$CaCO_3 \longrightarrow \underset{\text{Quicklime}}{CaO} + CO_2$$
Limestone

When this quicklime is mixed with water—or 'slaked'— calcium hydroxide or 'slaked lime' is produced:

$$CaO + H_2O \longrightarrow \underset{\text{Slaked lime}}{Ca(OH)_2} + heat$$

Slaking is accompanied by a considerable evolution of heat and an increase in volume. The slaked lime subsequently hardens slowly as it reacts with carbon dioxide from the atmosphere:

$$Ca(OH)_2 + CO_2 \xrightarrow{H_2O} CaCO_3 + H_2O$$

This reaction is accelerated when an excess of water is present in the cement. Since both carbon dioxide and calcium hydroxide are soluble in water a quicker reaction results than if a direct gas/solid reaction were involved. As $CaCO_3$ is much less water-soluble than $Ca(OH)_2$ it precipitates in crystalline form and the interlocking crystals of $CaCO_3$ convert the soft mass to a hard solid.

14.5.1 Lime Mortar   obtained by mixing slaked lime with clean sand and water was the principal building adhesive before the advent of Portland cement. It is still used in situations where a slow hardening process is admissible.

14.5.2 Portland Cement Lime-based cements were developed by the Romans by the addition of volcanic ash (pozzuolana) which produced a quick-hardening, strong hydraulic cement. The success of this cement is evident in the Colosseum and the various aqueducts still in excellent preservation. There was little further development in cement technology until 1756 when John Smeaton was engaged in building the Eddystone Lighthouse. He required a quick-setting cement which would not be attacked by sea water. By 'burning' a mixture of limestone and clay he produced a suitable hydraulic cement. This material was perfected in 1824 when Joseph Aspdin heated the limestone–clay mixture so that it formed a hard 'clinker'. This was then ground to a fine powder giving the product we now know as 'Portland Cement', so named by its inventor as 'a cement of superior quality resembling Portland stone' (a white limestone of the island of Portland). Portland cement is the principal building cement in use. It is essentially a calcined mixture of lime-bearing and clay-base materials. The lime-bearing material may be ordinary limestone or even dredged marine shells, whilst the clay-bearing material comprises suitable clays or shales. These materials are pulverised and then mixed in the correct proportions prior to being calcined in rotary kilns at about 1500°C. The resultant clinker is then ground with small amounts of gypsum and water and a greenish-grey powder—the well-known Portland cement—is the result.

Whilst the composition of this cement can vary, the most widely used formulation is given in Table 14.4.

Table 14.4

Constituent	Symbol*	Weight (%)
Tri-calcium silicate ($3CaO \cdot SiO_2$)	$C_3.S$	45
Di-calcium silicate ($2CaO \cdot SiO_2$)	$C_2.S$	28
Tri-calcium aluminate	$C_3.A$	10
Tetra-calcium alumino-ferrite ($4CaO \cdot Al_2O_3 \cdot Fe_2O_3$)	$C_4.A.F$	8
Gypsum ($CaSO_4$)	—	3
Magnesia (MgO)	M	3
Calcium, sodium and potassium oxides	—	Balance

*Cement scientists usually designate each constituent oxide by a single letter, i.e. CaO = C; SiO_2 = S; Al_2O_3 = A; Fe_2O_3 = F; H_2O = H. This sytem represents meaningful chemical 'shorthand'. In passing, note that an *empirical* formula written as Ca_3SiO_5, which would indicate the ratio of actual ions of each element in tricalcium silicate, would be meaningless to a professional chemist since it does not show the manner in which tricalcium silicate is constituted.

14.5.3 When mixed with water a workable paste is produced and the mixture remains in this state during a dormant period. It then gradually stiffens until it becomes unworkable as an *'initial set'* takes place. This stiffening process continues until a state of *'final set'* is reached. Setting and consequent hardening are achieved by a reaction between the cement and water. This reaction is usually termed *hydration*.

Tricalcium aluminate, C_3A, reacts more quickly with water than do the other constituents. Thus an initial fast reaction occurs:

$$C_3A + 6H \rightarrow C_3A.H_6 + \text{High heat of hydration}$$

If the above reaction were allowed to proceed without restraint the cement would set rapidly forming a structure of low strength. However, the reaction is controlled by the small quantity of gypsum present. This forms a compound which coats the existing C_3A grains and so slows down the hydration process thus regulating the setting rate. Subsequently the C_2S and C_3S constituents are slowly hydrated so that the strength of the cement gradually builds up. These chemical reactions are of a fairly complex nature but can be represented by:

$$2(C_3.S) + 6H \rightarrow C_3.S_2.H_3 + 3(C.H) \tag{1}$$

$$2(C_2.S) + 4H \rightarrow C_3.S_2.H_3 + C.H \tag{2}$$

In the above equations the formula 'C.H' represents calcium hydroxide, $Ca(OH)_2$, which may further react with atmospheric carbon dioxide to form calcium carbonate, producing a further gradual increase in strength. Whilst products based on cement of this type can be safely 'handled' within about a week they only attain some two-thirds of their maximum strength in about six months. In fact concrete based on Portland cement continues to harden for years.

The Structure of Portland Cement

14.6   When the irregular particles of cement powder are mixed with water the reaction:

$$C_3.A + 6H \rightarrow C_3.A.H_6 + \text{Heat} \tag{3}$$

begins and the *dormant* period of some 15 minutes follows. During this time the grains become coated with a gelatinous envelope of $C_3.A.H_6$. Since the grains become joined at points of contact (Fig. 14.4(ii)) by this gelatinous film there is some loss of plasticity due to the network of weak chemical bonds formed. After about two hours the second stage of hardening sets in as hydration of both $C_2.S$ and $C_3.S$ take place by the equations (1) and (2) above.

Thin 'needles' (actually, tubes) of $C_3.S_2.H_3$ begin to protrude from the gel coating. These grow and form densely-packed rods radiating from the original cement particles which are progressively used up. As hydration of the $C_2.S$ and $C_3.S$ continues the rods begin to fill the spaces between the original cement particles so that an interlocking framework of rods ultimately consolidates into a rigid body. Such a framework will penetrate into and so adhere to the rough porous surfaces of stone, brick or other concrete.

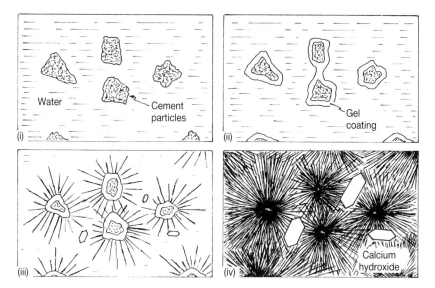

Fig. 14.4   The setting of Portland cement: (i) the cement particles are mixed with water; (ii) the particles become coated with a film of semi-permeable gel which forms weak bonds between some particles; (iii) 'needles' (tubes) of $C_3.S_2.H_3$ begin to grow leading to (iv) setting. This is followed by hardening which takes some days. The C.H (calcium hydroxide) separates out as hexagonal ⬡ crystals.

14.6.1   If, like the writer, you were entertained by your chemistry master during the limbo period following 'exams' you may recall his demonstration of a 'silica garden' in which crystals of various coloured inorganic salts were dropped into a solution of sodium silicate. Almost immediately the crystals were coated with a semi-permeable membrane through which coloured 'stems' burst and began to grow towards the surface of the silicate solution. The growth of 'needles' in setting cement follows a similar principle.

14.6.2   The gel coating of $C_3.A.H_6$ (equation 3) which covers the cement grains in the initial stages of setting (Fig. 14.5(ii)) is in fact a semi-permeable membrane. That is water molecules are drawn through it by osmosis which seeks to dilute the concentrated solution at the internal face of the membrane. This leads to an increase in pressure within the membrane coating (Fig. 14.5(iii)) until, at its weakest point, it bursts (Fig. 14.5(iv)) spouting a minute jet of a concentrated solution of $C_3.S$ and $C_2.S$ (dissolved from the cement particle) into the surrounding water. The outer surface of such a jet hydrates further on contact with water to form a tube consisting of $C_3.S_2.H_3$ (Fig. 14.5(v) and (vi)) (equations (1) and (2)).

Osmotic pressure within the original gel membranes continues to force solution through the tube so that the tube continues to grow in the needle-like form described, until the cement granule or the water is used up. CH i.e. calcium hydroxide, a by-product of equations (1) and (2)

Fig. 14.5 The formation of 'needles' (actually tubes) from a cement particle during setting.

remains in solution until its concentration is such that it recrystallises out (Fig. 14.4).

14.6.3 It should be appreciated that the foregoing is a much simplified description of cement chemistry. In fact other reactive constituents such as Fe_2O_3 and gypsum $(CaSO_4)$ are present in cement the latter also giving rise to a needle-like structure of a calcium sulpho-aluminate sometimes called *ettringite*. The setting process continues and the filling of pores in the structure increases the stiffness of the cement paste progressively. Since the hydrated product of the chemical process occupies a greater volume than did the 'dry' Portland cement it follows that expansion accompanies hydration and setting. Consequently cement—and concrete—give an excellent impression of the mould into which it is cast.

The setting of cement is another example of a chemical reaction which, like diffusion, is governed by the Arrhenius Law (5.8.6):

$$\text{Rate of setting} = A \cdot e^{-Q/RT}$$

Thus the rate of setting is controlled by the rate at which water molecules diffuse through the gel membrane which in turn depends on temperature, T.

14.6.4 *High alumina cement* consists of a very high proportion of C.A and a relatively small amount of C_2.S and C_3.S and in this respect it differs considerably from ordinary Portland cement. The high proportion of C.A confers a high hardening rate on this cement so that it reaches in a day the hardness achieved by Portland cement in about 20 days. Nevertheless there is a price to pay for this as the cement is particularly sensitive to both temperature and unsuitable environmental conditions. During the setting process in high-alumina cement two metastable hydration compounds are

358 . Composite Materials

formed: $C.A.H_{10}$ and $C_2.A.H_8$. The cement begins to harden as soon as these compounds begin to form but both eventually undergo a further change to form $Ca_3.A.H_6$. This reaction is known in the trade as 'conversion'. Since the relative density of $C_3.A.H_6$ is considerably higher than that of either $C.A.H_{10}$ or $C_2.A.H_8$ conversion is accompanied by a reduction in volume. This results in porosity usually leading to cracking and a big loss of strength. Conversion occurs very slowly in dry, cold conditions (below 5°C) and the defects may be unnoticed for years but in wet, warm conditions catastrophic failure may occur in a matter of months. (Above 50°C conversion is almost immediate.) Much high-alumina cement was used in the immediate post-war period because of its rapid hardening properties, but often under unsuitable conditions. We are still clearing up the mess!

14.6.5 A *rapid-hardening cement* which is in effect a form of ordinary Portland cement achieves an increased speed of hardening by raising the proportion of $C_3.S$. The cement is also ground to a much finer powder which increases the area of cement particle surface exposed to water. With this cement about twice the strength is achieved during the first day as compared with ordinary Portland cement. Nevertheless its use for large structures is inadvisable since the greater *rate* of heat output during the setting process may lead to the generation of dangerous residual stresses in the structure.

14.6.6 *Sulphate-resistant cements* are also necessary under some circumstances and contain no more than 6% $C_3.A$.

14.6.7 *Concrete* is produced from a mouldable mixture of stone 'aggregates', sand, Portland cement and water. When hard it has many of the attributes of natural stone and is extensively used in civil engineering and building. It is very easy to mould in the wet state and is one of the cheapest constructional materials available in Britain because of the availability of the necessary raw materials.

A variety of materials may be used as aggregate. Stone and gravel are the most widely used and also the most satisfactory. Other substances which may be available cheaply, e.g. broken brick, blast-furnace slag, can be used where a lower-quality product will suffice. Sand is also included in the aggregate and to obtain a dense product a correct stone/sand ratio is essential. By mixing various 'grades' (sizes) of aggregate in suitable proportions (Fig. 14.6) a relatively small quantity of cement is all that is necessary to provide an adherent film between the aggregate particles.

The proportion of cement to aggregate used is governed by the strength required in the product and can vary from 1:3 for a 'rich' mixture to 1:10 for a 'lean' one. The ratio commonly specified is 1:6 and this produces very satisfactory concrete if the aggregate material is sound, well-graded, properly mixed with the cement paste and, finally, adequately consolidated.

Other things being equal the greater the size of the large aggregate the

GRADES

Fig. 14.6 A satisfactory concrete structure.

smaller the cement ratio required to produce concrete of a given strength. This is to be expected since a smaller area of interface will need to be coated for a given volume of aggregate. Concrete made with Portland cement usually hardens in about a week but rapid curing methods can be used to shorten this time.

Formerly the *compressive* strength of concrete used in civil engineering was, typically, no more than 40 MPa. In recent years, however, serious attempts have been made to increase this value. With ordinary concrete sufficient water must be added, over and above that required for complete chemical hydration of the cement, to render the mixture sufficiently workable so that it can be moulded to shape. Unfortunately this extra water also reduces the ultimate strength of the concrete.

Hence the current tendency is to control the water/cement ratio more accurately so that a minimum excess of water is present. This is also made possible by the addition of 'superplasticisers' (polymer-type lubricants) which increase the workability, that is, allow particles to pack more closely thus reducing porosity. Milling the Portland cement more finely also reduces porosity. As a result of such simple devices concretes with compressive strengths of 100 MPa are now produced in North America whilst in Europe, Norwegian oil-production platforms containing concrete with strengths in excess of 80 MPa are in production.

14.6.8 *Plain concrete* is suitable for many purposes such as retaining walls, dams and structures which rely for their stability on great mass. Modern motorways, and foundations where large excavations have to be filled, are also examples where concrete is cast *in situ*.

At the other end of the scale small pre-cast parts can be manufactured by casting concrete into moulds, e.g. 're-constituted stone' blocks and various items of garden furniture.

Tarmacadam and Allied Substances

14.7 Modern methods of road making were originated in the early years of the nineteenth century by Scots engineer John McAdam. The method he originated, that of coating suitable aggregate with tar, is basically that used today. Two types of material are now used as the bituminous matrix: tars and asphalts. Tars used are the residues derived from the destructive distillation of coal. Some asphalts occur naturally, e.g. 'Trinidad Lake', but most are residues obtained during petroleum refining processes.

The bituminous material is mixed with a suitable aggregate such as blast-furnace slag for coarse foundation work and fine gravel for the finishing layer. Mixtures are designed to meet specific requirements such as stability, durability, skid resistance and resistance to penetration by water. Those mixtures using harder grades of asphalt are normally processed in a 'hot-mix' plant at about 135°C, whilst those containing the more fluid asphalts may be mixed at ambient temperatures.

The resultant mixture is tough and crack-resistant because of the bituminous matrix whilst it is hard-wearing because of the presence of protruding pieces of aggregate. In many ways this resembles the principles upon which a bearing metal is designed (14.4) and is a good example with which to finish a chapter discussing the diverse properties which can be brought together by the use of composite materials.

15

Fibre-reinforced Composite Materials

15.1 During a recent ramble through the forests of British Columbia I rested for a while in a grove of mighty Western Red Cedars and reflected that trees are the oldest living things on Earth and that some of those were sizeable saplins at the time when, far away in England, King John was signing *Magna Carta**.

15.1.1 During many millions of years of evolution trees have found it necessary to raise a complex biological structure high above the ground and to convert simple elements by photosynthesis into a material of high strength. Thus they have developed as tall slender columns with an array of subsidiary members attached cantilever fashion which have to carry appreciable loads of leaves—and often, snow—as well as to resist considerable wind forces. The materials from which these trunks and branches are constructed requires high strength and stiffness. In fact because of its low relative density wood has a specific strength higher than many man-made materials (Table 15.1).

15.1.2 The high specific strength—or strength/weight ratio as it was originally termed—of a wood such as pine is a result of its internal structure (Fig. 15.1), the major part of which consists of cells, known as *tracheids*, which are aligned vertically in the trunk and cemented together by an amorphous matrix composed mainly of the complex polymer, lignin. The tracheids are fibres about 2 to 4 mm long and 0.03 mm thick and since their strength is some five times that of wood as a whole, wood can be regarded as a relatively weak amorphous matrix successfully reinforced by the fibrous tracheids.

---

*My Editor tells me that further South in the arid uplands of Eastern California *Bristlecone Pines* eke out an existence for much longer—up to 5000 years in fact.

Table 15.1   *Properties of Some Fibre Materials*

Material	Tensile strength (GPa)	Young's modulus (GPa)	Relative density	Specific strength (GPa)	Specific modulus (GPa)	Fibre diameter (μm)
Polymer						
'Aramid' (Kevlar.29)	2.7	60	1.44	1.9	42	12
'Aramid' (Kevlar.49)	2.75	125	1.44	1.9	87	12
Glasses						
Silica (SiO_2)	5.9	74	2.2	2.7	34	10
E–glass (ordinary)	1.7	70	2.5	0.7	28	10
S–glass	4.6	84	2.6	1.77	32	10
$Fe_{(80)}B_{(20)}$ (metallic glass)	3.6 (Y.Pt.)	125	—	—	—	—
Crystalline ceramics						
Carbon (PAN)						
(high strength)	3.5	200	1.7	2.0	117	7–10
Carbon (high modulus)	2.5	400	2.0	1.25	200	7–10
Silicon carbide, SiC	7.0	400	3.2	2.2	125	10
Boron	3.9	400	2.6	1.5	154	130
α-alumina, Al_2O_3	1.7	380	3.95	0.43	96	20
Metals, etc. (for comparison)						
Steel	1.3	260	7.8	0.2	26 ⎫	
Aluminium	0.3	70	2.8	0.1	26 ⎬ Bulk	
Titanium	0.4	100	4.0	0.1	25 ⎪ materials	
Spruce (wood)	0.1	10	0.46	0.22	22 ⎭	

15.1.3   The internal structure of the tracheids is fairly complex but consists essentially of long *microfibrils* of cellulose molecules arranged in a crystalline pattern forming 'tubes', surrounded by a layer of *polysaccharides* built up of various sugars. Those polysaccharides adjacent to the crystalline cellulose core are themselves partially crystalline but become increasingly amorphous towards the surface of the tracheid. This means that between the crystalline cellulose core and the amorphous matrix there is no sharp boundary at which steep stress gradients could occur. It has been found that the avoidance of a well-defined crystalline-amorphous interface is important in developing satisfactory fibre-reinforced plastics materials.

Fig. 15.1   The composite nature of wood (white pine). The cellulose 'tubes' (tracheids) are cemented by a matrix of lignin.

As suggested above the concept of fibre-reinforced composites had its origin in nature. Thus the reinforcement of plastics materials by using glass fibre in the 1940s was based on the natural structure of bamboo, whilst the Ancients reinforced their air-dried bricks with straw—a practice one can still observe in some of the more remote regions of the Balkans.

15.2   Fibre-reinforced composites include:

(i)   matrix materials such as thermoplastic or thermosetting resins, and some metals; reinforced with carbon, glass, organic polymer and other fibres;
(ii)   plastics materials (usually thermosetting) reinforced with laminations of woven textiles;
(iii) vehicle tyres in which vulcanised rubber is reinforced with woven textile cords or steel wire;
(iv)   materials such as reinforced or pre-stressed concrete.

When a fibre-reinforced composite is stressed the matrix may begin to flow but in doing so will cause a force to be set up at the interface with the fibre. If the fibre is long enough this transmitted force will ultimately lead to its fracture and the fibre will therefore have contributed fully to the strength of the composite. Obviously strength will be directional, i.e. at a maximum value parallel to the direction of the fibres.

The nature of the interface between fibre and matrix influences the extent to which a force is transferred from the matrix to the fibre, and cohesion at the interface may be achieved by either mechanical, physical or chemical bonding (14.2.1).

15.2.1   When there is little or no chemical interaction between matrix and reinforcement material the relative density, ρ_c, of such composite materials can be calculated by the very simple 'rule of mixtures'. Then:

$$\rho_c = \rho_r V_r + \rho_m (1 - V_r)$$

where V_r is the volume of the reinforcement fraction and ρ_r and ρ_m the relative densities of reinforcement and matrix respectively.

Q   In a fibre-reinforced composite the continuously-aligned cylindrical fibres are packed as closely as is possible in a single direction. Given that the relative densities of fibre material and matrix material are 2.41 and 1.28 respectively calculate the maximum relative density of the resultant composite assuming pure ingredients.

A   The closest packing of cylindrical fibres is achieved in a hexagonal pattern as indicated in the following cross-sectional diagram:

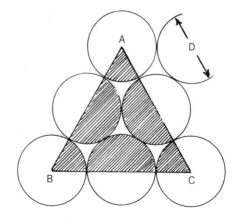

Consider the surface area of the cross-section within the triangular element, ABC.

The length of side of triangle ABC is:

$$D + 2\left(\frac{D}{2}\right) = 2D$$

Therefore the area is:

$$\frac{2D \times D\sqrt{3}}{2} = D^2\sqrt{3}$$

The number of circles within the triangle is:

$$(3 \times \tfrac{1}{2}) + (3 \times \tfrac{1}{6}) = 2 \qquad \text{radius } D/2$$

Packing density is given by:

$$\frac{\text{Area of 2 circles}}{\text{Area of } \triangle ABC} = \frac{2\pi D^2/4}{D^2\sqrt{3}}$$

$$= \frac{\pi}{2\sqrt{3}}$$

$$= \underline{0.907}$$

Let ρ_c be the relative density of the composite, then using $\rho_c = \rho_r V_r + \rho_m(1 - V_r)$:

$$\rho_c = (2.41 \times 0.907) + 1.28(1 - 0.907)$$

$$= 2.19 + 0.119 = \underline{2.31}$$

The relative density of the composite is 2.31.

15.2.2 Composites based on carbon or other advanced fibre materials in a thermosetting or thermoplastic matrix have a range of useful properties. Longitudinal strength and stiffness of the composite can be very high and, because of its low relative density, specific properties, i.e. Mechanical property/Relative density are excellent. The mechanical properties of unidirectional composites are of course very anisotropic for the same reason that wood is strong 'along the grain' but weak 'across the grain'. Such difficulties can be largely overcome by using a multi-directional fibre filling, e.g. woven fabrics and the like.

Functions of the Constituents in a Fibre Composite

15.3 The simplest composite consists of continuous aligned fibres which have been completely infiltrated by a matrix material which has bonded to the fibres and then hardened. The reinforcing fibres contribute the longitudinal mechanical properties such as strength, stiffness, creep and fatigue properties; whilst the matrix will generally be physically weaker than the fibre infilling though its functions are vital. Thus the matrix protects the fibre surface from damage and from the action of the environment, whilst keeping fibres in the desired position to receive stress distribution. Generally transverse properties such as shear and compressive strength, impact and toughness values are largely controlled by the matrix, though it may also give some protection against crack propagation perpendicular to the fibre direction by allowing separation to occur between fibre and matrix (Fig. 15.2).

Fig. 15.2   Delamination or separation of fibre from matrix which will arrest an advancing crack.

15.3.1 The interface region between fibre and matrix is generally about 0.1 μm or less in thickness and can be considered as a third phase in the composite. The strength of the fibre–resin bond may be between 1 and 100 MPa. Bond strength has an important influence on the overall mechanical properties of the composite. If too low then the fibre and matrix trend to operate independently leading to poor overall properties, but if too high the composite may be strong and stiff but also excessively brittle. In practice bond strength is usually controlled by the degree of treatment given to the surface of the fibre material in so far as this affects the degree of adhesion.

Reinforcement Mechanisms

15.4   We will consider the forces acting on a single fibre bonded in a bar of suitable matrix material, the strain at failure of which is greater than that of the fibre material (Fig. 15.3(i)). When a force is applied to the ends and

(i)   (ii)   x

Fig. 15.3   (i) A single short fibre in a matrix block; (ii) the stress distribution along a short fibre.

sides of the composite block different axial displacements in both fibre and matrix at either end of the fibre initiate shear stresses at the fibre surface. These stresses are transmitted to the fibre causing an increase in tensile stress within it. Assuming that a stress σ_x is produced at a distance x from the end of the fibre (radius r), then:

$$\text{Tensile force in fibre} = \pi r^2 . \sigma_x$$

Also:

$$\text{Force acting within the fibre-matrix interface} = 2\pi r . x . \tau$$

where τ is the shear strength of the interface material. Since these forces are in equilibrium in the composite:

$$\pi r^2 . \sigma_x = 2\pi r . x . \tau$$

or
$$\sigma_x = \frac{2x\tau}{r} \tag{1}$$

As the external force is increased the region of maximum shear stress advances along the fibre and tensile stress increases. If $2x$, the total length of the fibre is such that σ_x never exceeds the breaking stress of the fibre, then the fibre will not break if the composite is stressed further, i.e. failure will occur first at the interface. Stress distribution in the fibre is indicated in Fig. 15.3(ii) and the mean stress is $\sigma_x/2$.

15.4.1   We will now consider a long fibre, the length of which is such that failure can ultimately take place in it. If E_f is Young's Modulus for the fibre and ε is the overall strain then:

$$E_f = \frac{\sigma_x}{\varepsilon}$$

or $\sigma_x = E_f . \varepsilon$ $\hspace{2cm}$ (2)

Obviously the strain in the fibre cannot exceed that in the matrix and, before failure, stress in the fibre will increase with distance from either end until $\sigma_x = E_f . \varepsilon$ and then assumes a constant value. From equations (1) and (2) this occurs when:

$$x = \frac{E_f.\varepsilon.r}{2\tau}$$

As stress in the composite increases σ_x reaches the breaking stress σ_b and failure occurs. The value of x (from equation (1)) is then given by:

$$x = \frac{L_c}{2} = \frac{r.\sigma_b}{2\tau} \qquad (3)$$

where L_c is known as the *critical stress transfer length*, i.e. the fibre length which must be exceeded if failure of the fibre is to take place. The

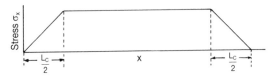

Fig. 15.4   The stress distribution along a long fibre.

distribution of stress along the fibre is as shown in Fig. 15.4 and the average stress level $\bar{\sigma}_x$ is given by:

$$\bar{\sigma}_x = \sigma_x\left(1 - \frac{L_c}{2x}\right) \qquad (4)$$

The longer the fibre $(2x)$ the nearer the overall stress approaches the constant value acting in the centre of the fibre and hence the more efficient is the use of the reinforcement.

If the tensile strength and the radius of the fibre are known along with the shear strength, τ, of the interface, then equations (1) and (2) can be used to estimate the fibre length necessary for optimum use of the fibre strength in a short-fibre composite and also the average stress level.

15.4.2   In the above it is assumed that the breaking stress of the fibre is less than that of the matrix but if the reverse obtains the matrix will fail first. The question of which component of the composite fails first depends upon the proportions and the stress/strain characteristics of the two components. In practice the fibre volume of a composite is usually between 40 and 70%—in a 'real' material a bar say 10 mm × 1 mm in cross-section would contain some 10^5 individual fibres, between 5 and 10 µm in diameter and between 0.1 and 1 µm apart.

Fibre failure generally leads to immediate failure of the composite but here we have been dealing with the simple case of a single fibre embedded in a matrix. In real multi-fibre composites the structure is much more complex and such variables as fibre direction relative to the direction in which stresses are acting must be considered.

If we make the assumption that the fibres are long, continuous and aligned parallel in a single direction and that there is complete bonding between fibres and matrix, then, the tensile strength, σ_c, of the composite is given by:

$$\sigma_c = \sigma_f V_f + \sigma_r (1 - V_f)$$

where σ_f is the fracture strength of the fibre material and σ_r the yield strength of the resin matrix, whilst V_f is the fibre volume fraction.

Similarly, in terms of the Elastic Modulus:

$$E_c = E_f V_f + E_r (1 - V_f)$$

15.4.3 The properties of these fibre-reinforced materials are extremely anisotropic so that the Elastic Modulus, E_{ct}, of the same material loaded transversely to the fibre direction would be given by:

$$E_{ct} = \left\{ \frac{V_f}{E_f} + \frac{(1 - V_f)}{E_r} \right\}^{-1}$$

which shows E_{ct} to be small compared with E_c.

Q A composite composed of long parallel fibres of *Kevlar* in an epoxy resin matrix, is required to have a tensile strength of 1600 MPa. If the strengths of *Kevlar* and epoxy resin are, respectively, 2560 MPa and 80 MPa, what volume fraction of *Kevlar* must be used in the composite?

A If σ_c is the tensile strength of the composite then:

$$\sigma_c = \sigma_f V_f + \sigma_r (1 - V_f)$$
$$\therefore \quad 1600 = 2560 V_f + 80(1 - V_f)$$

Rearranging: $\quad = 2560 V_f + 80 - 80 V_f$
$$1600 - 80 = 2480 V_f$$
$$\therefore \quad V_f = \underline{0.613}$$

The composite must contain 61.3% *Kevlar* by volume.

Fibres Used in Composites

15.5 Thermosetting polymers strengthened by woven textile fibres, e.g. *Tufnol*, were introduced many years ago for use in the electrical industries as tough insulating panels and the like but glass fibre was the first man-made fibre to be employed as the reinforcement in a strong *lightweight* composite. Since then glass fibre has been joined by a large number of materials designed for modern engineering uses. These include carbon, boron, ceramic and certain polymer fibres the necessary mechanical property requirements of which are high strength, stiffness and low relative density. Other important requirements are temperature tolerance, chemical resistance, thermal conductivity and—ultimately—reasonable cost.

Much modern research is aimed at producing materials of high *specific strength* for the aerospace industries and since the relative density of a substance usually follows the atomic masses of its constituent atoms many of these fibres are derived from compounds based on the low atomic mass elements: H, B, C, N, O, Al and Si. Moreover compounds based on these elements usually rely for their strength on the covalent bond, the strongest by far of all chemical bonds.

15.5.1 Organic Polymer Fibres   Polymers containing long-chain molecules generally have poor stiffness and strength because these molecules are coiled or folded in the normal bulk material. If, however, these molecules are stretched and straightened during a fibre manufacturing process, high strength and stiffness may be developed. Aromatic* polyamide (hence, *'Aramid'*) fibres are currently the most successful. These are based on the formula:

and are manufactured commercially under the name *Kevlar*. After being spun, the fibres are heated *under tension* in a nitrogen atmosphere at temperatures up to 550°C so that the 'stretched' properties are retained. These molecular chains are only held together by weak hydrogen bonds (Fig. 15.5) so that transverse and compressive strengths are low. However,

Fig. 15.5   Strong covalent bonding acts within aramid molecules but relatively weak hydrogen bonds operate between adjacent molecules. The aligned molecules form 'folded' crystals along the fibre direction.

*During the early classification of organic compounds substances were divided into two groups—*aliphatic* compounds in which carbon atoms are arranged in open 'chains' (as in alkane series, i.e. methane, ethane, etc.) and *aromatic* compounds in which the 'benzene ring' or a derivative such as phenol (Fig. 12.5) is involved. Not all 'aromatic' compounds have a pleasant aroma, whilst with many aliphatic compounds the scent is just as pleasant—or malodorous!

this in turn imparts a high capacity for absorbing energy so that aramid composites are useful for resisting impact damage. Consequently they are used in bullet-proof vests and the like, whilst their low relative density makes them valuable in the aerospace industries. They are sensitive to ultra-violet light from which they must be protected. A further disadvantage is that they absorb water easily making them unsuitable for use in some chemical industries.

Strong nylon fibres have been used for reinforcement for a considerable time and although polythene is not normally considered as a possible strong material it can be made strong provided polythene of a suitable molecular mass is chosen and if treatment is applied to obtain molecular alignment. Such fibres would be useful only at low temperatures since above 100°C the molecules acquire sufficient thermal energy to rearrange themselves randomly.

15.5.2 Carbon Fibres   Whilst polymer fibres are composed of aligned *linear* molecules carbon fibres are made up of *two-dimensional* planes or layers of carbon atoms arranged as in the graphite structure (Fig. 3.29). The graphite crystal is *anisotropic** in respect of many of its properties and this in turn is due to its layer structure. For example Young's Modulus measured in a direction parallel to the layers is 1030 GPa whilst perpendicular to the layers it is only 35 GPa. This is to be expected since individual carbon atoms in a layer are held to their neighbours by strong covalent bonds only 0.142 nm long, whereas only weak van der Waals forces operate between adjacent layers which are comparatively far apart (0.335 nm). Thus the bond energy due to the covalent bonds operating within the layers is about 600 kJ/gram-atom, whereas the interlayer binding energy is only 5.4 kJ/gram-atom. The relatively high binding energy within the layers gives rise to the high value of Young's Modulus since high elasticity is dependent on a large restoring force acting between atoms.

In practice carbon fibres are polycrystalline and consist of a large number of small crystallites. These are about 10^{-8} m thick and 2.5×10^{-8} m in diameter and consist of parallel planes of carbon atoms grouped together. The main difference between the structure of these crystallites and the ideal structure shown in Fig. 3.29 is that there is no regularity of the orientation of the parallel planes and the structure is said to be *turbostatic*. The structure of perfect graphite has been likened to a pack of playing cards which has been tapped on a table to square up individual cards (layer planes) whilst the structure in the turbostatic graphite of carbon fibres has been likened to the pack of cards collected together into a pile but not squared at the edges so that the orientation of individual cards (layer planes) is random. Examination by the electron microscope shows that the crystallites are grouped together in units known as *fibrils*, which are sub-units of the fibre. The fibre itself some 8×10^{-6} m in diameter consists of a network of closely packed fibrils each about 2.5×10^{-8} m in diameter (Fig. 15.6).

---

*Anisotropy is the characteristic of exhibiting different physical properties in different directions in a body of material.

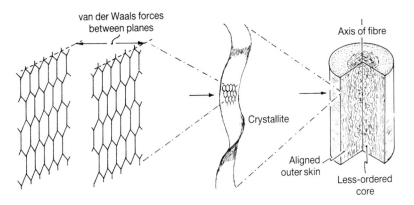

Fig. 15.6   The form of structure present in carbon fibre (after J. D. H. Hughes).

The bulk of carbon fibre is produced by the pyrolisis or heat-treatment of polyacrylonitrile filament. The process takes place in stages in an inert atmosphere:

(i)   a low temperature treatment at 220°C is first given. This promotes cross-linking (thermosetting) between molecules rendering the material infusible during the subsequent high-temperature process;

(ii)   the temperature is raised to 900°C to 'carbonise' the filament. Decomposition takes place and the polymer chains are stripped of all other atoms and groups of atoms leaving a 'skeleton' of carbon atoms in the graphitic structure;

(iii)   heat treatment is then applied which will produce the required structure and combination of properties. Lower temperatures (1300–1500°C) produce fibres of high strength and low modulus whilst higher temperatures (2000–3000°C) provide fibres of low strength but high modulus.

During the process the fibres are kept in tension to prevent curling and to encourage good alignment. Carbon fibres have diameters between 5 and 10 μm and are available in 'tows' containing 1, 3, 6, 12 or 20 thousand filaments.

Carbon fibre continues to be the most promising of composite reinforcements in terms of specific strength particularly since production costs are falling and are likely to fall further if current projects succeed in producing satisfactory fibre from pitch.

15.5.3 Glass Fibres   When in the form of fibre glass is considerably stronger than it is in any other form and as such enters the ranks of engineering materials. Silica glass fibre in the freshly-drawn condition, that is if tested within ten minutes of being drawn and 'untouched by human hand' so to speak, may have a tensile strength in the region of 15 GPa. The most minute of scratches, even such as would be produced by drawing a feather across the surface of the fibre, will cause fracture at considerably lower stress. Such minute cracks or scratches act as stress-raisers so that at the tip of the crack the stress is well above that necessary to cause fracture, whereas the overall average stress in the fibre is much lower than this. Consequently, the crack will extend and, as long as the stress concentration is maintained at the new tip of the crack, fracture will continue (Fig. 15.7(i)). A ductile material on the other hand can adjust to the situation by means of plastic strain so that the high concentration of stress is automatically reduced (Fig. 15.7(ii)).

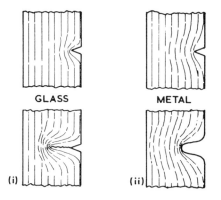

Fig. 15.7   The propagation of a crack in a brittle material (i) as compared with a ductile material (ii).

High quality glass fibre is produced from a melt containing high-purity materials. This is spun at a high temperature and cooled very rapidly before being given a protective coating termed 'size'. This size is a mixture of materials which protects the surface from damage and acts as a lubricant. It also contains a derivative of silane (SiH_4) which reacts with SiO– links in the glass fibre and is also compatible with chemical groups in the molecules of the polymer matrix, thus promoting fibre/matrix bonding.

Glass fibre was the first of the modern fibres to be used in composites. Now a number of different types are in production the best known of which are S–glass, a high-strength magnesia–aluminosilicate composition and E–glass, a non-alkaline borosilicate glass which was developed for its electrical insulation properties. Glass fibres are strong and cheap though not as stiff as carbon fibre. Their relative densities are also higher than that of carbon fibre.

15.5.4 Boron Fibre   has a low relative density and a very high Young's Modulus but is only moderately strong. Since it is difficult to fabricate it is very expensive. Nevertheless these fibres are used in the aerospace industry because of their high transverse strength which, along with large fibre diameter, enables them to withstand high compressive forces. Boron fibres are grown from the vapour phase on to tungsten filaments some

15 μm in diameter. The core filament is heated to 1300°C in a hydrogen atmosphere when volatile boron chloride is admitted:

$$2BCl_3 + 3H_2 \rightarrow 2B + 6HCl \uparrow$$

Boron is deposited on the surface of the tungsten core with which it reacts to form tungsten boride, W_3B_4. Pure boron then deposits on this core forming crystals with very few surface defects. The resulting fibres are

Core of W_3B_4 (17 μm)

100–200 μm

Boron crystal size approx. 7 μm × 2 μm

Fig. 15.8 A section through a boron fibre grown on to a tungsten filament (after J. D. H. Hughes).

100–200 μm in diameter. Boron fibre may also be grown on to a carbon filament some 30 μm in diameter but in this case there is no reaction between core and coating.

15.5.5 Other Fibres Mention must be made of those ceramic fibres which are now being developed for the production of metal matrix composites (MMCs). These are important for future developments in the aerospace industries where cast and extruded aluminium alloys can be further strengthened by fibre reinforcement. The most important fibres available in this field are those of silicon carbide and alumina, though the latter undergoes both phase changes and grain growth at high temperatures.

15.5.6 Extremely high strengths are attainable in materials produced in the form of 'whiskers'. These are hair-like single crystals, between 0.5×10^{-3} and 2.0×10^{-3} mm in diameter and as much as 20 mm long, grown under controlled conditions and generally having a single dislocation running along the central axis. This relative freedom from dislocations means that their yield strength is nearer to the theoretical for the material (Table 15.2). Thus a graphite whisker may have a strength of as much as 21 GPa whilst the strength of carbon fibre is of the order of only 3 GPa. Unfortunately whiskers are normally produced only under laboratory conditions and their great strength is difficult to utilise under ordinary production processes. Thus the high cost of forming a whisker composite generally rules out its use at present and ordinary fibre composites are used instead.

Table 15.2 *Peroperties of Some 'Whiskers'*

Material	Tensile strength (GPa)	Relative density	Young's modulus (GPa)	Specific strength (GPa)	Specific modulus (GPa)
Alumina	21	3.96	430	5.3	110
Silicon carbide	21	3.21	490	6.5	150
Graphite	20	1.66	710	12.0	430
Boron carbide	14	2.52	490	5.6	190
Silicon nitride	14	3.18	380	4.4	120

Matrix Materials for Fibre Composites

15.6 Matrix materials used in fibre composites may be of polymer, ceramic or metal though by far the greatest output to date is of composites employing thermosetting resin matrices. To produce a successful composite the matrix material must form either a frictional, or more often, a chemical bond with the fibre surface. In the latter case the chemical reaction between the two must not be so vigorous as to damage the fibre. The matrix material must be sufficiently mobile to infiltrate among individual fibres at as low a temperature and pressure as possible during moulding and then solidify (or reach T_g) quickly. Fibres should not be damaged physically during the moulding process.

Apart from compatibility with the fibre material the choice of matrix will be governed by service requirements. The resistance of the composite to temperature, chemical environment and the effects of moisture are determined by the matrix, though the fibre reinforcement must be stable to a temperature above that at which the matrix begins to deteriorate.

15.6.1 Thermosetting Resins These are polymers which cross-link (12.3) during 'curing' to form a glassy hard solid (polyesters, vinyl esters and epoxies). Polyesters were the first thermosetting resins used (with glass-fibre reinforcement) to make protective housings for aircraft radar antenna during the Second World War, the resultant composite being transparent to radio transmissions. These resins still account for a high proportion of composite manufacture since they are cheap, easy to work with and can be 'cured' at relatively low temperatures. The main disadvantage of polyesters is that they do not bond well with fibres so that strength of the composite depends largely on relatively weak van der Waals forces. Polyesters also have a relatively high shrinkage coefficient which further impairs adhesion to the fibres. For these reasons polyesters are not generally used for high-performance composites.

Epoxides are the most extensively used matrix materials for those composites based on continuous carbon or aramid fibres. The resin is mixed with a 'hardener' (the reader will be familiar with DIY epoxide glues) and heated for up to eight hours at 160–180°C to produce a hard, insoluble cross-linked structure* with a tensile strength of 80–100 MPa and

---

*The reaction involved is of the type described in 12.3.1 though here we are dealing with cold-setting polymers. A much stronger harder bond can usually be obtained with DIY glues if the joint is warmed at 100–150°C—a greater degree of cross-linking is obtained.

a Young's Modulus of 2–4 GPa. Most of the resins in this class are based on molecules having two or four epoxide groups. This introduces a greater density of cross-linking and consequently a higher temperature tolerance which is usually specified in terms of T_g but in practice a temperature of 160°C is probably the maximum at which such a load-bearing composite can work continuously. A greater degree of cross-linking will of course mean increased hardness—and brittleness.

Other useful thermosets include well-known phenolics (12.3) which are inexpensive, fire-resistant but tend to be brittle and suffer from high shrinkage. High-performance resins include polyimides which may be either thermoplastic or thermosetting materials and can withstand temperatures of 280–315°C continuously. Unfortunately they are very expensive and require long curing cycles following the fibre-impregnation stage.

15.6.2 Thermoplastic Polymers These are generally high-molecular mass, long-chain molecules which develop strength at ambient temperature either by remaining amorphous (but becoming entangled), e.g. polycarbonates; or by crystallising partially, e.g. nylon. For many applications thermoplastics have the advantage over thermosets of greater toughness and shorter fabrication cycles, but are at a disadvantage in respect of their high viscosity at the moulding temperature, making fibre impregnation difficult.

Best known among the many thermoplastics used as matrices are Nylon 66, polyethylene terephthalate (PET), polysulphone (PS: 150–170°C), polyphenylene sulphide (PPS: 200–220°C), polyamide imide (PAI) and polyether ether ketone (PEEK: 250°C). (Figures represent maximum service temperatures.) PEEK-based composites have excellent resistance to failure under impact, good strength and a work-to-fracture value up to 13 times that of a comparable epoxide composite.

15.6.3 Metal Matrices Since the general aim in producing a composite is to develop a material of high specific strength, suitable metals will be those of low relative density, i.e. aluminium and its alloys, magnesium and titanium and its alloys. Of the latter the alloy: Ti–6Al–4V is the most used. Metal matrix composites are of interest because of the higher temperature (around 500°C) at which some of them can operate, though other valuable properties include higher transverse and shear properties, good impact resistance and the fact that they can be joined by orthodox metallurgical techniques. The main disadvantages are high relative density and the general difficulty of fabrication as compared with polymer matrices.

The Manufacture of Fibre Composites

15.7 The commercial forms in which the fibre materials are supplied include:

(i) *rovings*   this is the basic material, a roving consisting of a number of strands or bundles of filaments wound into a package or 'creel'. The length of the roving may be several kilometres and the term is usually applied to glass fibres whereas bundles of continuous carbon fibres are known as 'tows'. A 'strand' of glass fibres contains typically some 200 filaments each about 10 μm in diameter;

(ii)   *woven fabrics*   in which various types of weave styles are used;

(iii)   *chopped fibres*   from about 1 mm to 50 mm in length.

Composites are manufactured either as individual single components or in the form of sheets or sections in any required length. In the former case the fibre material may be woven into a 'preform' which roughly fits the die cavity into which the resin is injected. Of a number of processes available the following are examples.

15.7.1 Hand and Spray Placement   These processes account for the greatest proportion of glass-reinforced plastics (GRP) produced currently in Britain. In the simplest form woven rovings are laid over a mould surface which has been coated with a suitable non-stick agent. The liquid resin is then worked into the reinforcement by hand using a brush or roller. Polyester resins are most commonly used with glass fibre giving a combination of low cost and good chemical resistance. This process—or at least its results—will be familiar to those who go 'messing about in boats' since it is the foundation of the manufacture of many modern pleasure craft, though large vessels such as mine sweepers up to 60 m long have been 'hand layered'.

15.7.2 Press Moulding   the most widely used process for mass producing reinforced plastics components. The process is fundamentally similar to that in Fig. 6.29, the charge to the mould cavity consisting of fibre and resin.

15.7.3 Resin Transfer Moulding   similar in principle to the process shown in Fig. 6.30. The fibre, usually as a woven pre-form, is placed in the die cavity and the resin then injected.

15.7.4 Pultrusion   a process whereby profiles of fixed cross-section are manufactured in continuous lengths. The process resembles that used in the extrusion of metals and plastics except that in *pul*trusion the reinforcing fibres pull the composite through the die system (Fig. 15.9).

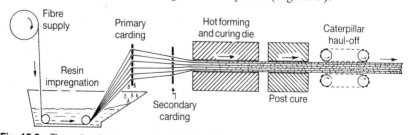

Fig. 15.9   The pultrusion process in which the reinforcing fibres pull the composite through the die system. The finished pultrusion is cut to length by a 'flying saw'.

Most pultrusions are of glass-fibre/polyester composite giving good corrosion resistance at low cost. Rod, sheet, tube and channel profiles are typical products but building panels, window frames, railings and walkways in chemical plant are also made by the basic process.

15.7.5 Metal Matrix Composites (MMCs)   these are difficult to produce because relatively high temperatures and/or pressures need to be used during the process and this in turn may cause damage to the fibre reinforcement. Most MMCs are based on aluminium alloy matrices which are reinforced with particles or fibres of alumina, silicon carbide, carbon or boron.

Low melting point metals may be gravity cast to form the matrix but in order to achieve adequate infiltration the use of pressure is generally necessary. In *squeeze casting* mechanical pressure is used to force molten metal amongst the reinforcement material but in *liquid pressure forming* a fibre 'pre-form' is first placed in the die. The die is then evacuated to reduce the incidence of porosity and assist infiltration. Molten metal is then forced in. The required pressure is lower than with squeeze casting and this offers greater flexibility in terms of size and shape of the castings produced.

Two methods are available for producing MMCs from solid metals. In the *powder metallurgy* process a metal powder is mixed with particles or chopped fibres, pressed into shape and then sintered using the basic method described in 6.5.1. *Diffusion bonding*, however, uses metal foil instead of powder. Alternate layers of metal foil and fibres are pressed whilst being heated so that sufficient diffusion and grain growth occurs across the metal–metal interfaces to bond the foils together.

Mechanical Properties of Fibre-reinforced Composites

15.8   A composite made by simple pultrusion as described above will be extremely anisotropic with regard to its mechanical properties. In fact its strength measured transversely to the direction of pultrusion may be only 2 to 5% of that measured longitudinally, i.e. along the direction of the fibres. In most cases the transverse strength of the composite is less than that of the matrix resin due to the presence of voids and weak interfacial bonds. Nevertheless we must not forget that many of these fibre-reinforced composites offer higher values in terms of *specific* strength and *specific* modulus than can be found in any other materials. This above all makes them pre-eminently useful in the aerospace industries where engineering design is sufficiently sophisticated to accommodate the distinctive characteristics of these materials.

In moulding or casting single components it is possible to arrange fibre direction by the placement of a fibre pre-form in the die or mould so that expected stresses are catered for.

Table 15.3 indicates values for uni-directional fibre-reinforced compo-

Table 15.3 Mechanical Properties of Some Uni-directional Fibre Composites

Material (60% by volume fibre in epoxy matrix)	Relative density	Tensile strength (MPa)		Young's modulus (longitudinal) (GPa)	Specific strength (longitudinal) (MPa)	Specific modulus (longitudinal) (GPa)
		Longitudinal	Transverse			
Aramid	1.35	1600	30	80	1185	59
E–glass	2.0	1150	40	44	575	22
S–glass	2.0	1800	40	55	900	28
High-strength carbon	1.55	1770	50	174	1140	112
High-modulus carbon	1.63	1100	21	272	675	167
Boron	2.2	1240	70	210	564	95
Metals (in bulk) for comparison						
Duralumin	2.7	460	—	69	170	26
Titanium alloy (6Al–4V)	4.4	1340	—	115	304	26
High-tensile alloy steel (Ni-Cr-Mo)	7.83	2000	—	210	255	27

sites and although quoted strengths and stiffnesses of composites are not very different from those of metals, their relative densities are much lower. Hence these composites have greatly improved *specific* properties. The weight savings obtained in practice are not so great as Table 15.3 might suggest because fibre composites are anisotropic and this must be allowed for in design. Nevertheless savings in weight of some 30% are achieved over aluminium alloys.

Uses of Fibre-reinforced Composites

15.9 The most important of these commercially are polymer matrix composites reinforced with either glass, carbon or aramid fibres, and of these by far the greatest volume consists of GRP principally because of their relatively lower cost.

Many commercial airliners fly with some parts of their primary structure and a large proportion of the interior fittings and trim made from such composites; whilst sporting and leisure goods—boats, gliders, sailboards, racquets and skis—are made largely from fibre-reinforced composites. Domestically, most plastics-bodied appliances use reinforcement in the form of short, chopped fibres incorporated during moulding of the product.

15.9.1   The following characteristics of composites dictate their uses:

 (i)   low relative density and hence high *specific* strength and modulus of elasticity;
 (ii)   good resistance to corrosion in many environments;
 (iii)   good fatigue resistance, particularly along the fibre direction;
 (iv)   the possibility of economising on material bulk by designing to make use of the anisotropic properties;
 (v)   the ability to reduce production costs by 'parts integration', e.g. by using a 'one-shot' moulding instead of the assembly of several separate parts;
 (vi)   the generally low thermal expansion, particularly with carbon or aramid fibres.

15.9.2   The building of non-magnetic GRP mine sweepers was mentioned earlier but on a more general scale this low-density, corrosion-resistant composite has revolutionised the leisure and small work boat-building industries because of its ability to mass produce complete hull and deck mouldings. Similarly in the building industry mouldings of decorative texture and colour are used in cladding prestige buildings, whilst GRP flat and corrugated translucent sheetings are used for roofing.

15.9.3   Large quantities of GRP are used in land transport vehicles because of economies in manufacturing where appropriate, and also in the lower fuel costs which result. In motor car manufacture the use of fibre-reinforced plastics is growing steadily in the form of panels, bumpers and interior fittings; whilst leaf springs made in GRP have excellent fatigue properties, offering a large saving in weight and also a decrease in noise transmitted to the driving seat.

E–glass, the most common type of glass fibre used for reinforcement work, was originally developed as an electrical insulation material. Hence it is not surprising that, when combined with polymer resins, a group of composites is available which can be moulded into complex shapes possessing excellent insulating properties. Switch casings, cable and distribution cabinets, junction boxes, etc. are produced in GRP. Similarly in terms of corrosion resistance these materials are suitable for use in the chemical industries for storing potentially aggressive chemicals as well as in vats, silos, wine vats and pipelines for water and sewage.

In more sophisticated applications where maximum specific strength and specific modulus are required then plastics material reinforced with either carbon fibre or aramid fibre is used. Aerospace and military aircraft design usually employ carbon fibre-reinforced plastics (CFRP), whilst most helicopter blades are of this material. In the leisure industries some of the advanced materials developed for aerospace have been used to produce strong rigid hulls in racing yachts, whilst on land many racing cars employ a composite chassis consisting of CFRP skins bonded to a honeycomb core.

Concrete Products Reinforced with Steel

15.10   Whilst plain concrete (14.6.7) has a reasonably high compressive strength it is weak in tension. This disadvantage can be overcome for some purposes by reinforcement with steel rods or cables the product then bearing similarity with other fibre-reinforced composites.

15.10.1 Reinforced Concrete   Most ceramic materials are relatively strong in compression but relatively weak in tension and concrete is no exception to this rule. For structural members such as beams the use of an adequate section of plain concrete would be uneconomic and often unsuitably cumbersome (Fig. 15.10(ii)). By using steel reinforcement rods (Fig. 15.10(iii)) in that part of the beam which will be in tension, full advantage can be taken of the high compressive strength of the concrete and the cross-section reduced accordingly. The ends of the rods are so shaped that they are firmly gripped by the rigid concrete.

15.10.2 Pre-stressed Concrete   A source of weakness in reinforced concrete is indicated in Fig. 15.10(iii) which shows the formation of small cracks in the concrete on that side of the beam which is in tension. Since steel has a much greater elasticity than concrete the steel rods will still be stretching within their elastic limit when the concrete has already begun to crack. The presence of small cracks which absorb moisture would lead to

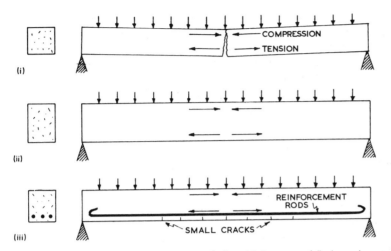

Fig. 15.10 The principle of reinforced concrete. In (i) a plain beam can fail when micro-cracks, acting as stress raisers, are formed in the edge which is in tension. A plain beam which is strong enough (ii) will be uneconomically bulky. In (iii) steel reinforcement-rods support that section of the beam which is in tension. Small cracks may, however, form in the concrete (15.10.2).

the gradual exfoliation of the concrete due to the action of frost. One method by which the situation can be met is to use the elastic properties of steel to apply an initial *compression* to the concrete, that is, to 'pre-stress' it.

A steel cable is stretched within its elastic limit and whilst it is in tension a concrete beam is cast around it (Fig. 15.11). When the concrete has completely hardened the externally applied tension in the steel cable is released (Fig. 15.11(ii)). As the cable attempts to contract it exerts

Fig. 15.11 The principles of prestressed concrete.

compressive forces on the concrete. Assuming that the cable is in the correct position relative to the cross-section of the beam these compressive forces in the concrete will exactly balance the tensile forces caused by the load applied to the beam during service (Fig. 15.11(iii)). Since there is no resultant tension in the concrete cracks will not develop in it.

This method obviously depends upon the formation of a good concrete/steel bond without which the steel members would tend to slip through the concrete when the external tensioning force was released. Bonding between steel and concrete is not of a chemical nature but is more likely to be due to surface irregularities in the reinforcement steel. This bonding is improved by a thin layer of rust on the steel and assisted by the shrinkage of the concrete which occurs during hardening.

16

The Failure of Materials in Stress

16.1 Engineering components must in most cases be designed in such a manner that any stresses involved during service are insufficient to give rise to plastic deformation. Whilst in tensile or compressive tests the specimen is subjected only to uni-axial stress, that is, stress in one direction, in practice a multi-axial stress system must be allowed for. Moreover, it is not realistic to use the simple yield stress as the criterion in engineering design and practical engineers have long been in the habit of using a suitable *factor of safety*, *f*, in order to arrive at a maximum allowable stress. Generally this maximum allowable stress is given by Yield Stress/*f*, though in some cases the value, Ultimate Stress/*f*, is used.

Be that as it may, many components which have been designed such that the stresses arising from applied loads are insufficient to lead to plastic deformation, nevertheless fail in service. There are many causes for such failure. These are now generally recognised by designers and are considered, as far as is possible, in subsequent design. Amongst the more important potential causes of failure which can be examined quantitatively are brittleness, creep, fatigue and the chemical influence of the environment in which the component is operating. The first three of these will be considered briefly in this chapter whilst the subject of corrosion is discussed in Chapter Nineteen.

As will be apparent from a study of Chapter Four, the results associated with such different tests as the simple tensile test and the impact test are not comparable. Thus strength must not be confused with toughness since the two values are not interrelated. The Izod impact test assesses different properties from those examined in the tensile test which is in any case a relatively 'long-time' method of loading when compared with the impact test. Even longer periods of static loading, particularly at high temperatures, may involve creep (or gradual extension) which leads to failure; whilst the presence of stresses which are continually varying in magnitude or direction (though well within the yield stress) may cause failure by fatigue.

Fracture

16.2   The application of stress to any material will lead to the production of elastic and/or plastic strain and if the stress is increased progressively fracture will ultimately occur. This fracture can be classified as either 'brittle' or 'ductile' and the mode of fracture produced is governed by the stress at which it occurs in relation to the elastic/plastic properties of the material (Fig. 16.1).

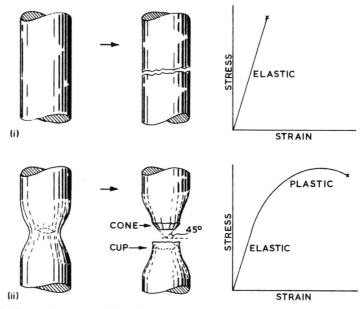

Fig. 16.1   Types of fracture: (i) brittle; (ii) ductile.

In the case of brittle materials failure occurs *before* any appreciable plastic deformation can take place by shear. This is typified in materials like cast iron, glass and concrete. Fracture of this type—also termed 'cleavage fracture'—follows paths between adjacent crystallographic planes which are generally those with the weakest atomic bonding. In addition to the materials mentioned, cleavage fracture is prevalent in all metals with BCC and CPH structures rather than in those which are FCC. Brittle fractures generally appear bright and 'granular' because of the reflection of light from the sections of individual crystals.

Ductile fracture takes place at some stress *above* the shear strength of the material so that some plastic flow precedes fracture. As the neck begins to form (Fig. 16.1(ii)) the stress within the necked region obviously increases and reaches a level where small cavities begin to propagate in the central region of the section. Such cavities will nucleate more easily if a second phase or particles of some impurity are present for which reason very pure metals are more ductile than impure ones. The cavities which have formed join up to form a crack which is roughly normal to the axis of stress of the test piece. Final fracture then occurs by simple shear at an angle of 45° to the axis of stress, shear stress being at a maximum in this direction. The resultant fracture is of the well-known cup-and-cone variety and occurs particularly in FCC metals.

The type of fracture produced is dependent mainly on the nature of the material and in particular on its lattice structure. It is also affected by other factors such as the condition of the material in respect of the amount of

(*Courtesy of J. A. Nicholls Esq., W.B.C.C.T.*)

Fig. 16.2   A typical 'cup-and-cone' fracture in a ductile carbon steel, obtained during a normal tensile test. (The specimen was a 0.1% carbon steel, normalised at 900°C.)

cold-work it may have received; the quality of its surface finish; its environment and temperature; and the rate of application of stress. The latter factor is illustrated by the behaviour of a material in a simple tensile test as compared with its behaviour during the Izod test where impact loading allows insufficient time for the adequate movement of dislocations to take place, so that brittle failure tends to result under some circumstances.

16.2.1 Griffith's Crack Theory   The fact that the real strength of a metal is but a very small fraction of the theoretical value (derived from a knowledge of inter-atomic forces) has already been noted (5.2.3). This has been explained in terms of the movement of dislocations the theory of which was developed more than half a century ago. Nevertheless the fracture of brittle materials was studied a decade or so earlier by A. A. Griffith who, in 1920 suggested that in any *brittle* non-metallic substance such as glass minute cracks or fissures, particularly occurring at the surface, act as stress raisers by concentrating stresses at their tips. This is a concept very familiar to engineers.

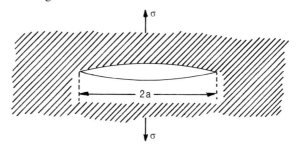

Fig. 16.3

For a small crack of elliptical section (Fig. 16.3) it can be shown that for a stress σ applied perpendicular to the major axis of the crack, length 2*a* (m), then:

$$\sigma = \sqrt{\frac{2\gamma E}{\pi a}}$$

where E is Young's Modulus for the material and γ is the *surface energy per unit area* (J/m^2).

The concept of 'surface energy' is perhaps best understood if we imagine a block of plastics material cut into two. The chemical bonds which formerly crossed the boundary between the two pieces of material will now transfer their forces to atoms within each of the newly created surfaces. The result of these atomic forces acting amongst themselves along the surface constitutes the surface energy. In fact surface energy is equivalent to the work done when bonds are broken during the process of creating new surfaces. Another manifestation of surface energy is in the *surface tension* of liquids. Globules of 'quicksilver' (Hg) are almost perfectly spherical in form, the forces acting along the surface of the metal being so great that the maximum volume of metal is contained within the minimum area of surface—a sphere. To subdivide a globule requires some dexterity since to produce extra new surfaces requires an input of energy. Similarly when a crack is produced in a brittle solid new free surfaces are generated and energy must be supplied to achieve this, for example, in the form of strain energy.

Thus as soon as the applied stress reaches the value σ, small cracks present can begin to propagate. As *a* increases σ must decrease and this can only be achieved by the rapid spread of cracks ending in catastrophic failure.

Griffith's Theory has been verified experimentally on glasses and brittle ceramic materials. In fact very fine fibres freshly drawn from a melt have strengths near to the theoretical value. However if these fibres come into contact with any object, or even with the atmosphere for short periods, these strengths are considerably reduced. This suggests that the strength of the fibre is very dependent on surface perfection and that anything which may initiate even the most minute surface irregularities will weaken it.

Q Examination of a ceramic material revealed a surface flaw 90 μm long lying normal to the direction in which stress was acting. Given that Young's Modulus for the material was 350 GPa and that the surface energy was 24 J/m² calculate the fracture stress for the material.

A Using Griffith's equation: $\sigma = \sqrt{(2\gamma E/\pi a)}$. In this case:

$a = 90/2 \times 10^{-6}$ m (i.e. half the flaw length in metres)
$E = 350$ GPa $= 350 \times 10^9$ N/m²
$\gamma = 24$ J/m²

Hence:

$$\sigma = \sqrt{\frac{2 \times 24 \times 350 \times 10^9}{3.142 \times 45 \times 10^{-6}}} = \sqrt{118.8 \times 10^{15}}$$

$$= 34.47 \times 10^7 \text{ N/m}^2 = \underline{344.7 \text{ MPa}}$$

16.2.2 Factors Leading to Crack Formation   It is reasonable to suppose that crack initiation in metals occurs most frequently in the vicinity of an included particle. When slip in a metal takes place past a rigid inclusion dislocations will pile up near to the inclusion/metal interface. Assuming that the inclusion itself does not shear, a minute fissure will begin to develop at this interface (Fig. 16.4). Obviously fissures can develop most easily when there is little adhesion between the inclusion and the metal containing it. Thus fissures can form more easily at a copper/cuprous oxide interface than at an aluminium/aluminium oxide interface since the degree of adhesion between metal and particles of its oxide is much greater in the latter case.

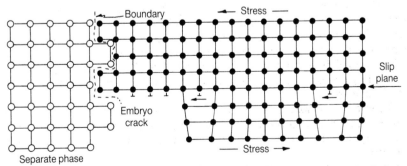

Fig. 16.4   A 'pile-up' of dislocations at a separate non-cohesive phase, e.g. an inclusion, leading to the formation of an embryo crack which will be propagated as stress increases.

Other barriers to the movement of dislocations can initiate micro-cracks in this manner. Thus a grain boundary can act as a barrier to the passage of dislocations so that a pile-up occurs and they become so closely packed that a micro-crack is nucleated. Such a crack grows by the addition of further dislocations which move towards it.

Crack formation also takes place by the movement of dislocations along close-packed or other slip planes within a crystal (Fig. 16.5(i)). These dislocations pile up at the intersection of the slip planes and so initiate a micro-crack (Fig. 16.5(ii)).

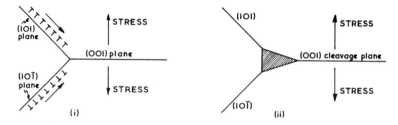

Fig. 16.5   The initiation of a micro-crack by the running together of dislocations (after Cottrell).

In some cases the presence of a discontinuity can arrest the propagation of a crack by reducing the stress concentration which is always present at the tip of a crack. Thus the propagation of a crack in a plate-glass window can be stopped by carefully drilling a hole in front of the advancing crack. Similarly brittle fracture in welded-steel ships (16.3) was arrested by including dummy rivet holes to arrest the spread of cracks. Some fibrous materials—notably the now-obsolete wrought iron—are tough along the direction of the fibres because any crack formed at right angles to the stress will then have to change direction on meeting a fibre barrier. If the interface between fibre and metal is weak this can actually *increase* the overall strength of the composite *in the direction of the fibres*. The weak interface reduces the stress concentration at the crack tip because the radius of the tip increases. These examples are governed by the approximate relationship:

$$\text{Stress concentration} = \sqrt{\frac{\text{Crack length}}{\text{Crack-tip radius}}}$$

i.e. as the crack-tip radius increases the stress concentration will decrease.

16.2.3 Fracture Toughness   In all cases of brittle fracture the mechanics of the process are similar. Resistance to failure is related to the energy which must be used to separate particles in the region of the fracture and here we are virtually describing the surface energy, γ, of the material. Catastrophic failure is likely to occur in metals which contain small cracks but which in other situations might deform plastically before failing. A 'coarse grey' cast iron consists largely of ductile ferrite but is brittle because of the presence of flake graphite (Fig. 10.10) which behaves as fissures in so far as lack of strength is concerned.

When dealing with engineering problems it is usual to combine γ with Young's Modulus in the form:

$$K_c = \sqrt{\frac{2\gamma E}{\pi}}$$

Here K_c is known as the *fracture toughness* or, alternatively, as the *critical stress intensity factor* of the material and is important in engineering design. It is measured in units $N/m^{1.5}$.

If σ_f is the stress necessary to cause catastrophic failure then:

$$\sigma_f = \frac{K_c}{\sqrt{\pi a}}$$

or $K_c = \sigma_f \sqrt{\pi a}$

where a is one half of the crack length (in metres). In situations where a is very small σ_f will be greater than the yield strength of the material and some plastic deformation will occur but when the crack is of such a size that $\sigma_f > \sigma_y$ then catastrophic failure occurs, that is, without any plastic deformation.

In hard, brittle substances such as ceramics the design strength is determined by the length of the microcracks present, i.e. if the longest microcrack present is of length $2a$ then the *tensile strength* is equivalent to $\sigma_{ts} = (K_c/\sqrt{\pi a})$.

Q A 1% nickel steel has a fracture toughness of 32 MPam$^{0.5}$ along with a yield strength of 494 MPa. What will be the limiting design stress if the maximum allowable quench crack present is 2.4 mm long, assuming that no plastic deformation occurs?

A The stress required to cause catastrophic failure is given by:

$$\sigma_f = \frac{K_c}{\sqrt{\pi a}}$$

where K_c is the fracture toughness (Pam$^{0.5}$ or N/m$^{1.5}$) and a is one half of the permissible crack length (m). Therefore:

$$\sigma_f = \frac{32 \times 10^6}{\sqrt{\pi \times 1.2 \times 10^{-3}}} = \underline{521.2 \text{ MPa}}$$

This indicates that σ_f is greater than σ_y (494 MPa) and since no plastic deformation is permitted for design purposes the limiting stress will be the yield stress, i.e. 494 MPa. However, if the specification also requires a Factor of Safety of 2 to be used the maximum design stress would be $494 \div 2 = 247$ MPa.

Table 16.1 *Typical Fracture Toughness Values for Some Common Engineering Materials*

Material	Fracture toughness (MPam$^{1/2}$)	Material	Fracture toughness (MPam$^{1/2}$)
Alloy steels	50–170	Nylons	3–5
Copper alloys	30–125	Polypropylene	3.5
Aluminium alloys	30–115	Scots pine	0.3 parallel to grain
Mild steel	20–50		6.0 perpendicular to grain
CFRP (epoxy)	30–50	Polythene (low density)	1–2
Cast iron	5–20	Soda glass	0.7
Zirconia	5–12	Concrete	0.2
Sialons	6		

Ductile-brittle Fracture and Transition Temperature

16.3   Some metals which are tough and ductile at ambient temperatures and above become very brittle at low temperatures. The appearance of their fractured surfaces varies accordingly. Low-carbon steel tends to become increasingly brittle as sub-zero temperatures are approached and it is possible that Scott's 1912 Antarctic Expedition would have met a less disastrous end had not brittle failure in the snow tractors in the early stages put the burden of transport solely on horses, dogs and, later, men alone. Little serious attention was paid to the problems of brittle fracture at low temperatures prior to the catastrophic failure of some of the all-welded 'Liberty' ships during the Second World War.

Under normal circumstances the stress required to cause cleavage is higher than that necessary to produce slip, but if for some reason slip is suppressed, brittle fracture will occur when internal stresses increase to the value necessary to produce failure. Tri-axial stresses acting in the material may be residual from some previous treatments and the presence of points of stress concentration may aggravate the situation. The 'Liberty' ships, mentioned above, were fabricated by welding together steel plates to form a continuous hull. Cracks often, though not always, started at sharp corners or arc-weld spots and propagated right round the hull so that the ship broke in half. A completely welded structure such as a ship or a large pressure vessel is in truth a single continuous piece of metal and so the uninterrupted propagation of these cracks was very rapid. Similar failure does not occur in riveted ships because any crack which begins to run will terminate in the first rivet hole it reaches. As mentioned above (16.2.2):

$$\text{Stress concentration at a crack} = \sqrt{\frac{\text{Crack length}}{\text{Crack-tip radius}}}$$

and since the crack-tip radius has been replaced by the radius of the rivet hole, which is extremely large by comparison, the stress concentration at the hole decreases almost to zero.

Though brittle fracture often starts at some obvious fault which acts as a stress-raiser this is not always the case and it seems likely that one or more of a number of possible contributory causes lead to this type of failure. It seems likely, however, that sluggishness in the movement of dislocations at low temperatures is common to all cases of brittle failures. Plastic flow depends upon the movement of dislocations and this requires some finite time. If stress increases very rapidly it is possible that it cannot be relieved adequately by slip and a momentary increase in stress to a value above the shear strength can cause failure.

As temperature decreases the movement of dislocations becomes more difficult so that internal stresses may exceed the shear strength at some instant. Brittle fracture is therefore more common at low temperatures, a fact supported by the failure of the 'Liberty' ships during service in the cold North Atlantic where high mechanical stresses due to rough weather combined with the effects of low temperature. In some instances, however, these ships broke up whilst in dock, the early morning sun having warmed the deck and so generated thermal stress. Such catastrophic fracture moved at a speed of 2 km/s.

16.3.1 Metals with FCC structures are generally more ductile than those with either BCC or CPH structures because of the greater number of available slip directions associated with the FCC structure (5.2.1). At low temperatures this becomes even more apparent so that BCC metals like W, Mo, Cr, Nb, Ta and of course Fe all exhibit low-temperature brittleness as do CPH metals like Zn and Be. BCC ferrite, the principal constituent of low-carbon steels, is particularly susceptible to brittle fracture which follows a transcrystalline path along the (100) planes. As indicated in Fig. 16.6 this occurs at a fairly low temperature and the temperature at

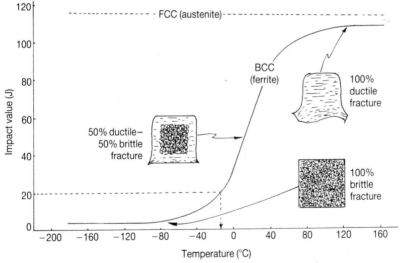

Fig. 16.6 The relationship between brittle fracture and temperature for ferritic and austenitic steels. Here a Charpy value of 20 J is taken as an alternative indication of the temperature of transition to a 50% ductile—50% brittle fracture of a ferritic steel.

which ductile fracture is replaced by brittle fracture is termed the *transition temperature*. However, as Fig. 16.6 shows the complete ductile→brittle transformation takes place over a *range* of temperatures rather than at a single temperature and it is more satisfactory to determine some temperature at which, say, 50% of the failed surface exhibits a ductile fracture with the remaining 50% showing brittle fracture. Alternatively the temperature corresponding to some arbitrary value of J(Charpy) can be judged as coinciding with the transition temperature. In Fig. 16.6 a notional value of 20 J has been taken and this would coincide with a representative transition temperature of −14°C.

Some elements, notably phosphorus, carbon and nitrogen raise the transition temperature in mild steel for which reason plates used in the building of ships' hulls must be low in phosphorus. Other elements, but in particular manganese and nickel, depress the transition temperature of mild steel. Thus the least expensive way of reducing its transition temperature is to increase the manganese/carbon ratio. A suitable steel, in which the transition temperature has been reduced to a safe low limit, contains 0.14% C and 1.3% Mn. When lower temperatures are involved a low-nickel steel must be used.

Creep

16.4 'Creep' describes the continuing slow deformation with passage of time in materials subjected to steady persistent stress. This deformation is plastic and occurs even though the acting stress is *below the yield stress* of the material. At temperatures below $0.4 \, T_m$ (on the Kelvin scale) the rate of creep is very small but at temperatures higher than this it becomes increasingly important. For this reason creep is commonly regarded as being a high temperature phenomenon associated with steam plant and gas-turbine technology. Nevertheless with some of the soft, low-melting point metals and alloys creep will occur significantly at ambient temperatures. It has been reported that some ancient lead roofs are measurably thicker at the eaves than at the apex, the lead having crept over the centuries even under its own weight. Creep in polymer materials below T_g follows roughly the same pattern as it does in metals (12.5.6).

16.4.1 When a metal is suitably stressed it undergoes an instantaneous elastic deformation (Fig. 16.7) but this is followed by a relatively low rate of plastic strain which is both *time and temperature dependent* as well as being dependent upon the applied stress. This plastic strain occurs in three distinct stages (Fig. 16.7):

(i)   primary, or transient, creep, OP, beginning at a fairly rapid rate which then decreases with time as strain hardening sets in;

(ii) secondary or steady-state creep, PS, in which the rate of strain is uniform and at its lowest value. In this stage the rate of work hardening is balanced by the rate of recovery.

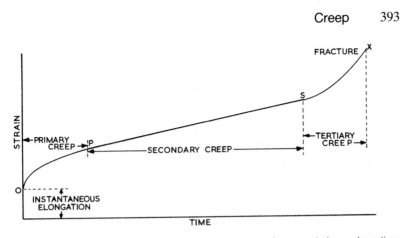

Fig. 16.7 Typical creep curve showing the three stages of creep during a long-time, high-temperature creep test.

$$\varepsilon = Ct \qquad (1)$$

where t is the time and ε the resultant strain due to creep. The creep rate, $d\varepsilon/dt = C$, where C is a constant for a given set of conditions of temperature and stress and is equivalent to the slope of the section PS of the creep curve. C will vary with both temperature and applied stress since recovery processes are both temperature and stress dependent. Temperature dependence is a consequence of the *diffusion* of dislocations caused by thermal activation and can therefore be represented by an Arrhenius-type equation:

$$\frac{d\varepsilon}{dt} = A \exp(-H/RT) \qquad (2)$$

where $d\varepsilon/dt$ is the secondary creep rate, H is the activation energy for creep, R is the universal gas constant, T is the absolute temperature and A is a constant.

The secondary creep rate also increases with increase of stress, the relationship being generally expressed by:

$$\frac{d\varepsilon}{dt} = \beta\sigma^n \qquad (3)$$

where β and n are constants, n usually varying between 3 and 8 for metals and between 1 and 2 for polymer materials.

Expressions (2) and (3) can be combined to give:

$$\frac{d\varepsilon}{dt} = K\sigma^n \exp(-H/RT)$$

(iii) tertiary creep, SX, in which the rate of strain increases rapidly so that fracture occurs at X. This stage coincides with the formation of microcracks at crystal boundaries followed by necking and rapid failure.

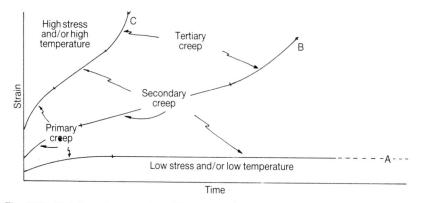

Fig. 16.8 Variation of creep rates with stress and temperature. In curve A creep ultimately becomes negligible, due presumably to work hardening. In curve C the secondary creep rate is higher than in curve B because of the use of a higher stress and/or a high temperature.

The relationships between stress, temperature and the resultant rate of creep described above are illustrated in Fig. 16.8. At a low stress and/or a low temperature some primary creep may occur but this falls to a negligible value in the secondary stage due presumably to strain hardening of the material. With increased stresses and/or temperatures (curves B and C) the rate of secondary creep also increases leading to tertiary creep and inevitable failure.

This set of curves (Fig. 16.8) relates the increase in strain with time; whilst temperature and stress are maintained at fixed values. Creep data may also be presented in other forms. For example, *isometric* stress/strain diagrams can be plotted (Fig. 16.9(i)) in which stress, strain and time are related for some single material X at a single fixed temperature *T*. Alternatively values of time, stress and temperature may be related to some fixed specific value of strain in what are termed *isochronous* curves (Fig. 16.9(ii)).

16.4.2 In creep testing a test piece in form similar to a tensile-test piece, is enclosed in a thermostatically controlled electric tube furnace which can be maintained accurately over long periods at the required temperature. A

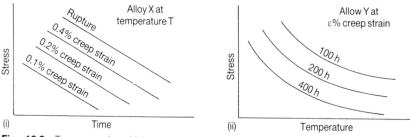

Fig. 16.9 Two ways in which creep measurements can be presented: (i) an isometric stress/time diagram; (ii) isochronous stress/temperature curves.

suitable tensile force is applied and the extension of the test piece determined at suitable time intervals by means of some form of sensitive extensometer. Generally a set of identical test pieces will be tested at the same temperature but at different stresses and from a study of the results it will be possible to arrive at a maximum stress which can be applied and which will result in no *measurable* creep taking place. This stress is known as the *limiting creep stress* or *creep limit*. Obviously this is not a very satisfactory measurement since it is dependent largely upon the sensitivity of the extensometer equipment. Moreover, formal creep tests can occupy long periods of time running into weeks or months depending upon the combination of stress and temperature involved and such tests are consequently very expensive to carry out.

For reason such as these it is generally more satisfactory to use some form of short-term test and then derive acceptable and usable information by extrapolation. Such information may not always predict when the tertiary or catastrophic stage of creep is likely to be reached and to deal with such a situation some authorities have proposed methods which involve accelerated creep tests in which a higher test temperature is used than will be experienced in service, the material being tested to rupture at this higher temperature. When fracture—often referred to as *creep rupture*—is specified then material selection would involve extrapolation of creep–rupture time with temperature and with stress. These higher temperature results are then used to predict the rate of creep over a long service life at a lower temperature.

One of the best-known methods is that of *Larson and Miller* and is based on the Arrhenius equation, modification of which gives:

$$\log_{10} . t_r + C = \frac{m}{T}$$

or $T(\log_{10} . t_r + C) = m$

where t_r = creep rupture time, T = temperature (K) and m = a function of stress which can be written as:

$P = f(\sigma)$ P being the Larson-Miller parameter

$P = T(\log_{10} . t_r + C)$

C is a constant the value of which is determined by plotting $\log_{10} . t_r$ against $1/T$ and measuring the intercept with the axis.

By conducting a creep test using certain values of stress and temperature the value of t_r can be determined. If this is repeated with other sets of values of stress and temperature the results can be plotted on a master curve (Fig. 16.10).

The value of parameter P holds good for a wide range of combinations of t_r and T, i.e. between short time–high temperature (test conditions) and long time–low temperature (service conditions). A number of other parameters, devised by other workers but similar in methods of application, are also in use.

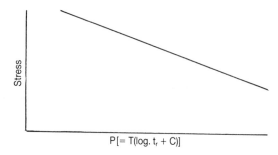

$$P[= T(\log. t_r + C)]$$

Fig. 16.10 A Larson-Miller master curve.

Q | A series of creep tests carried out at 500°C on a low molybdenum–vanadium steel gave the following results:

Stress (σ) (MPa)	Creep rate (ε) (%/10 000 h)
105	0.1
178	0.5
227	1.0

Calculate the maximum load which can be applied axially to a rod of similar steel operating at 500°C if creep is not to exceed 0.35 mm in 10 000 hours. The rod is 160 mm long and 500 mm² in cross-sectional area.

A | Permitted creep strain = 0.35/160 = 0.21875% in 10 000 hours. Steady secondary creep is related to stress by the equation:

$$\varepsilon = \beta . \sigma^n$$

To solve plot a graph of ln ε against ln σ:

σ	ln σ	ε	ln ε
105	4.654	0.1	−2.303
178	5.182	0.5	−0.693
227	5.425	1.0	0.000

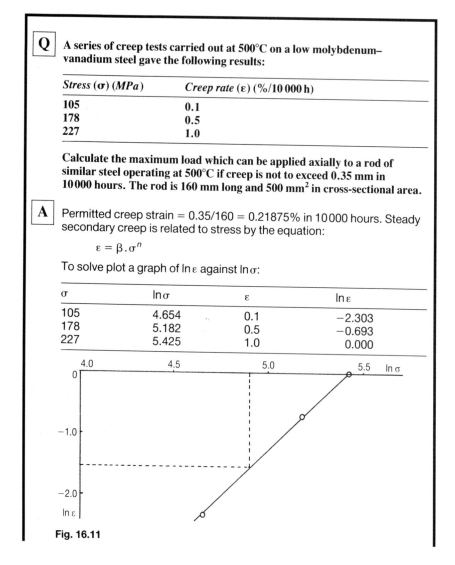

Fig. 16.11

For a creep strain rate of 0.21875%, $\ln \varepsilon = -1.520$. From the graph the corresponding value of $\ln \sigma = 4.895$. Therefore:

$$\sigma = 133.6 \text{ MPa} = 133.6 \text{ N/mm}^2$$

Since cross-sectional area of test piece is 500 mm², load is equal to:

$$133.6 \times 500 \text{ N} = \underline{66.8 \text{ kN}}$$

16.4.3 The Mechanism of Creep   Early study of the nature of creep showed that metals of small grain size were more prone to creep than were those of coarse grain size of similar composition and when tested under conditions of similar temperature and stress. Experiments show that creep rate is proportional to $1/d^2$ where d is the average grain diameter of the metal or alloy. This led to the realisation that creep is a phenomenon associated largely with the grain boundary regions since fine-grained metals contain a greater proportion of grain boundary material per unit volume than do coarse-grained metals. Moreover at low temperatures the grain-boundary material is harder and stronger than the material away from the boundaries but at high temperatures the position is reversed and the grain boundaries become weaker. At some intermediate temperature—appropriately called the *equi-cohesive temperature*—both grain boundary and crystal core are equally strong. Such a temperature is difficult to measure and is of little consequence in practice.

It is suggested that creep involves two separate modes of plastic deformation:

(i) that due to the normal migration of dislocations which occurs in crystalline materials;
(ii) that which is of a viscous character and is associated with sliding in the non-crystalline grain boundary regions.

In the primary stage of creep dislocation movements are initially rapid and various barriers such as solute atoms are easily overcome. New barriers are introduced as dislocation pile-ups occur at grain boundaries and elsewhere but thermal activation enables dislocations to surmount such barriers, though at a decreasing rate, so that the creep rate is reduced. Thus, as deformation continues, so does work-hardening, but since the temperature is high, recovery processes are also actively assisted by dislocation climb (5.8) and cross-slip which is in the motion of a dislocation from one slip plane into another intersecting plane. Both of these dislocation motions will oppose work hardening. Hence creep is essentially a process in which work-hardening is balanced by thermal softening. At low temperatures (curve A in Fig. 16.8) recovery does not take place and so unrelieved work-hardening leads to a reduction of the creep rate virtually to zero.

During the secondary creep stage the fact that work-hardening and recovery apparently balance each other gives rise to a constant rate of deformation by creep. It is generally believed that the climb of edge dislocations is the process which controls the rate of creep in this stage since it requires the higher activation energy. Other phenomena such as the annihilation of dislocations (5.4.3) will also assist the movement of other dislocations and restore the steady creep rate as the second stage proceeds.

The tertiary stage of creep coincides with the initiation of micro-cracks at grain boundaries. Some of these may form as the result of migration of vacancies so that they coalesce at points of high stress concentration. Other

Fig. 16.12   Methods by which micro-cracks may form due to slip at crystal boundaries.

voids may be generated by grain boundary sliding (Fig. 16.12) but by whichever method these micro-cracks are formed they will lead to necking and the consequent rapid failure of the member.

16.4.4 Creep Resistance   Because creep rate is low at temperatures below $0.4\ T_m$ it follows that metals with high melting point are likely to be useful where high creep resistance is required. Unfortunately many such metals— for example, tungsten ($T_m = 3380°C$)—are not only expensive but, because of the very 'stiffness' which contributes to their desirable creep resistance, also difficult to fabricate.

Close packing of the atoms in a metallic structure impedes the move-ment of dislocations and in consequence FCC and CPH structures tend to be more resistant to creep than do BCC structures. Thus nickel ($T_m = 1453°C$; FCC) and cobalt ($T_m = 1492°C$; CPH $\underset{}{\overset{400°C}{\rightleftharpoons}}$ FCC), both metals of intermediate values of T_m, are very useful in terms of their creep-resistant properties.

Since creep takes place by the movement of dislocations and by grain-boundary sliding, creep resistance will be increased if steps are taken to 'peg' dislocations in order to limit their movement and also to render grain boundary sliding more difficult. Creep resistance is therefore im-proved in the following ways:

(i)   by the addition of alloying elements which form a solid solution with the parent metal. This will only be really effective if the solute atoms have a *low mobility*. If on the other hand they diffuse freely with thermal activation they will also allow dislocations to travel so that recovery—and hence further creep—can proceed. Generally the greater the difference in size between solvent and solute atoms the greater the resistance to dislocation movement, but this is also likely to limit the extent of solubility. Nevertheless the addition of large amounts of alloying elements will probably lower the melting range of the alloy to the extent that it will be useless for high-temperature applications;

(ii) by the inclusion of alloying elements which will give rise to dispersion hardening. Coherent and small non-coherent precipitates are generally produced by precipitation treatment and it is essential that at the service temperature such particles should remain finely dispersed and not coagulate. Finely dispersed precipitates form effective barriers to the movement of dislocations, as in the Nimonic series of alloys (11.5.2) and other alloys containing finely dispersed carbides and oxides. Here it is the small dispersed particles which act as barriers to grain-boundary sliding and also to the movement of dislocations (14.1.5). Dislocations can only 'circumnavigate' such barriers if helped by a degree of diffusion assistance, so that metals with a low rate of self diffusion are superior in creep resistance;

(iii) by the promotion of a large grain size where this is appropriate. This reduces the proportion of grain-boundary material per unit volume which, in turn, limits the formation of vacancies and the density of grain-boundary slide. Since creep is a phenomenon associated with grain boundaries it would seem desirable to get rid of them completely! In fact this has been achieved with the high-temperature creep-resistant turbine blades of some jet engines such as the famous RB211—these components being carefully grown as *single* crystals.

Fatigue

16.5 Practising engineers have long been aware that 'live' loads and alternating stresses of relatively small magnitude can cause failure in a member which could carry a much greater 'dead' load. Under the action of changing stresses a material may become *fatigued*. So, whilst creep is a phenomenon associated with extension under a steady force acting for a long period, fatigue refers to the failure of material under repeated fluctuating stresses.

16.5.1 Early investigation into fatigue was carried out in 1861 by Sir William Fairbairn. He found that a wrought-iron girder which could safely sustain a static mass of 12 tonnes, nevertheless broke if a mass of only 3 tonnes was raised and lowered on to it some 3×10^6 times. From further investigation he concluded that there was some mass below 3 tonnes which could be raised and lowered an *infinite* number of times on to the girder without causing failure. Some years later the German engineer Wöhler did further work in this direction and developed the useful fatigue-testing machine which bears his name.

16.5.2 Many formal fatigue testing machines are based on the original Wöhler design. Here a standard test piece is rotated at high speed and carries at its end, cantilever fashion, a load transmitted by a ball-race

Fig. 16.13 This represents a test piece suffering alternating stress of range 2σ and a mean range of zero.

system (Fig. 16.14). With each revolution of the test piece the *amplitude* of the stress passes from σ_a in one direction to σ_a in the opposite direction so that the stress range is $2\sigma_a$ whilst the mean stress is zero. Testing is carried out until the test piece fails or withstands an 'infinite' number of reversals of stress—usually 20×10^6 in practice since an 'infinite number' would take too long a time! The number of reversals endured is recorded. This

Fig. 16.14 The principle of the Wöhler fatigue-testing machine. In this case the applied load is W (from which σ_a is derived) and the load range is 2W.

procedure is repeated for a series of applied loads applied to similar test pieces and the resultant stresses, σ_a, plotted versus the number of reversals, N, sustained giving what is usually termed a *S–N* curve (Fig. 16.15). For many steels the curve levels out at a horizontal asymptote (i) indicating that at any stress below σ_D the material *will not fail* by fatigue no matter how many reversals of stress are applied. The value σ_D is called the *fatigue limit* of the material.

Many non-ferrous metals and alloys, and also some steels operating under corrosive conditions, give *S–N* curves of the type shown in Fig.

Fig. 16.15 (i) A typical *S–N* curve for many steels; (ii) a typical *S–N* curve for many non-ferrous metals and alloys.

16.15(ii). Here there is no fatigue limit as such and any member will fail if it is subjected to the appropriate number of stress reversals even at extremely small stresses. In the case of these materials which show no fatigue limit an *endurance limit*, S_N, is quoted instead. This is defined as the maximum stress which can be sustained for N cycles of stress. Components in materials of this type must of course be designed with some specific life in mind and 'junked', as Americans so descriptively put it, before the number of cycles N (for stress S_N) has been attained.

16.5.3 Many fatigue testing machines are of the Wöhler type since it offers a simple and rapid means of testing. However, other forms of testing machine are in use in which stress can be applied in a more appropriate fashion, for example:

(i)   in which a push-pull force is applied axially;
(ii) in which the test piece is stressed torsionally by alternating torque;
(iii) in which reversed bending is applied to a test piece of rectangular cross-section by an alternating bending moment.

In some cases a specialised test procedure has to be developed as in the fatigue testing of complete aircraft frames.

16.5.4 The Wöhler form of test uses a system of loading in which the stress amplitude is σ_a but the mean stress σ_m is zero (Fig. 16.16(i)). In

(i)   (ii)

Fig. 16.16

practical engineering this will rarely be the case and whilst the amplitude of the applied stress may well be σ_a, the mean stress is not zero (Fig. 16.16(ii)). Experiments show that as σ_m increases the fatigure life of a metal decreases and so it has become necessary to devise modifications which take into account the effects of tensile *mean stress*, σ_m, in order to adapt results of the Wöhler test for use in engineering practice.
    A number of empirical equations named after their originators are available in order to relate σ_a and σ_m with σ_t the tensile strength, σ_y the yield strength and σ_D the fatigue limit (or σ_N the endurance limit where applicable). The most popular of these equations are:

$$\sigma_a = \sigma_D[1 - (\sigma_m/\sigma_t)] \quad \text{Goodman}$$

$$\sigma_a = \sigma_D[1 - (\sigma_m/\sigma_y)] \quad \text{Soderberg}$$

$$\sigma_a = \sigma_D[1 - (\sigma_m/\sigma_t)^2] \quad \text{Gerber}$$

Fig. 16.17 The relationships between stress amplitude, σ_a, and mean stress, σ_m: (i) the Goodman equation where any point within the hatched area represents conditions of σ_a and σ_m which will not cause failure due to fatigue; (ii) comparisons between Goodman, Gerber and Soderberg equations on predictions of failure.

The equations can be represented in graphical form (Fig. 16.17). Referring to the Goodman equation (Fig. 16.17(i)) any point representing the working conditions in terms of σ_a and σ_m which falls below the line connecting σ_D and σ_t (i.e. within the hatched area) indicates that failure will not occur due to fatigue. Similarly points below equivalent lines representing the Gerber and Soderberg equations (Fig. 16.17(ii)) will also indicate safe working conditions with respect to those equations; thus Gerber gives the most optimistic forecast.

It is found that the greater proportion of fatigue test results fall within the area between the two lines representing the Goodman and Gerber equations (Fig. 16.17(ii)). Nevertheless the Soderberg equation is very popular because it offers a greater margin of safety, working as it does on stresses below the yield stress, σ_y. This margin is often increased still further by applying a factor of safety, F, to *both* the alternating stress σ_a, and to the mean stress σ_m, so that the Soderberg equation becomes:

$$\sigma_a = \frac{\sigma_D}{F}\left[1 - \left(\frac{\sigma_{m.F}}{\sigma_y}\right)\right]$$

Q A steel bolt 10 mm in radius is subjected to a static mean force (P) of 19.2 kN. What is the maximum value of reversed direct fatigue load which the bolt will withstand assuming that the fatigue limit (under mean stress conditions of zero) is 287 MPa? Use the Soderberg equation and assume that the yield stress of the steel is 364 MPa.

A The static mean stress acting on the bolt is given by:

$$\sigma_m = \frac{P(\text{MN})}{A(\text{m}^2)}$$

$$\boxed{A = \pi \times (0.01)^2 \ (\text{m}^2)}$$

$$= \frac{19.2 \times 10^{-3}(\text{MN})}{\pi \times (0.01)^2 (\text{m}^2)}$$

$$= 61.12 \ \text{MN/m}^2 \ (\text{MPa})$$

The Soderberg equation:

$$\sigma_a = \sigma_D[1 - (\sigma_m/\sigma_y)]$$

$$= 287\left[1 - \left(\frac{61.12}{364}\right)\right]$$

$\sigma_D = 287$ MPa
$\sigma_m = 61.12$ MPa
$\sigma_y = 364$ MPa

$$= 238.8 \text{ MN/m}^2$$

The load is given by $\sigma_a \times A$, i.e.

$$238.8 \times \pi \times (0.01)^2 = 0.075 \text{ MN} = \underline{75 \text{ kN}}$$

16.5.5 Cumulative Damage   The effects of fatigue are cumulative and it is difficult to predict the fatigue life of components which work under varying conditions of stress. For example the flight of an aircraft through stormy weather will result in a greater reduction in its fatigue life than a similar flight in calm conditions. The assessment of such damage becomes particularly important when we are dealing with materials in which the fatigue properties are measured by a finite endurance limit rather than by a fatigue limit. Various attempts have been made to assess the fatigue life of components subjected to such variable stresses. Many such rules are complex and possibly the simplest and most widely used is known as *Miner's Law*. Miner suggested that in a component which was being fatigued, cumulative internal damage was taking place and that the extent of this internal damage was directly proportional to the number of cycles for a specific stress level. Thus, suppose a component undergoes n_1 cycles in a cyclic loading situation which would lead to failure after N_1 cycles, then n_1/N_1 is the proportion of damage which has occurred. If the stress amplitude now changes for a further n_2 cycles and at that amplitude failure would have occurred after N_2 cycles then n_2/N_2 is a measure of the proportion of damage caused during that period. Miner's Law states that the component will fail when the sum of all these 'cyclic ratios' equals unity, i.e.:

$$\sum_{N}^{n} = 1$$

or $$\frac{n_1}{N_1} + \frac{n_2}{N_2} + \frac{n_3}{N_3} + \ldots = 1$$

In practice it is found that $\sum_{N}^{n}$ yields a value somewhere between 0.5 and 2.0 and that this value is governed by the order in which the variations in σ_a are applied. Thus large values of σ_a followed by smaller values will give a shorter fatigue life than the reverse situation, that is, small values of σ_a followed by larger ones.

Q The following values were obtained during a series of endurance tests on an aluminium alloy.

Stress Amplitude, σ_a (MPa)	Reversals of Stress, N_f
141	2010
127	10500
121	25020
113	42100
112	60500
104	101000
95	180400

How many further reversals of stress would a similar specimen of the same material withstand at a σ_a value of 121 MPa given that it remained unbroken after a succession of stress reversals at the stated values of σ_a and N_f: 4100 at 127 MPa; 14300 at 112 MPa; 23200 at 104 MPa and 16450 at 95 MPa?

A If X is the required number of stress reversals then, using Miner's Law:

$$\frac{n_1}{N_1} + \frac{n_2}{N_2} + \frac{n_3}{N_3} + \frac{n_4}{N_4} + \frac{n_5}{N_5} + \frac{n_6}{N_6} + \frac{n_7}{N_7} = 1$$

so $$\frac{4100}{10500} + \frac{14300}{60500} + \frac{23200}{101000} + \frac{16450}{180400} + \frac{x}{25020} = 1$$

and $$0.3905 + 0.2364 + 0.2297 + 0.0912 + \frac{x}{25020} = 1$$

so: $$0.9478 + \frac{x}{25020} = 1$$

and: $$x = 1306 \text{ cycles}$$

16.5.6 Some Causes of Fatigue Failure   The fatigue characteristics of most engineering materials are now well known or can be measured with some accuracy. It would therefore be reasonable to expect that fatigue failure would not arise due to inadequate design on the part of the engineer! Yet fatigue failure *does* occur and such failure can usually be traced to:

(i)   the action of unforeseen factors in the working environment;
(ii) surface or subcutaneous flaws—the presence of either microscopical faults in the original material or a rough surface finish due to poor machining practice.

As an example of the former I was once called out to a factory site to investigate the failure by fracture of a stout copper tube which had carried, cantilever fashion, a small pressure gauge of but a few grams in mass. After a few days' service the tube had broken without warning. The fracture was of the characteristic fatigue type and when I visited the scene of failure the reason for this was immediately apparent. The gauge had been attached by means of the copper tube to an air compressor the vibrations from which could be experienced from a distance of several hundred metres. Since copper is represented by an $S-N$ curve of the type shown in Fig. 16.14(ii) it is obvious that even very small stresses will cause failure when the number of stress cycles reaches N. Here an unforeseen influence in the environment viz. high-frequency vibrations carrying stress reversals, was at fault but it was some time before I could convince the engineer-in-charge that the copper from which the tube had been manufactured was of perfectly good quality and not contaminated by some mysterious but highly undesirable impurities.

The disasters to the early Comet airliners followed a similar pattern in that fluctuations in cabin pressures in this case caused fatigue failure. Many causes of fatigue failure are due to unforeseen and sometimes undetected vibrations giving rise to reversals of stress at values *above* S_D (or S_N). Such vibrations are often sympathetic.

If we begin with the assumption that the engineering design has been impeccable then we must look elsewhere for the cause of failure. Small cavities at or near the surface will of course act as stress raisers and the same may be true of small inclusions of impurities, particularly when these are weak and brittle and consequently have a similar effect on properties to a cavity. Such faults may raise the local stress above σ_D (or σ_N) and so initiate a crack. However surface quality is probably of greater importance since poor surface finish, e.g. the presence of *machining tool marks* will tend to nucleate fatigue cracks.

If quench cracks occur in components such as heat-treated steel shafts they are certain to develop as fatigue cracks; whilst corrosive conditions in the environment particularly when stress corrosion (19.5) is involved, can also generate stress raisers in the form of etch pits and grain boundary fissures which can nucleate fatigue cracks.

16.5.7 The Development and Growth of Fatigue Cracks   A fatigue crack is likely to be generated by stress concentrations arising from any of the surface or subcutaneous defects enumerated above. Nevertheless in a material which is free of blemishes continual cyclic stressing produces a state of plastic flow even in regions where the stress range is still *below* the normal static yield stress. Such slip bands as do appear on the surface are either *intrusive* or *extrusive* (Fig. 16.18(i) and (ii)). Although such an intrusion is very small, being of the order of 1 μm long and 0.5 pm thick, it can act as a stress raiser and initiate a fatigue crack.

A fatigue fracture is generally considered to develop in three stages— nucleation, crack growth and ultimate failure(Fig. 16.19). Following the initiation of a crack by one of the procedures outlined above, the second stage, that of crack propagation, begins. As indicated in Fig. 16.18(iii) the crack emanating from the initial intrusion (ii) first changes course and becomes normal to the direction in which stress is acting. The crack tip

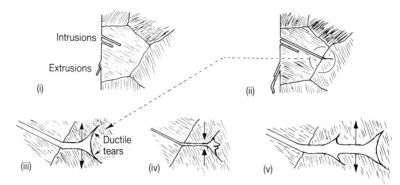

Fig. 16.18   The initiation and development of a fatigue crack. (i) Intrusions and extrusions may form at the surface due to slip along suitable crystallographic planes; (ii) intrusions progress, leading to stress concentrations and the formation of incipient cracks; (iii)–(v) alternating stresses lead to alternating tension and compression at the crack tip. This results in the progressive spread of the crack and the appearance of striations or 'ripples' on the fractured surface when this becomes revealed on ultimate failure.

then comes under the influence of a succession of tensile and compressive stresses. As compression changes to tension the crack opens and a system of shear stresses begins to operate at the crack tip. As the tensile stress increases plastic deformation and tearing takes place leading to crack growth, followed by blunting of the crack tip. When the cycle enters on its compressive phase the shear stresses are reversed and as a state of maximum compression almost closes the crack plastic flow also reverses. This causes a new striation to form and also leads to the re-opening of the crack tip. A succession of striations is generated in this way on the fractured surfaces. As the crack propagates *slowly* from the source the fractured surfaces rub together due to the pulsating nature of the stress and so the surfaces become burnished. Often conchoidal markings are present showing the direction of spread of the fatigue crack.

During this second stage of crack propagation the rate of crack growth da/dN is governed by the *Paris-Erdogan* Law:

Fig. 16.19   The stages of fatigue failure. A fatigue fracture is generally easy to identify as the crack-growth region is burnished by the mating surfaces rubbing together as stresses alternate. The ultimate fracture is crystalline.

$$\frac{da}{dN} = C(\Delta K)^m$$

where C and m are constants associated with the material (m usually being in the region of 2 to 5 for most ductile materials). ΔK is the range of stress intensity factor (K) and there is a value of ΔK below which the crack will not propagate.

Stage three of crack propagation is sudden and catastrophic and describes the ultimate fracture of the section when the unbroken portion is no longer able to sustain the load. This portion has the typical 'crystalline' appearance of a brittle fracture.

17

Methods of Joining Materials

17.1 At about the time that Diogenes, the original Cynic, was living in his tub, a fellow Greek, Archytas of Tarentum, is believed to have invented the screw. By the second century B.C. this device was being used extensively by the Greeks in their wine presses, whilst a less exotic use was made of the screw by Archimedes in his machine for raising irrigation water. Strangely, no one seems to know when the screw was first used as a fastening device.

Primitive man experienced great difficulty in joining his materials. He used sinews and thongs of hide to tie together the framework of his huts and also to tie on the heads of his stone axes. By 3000 B.C., however, the coppersmiths of Egypt were producing nails whilst, a little later, the woodworkers of that region were using quite sophisticated six-layer plywoods cemented by animal glues. Almost certainly by 3000 B.C. the smith had mastered the art of pressure welding using hammer and anvil but it was less than a century ago that the development of welding technology, as we now understand it, really began.

Adhesion and Adhesives

17.2 When the surfaces of two solids are brought sufficiently close together the interaction of surface forces will give rise to some degree of adhesion where the two surfaces are within roughly atomic range. This adhesion often manifests itself as friction if the two surfaces are made to slide relative to each other and the heat energy produced is a measure of the work done against these inter-surface forces of attraction. Rubber tyres do not skid if they can make contact with a dry road surface.

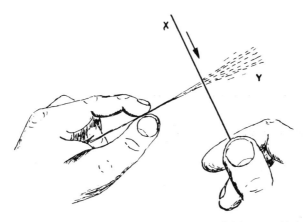

Fig. 17.1 Adhesion between *clean* glass fibres. If X is drawn lightly across Y, the fibres adhere causing vibration of Y due to the pulling action of X.

17.2.1   Two *freshly drawn* glass fibres will adhere to each other if they are brought into contact. This can be demonstrated if the fibres are held between the finger and thumb of each hand (Fig. 17.1) and the upper fibre (X) then gently drawn across the lower (Y). The motion will be found to be 'snatchy', indicated by the deflection and jerking of the lower fibre. If the two fibres are now drawn through the fingers and the experiment repeated the snatching motion is reduced almost to zero. Presumably grease from the fingers produces a lubricating film spacing the glass surfaces apart so that adhesion is no longer possible. Similarly if the freshly drawn fibres are allowed to remain in contact with the atmosphere for some time they will no longer adhere, apparently due to contamination of the surfaces with dust, moisture and fumes.

17.2.2   In practical terms an adhesive may be regarded as any substance capable of holding materials together by surface attachment. Typical adhesives include glue, mucilage, paste and a wide range of polymeric cements. Generally the mechanism by which surface attachment occurs is:

(i)   *specific adhesion* which depends principally upon the action of van der Waals forces (2.4) between molecules in the adhesive and the surface layers of the materials (*adherands*) being joined;
(ii) *mechanical adhesion* in which the adherands are to some extent keyed together by molecular filaments of the adhesive which penetrate into the substrates* of the adherands.

Since materials like glass and metal can be successfully bonded by adhesives it is reasonable to suppose that specific adhesion is the more important of these two possibilities. Generally speaking the larger the molecules involved the greater the adhesion since van der Waals forces will

* This term refers to the surfaces and those regions near to the surfaces of the adherands.

then be greater. Consequently most adhesives are organic compounds of very large and complex molecules having a high density of polar groups. The more common polar groups met with in adhesives include hydroxyl, carbonyl, nitro, nitrile, amide, sulphonic and some others. It is reasonable to suggest that the stickiness of sugar is due to the presence of hydroxyl groups in sugar molecules.

The strength of the adhesive bond is derived from the sum of all the van der Waals attractions between the molecules of the adhesives and neighbouring molecules in the substrate of the adherand. The molecules themselves will be strong due to the covalent bonds operating within them. However, not all large polymer molecules provide good adhesion. For example the fluoro-carbon grouping is associated with such poor adhesive properties that PTFE (Table 12.2) is used as a non-stick coating and also as a bearing material. Needless to say *cohesion* within the adhesive itself is just as important as *adhesion* between the adhesive and the adherand, otherwise failure in the adhesive layer will be likely. For this reason mobile liquids such as alcohol and nitrobenzene, whilst having suitable polar groups, are useless as adhesives since there is little cohesion between their relatively small molecules. In adhesives molecules are often so large as to have molecular masses of several hundred thousands or even millions. The sum total of polar bonds each molecule then provides is very great.

Another important feature is that the adhesive must mould itself perfectly to the surfaces being joined in order that its molecules will remain as close as possible to the molecules in these surfaces. For this reason ordinary glue must be applied hot whilst it is sufficiently mobile. Many polymeric adhesives are of the cold-setting type in which two mobile liquids are mixed together and then proceed to harden by polymerisation in the manner of cold-setting plastics materials (12.3.2) after the joint has been made.

17.2.3 So far only the level of adhesion between adhesive and adherand has been considered. The mechanical properties of the hardened adhesive layer are obviously important. As a modern joining method in engineering practice an adhesive is useless if it produces a brittle joint so that high toughness as well as shear strength is essential. Modern polymeric adhesives can be designed to give different combinations of these properties as indicated in Fig. 17.2. Here, curve B represents the combination of properties in a modern high-performance adhesive in which the elastic shear stress limit is approximately 40 MPa and the maximum shear stress 50 MPa. The more extreme values represented by curves A and C can be obtained by increasing the degree of cross linking during setting of the polymer; curve A in which maximum stiffness would give good high-temperature properties; or by decreasing the degree of cross linking, curve C, the latter being typical of the many acrylic-based adhesives. The length of the near-horizontal parts of A, B or C is a measure of the 'toughness' of the adhesive. In some a shear strain of 100% can be obtained with an adhesive of type C.

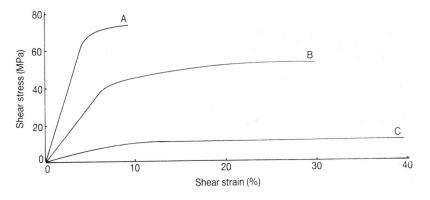

Fig. 17.2

17.2.4 Adhesives based on epoxy-type resins will be familiar to most DIY readers. In these the speed of setting—an important factor in industrial production—can be raised by formulation, as can the value of T_g in relation to the specified working temperature.

Table 17.1 *Adhesives*

Group	Adhesive substance (or source)	Materials joined
Animal glues	Animal hides or bones Casein (from milk) } protein materials Blood albumen	Wood, paper, leather and fabrics
Vegetable glues	Starch Dextrines } carbohydrates	Paper and fabrics
	Soya beans—protein	'Sizing' of paper
Elastomer materials	Natural rubber and its derivatives—latex	Rubber, sealing strips (automobiles), gaskets, weatherproofing, veneers, linoleum tiles, footwear industries
	Synthetic rubbers—styrene/butadiene, butylene, nitrile, neoprene and carbonylic elastomers	Nylon, footwear, metal/metal, adhesive tapes, tyre construction, cardboard industries, aircraft industries, bonding of abrasives in grinding wheels
Thermoplastic resins	Vinyls, e.g. polyvinyl acetate	Wood furniture assemblies, paper industry
	Cellulose derivatives, e.g. cellulose nitrate	Glass, balsa wood, paper, china
	Acrylics, e.g. polymethyl methacrylate	Optical elements
Thermosetting resins	Phenol formaldehyde, resorcinol formaldehyde	Weather-proof plywoods, cellulose acetate, nylon
	Epoxies	Optics, glass, polystyrene, nylon, metals—very versatile adhesives for dissimilar adherands

A more detailed list of engineering adhesives is given in Table 17.2 near the end of this chapter.

Soldering and Brazing

17.3   These processes differ from fusion welding in that the solder which is used melts at a *lower* temperature than the metal parts being joined so that no direct fusion of the latter occurs. It is essential, however, that a solder shall 'wet' the metallic surfaces it is meant to join, that is form a liquid solution with them. On solidification this liquid solution should be replaced by a solid solution if maximum toughness of the joint is to be attained, together wtith crystalline continuity across the interface. The formation of an intermetallic compound on the other hand will lead to brittleness. For example, the use of a tin-base solder to join copper may produce a brittle joint since at some point between the pure copper parts and the centre of the solder joining them, the metals copper and tin will be present in the correct proportions to form very brittle compounds such as $Cu_{31}Sn_8$ and Cu_6Sn_5. If such a joint is torn apart it will show a bluish fracture due to the presence of $Cu_{31}Sn_8$ in the surface through which failure has occurred. For this reason copper is soldered with a lead–silver solder in high-class work. Copper does not form intermetallic compounds with either lead or silver.

During soldering and brazing processes oxide films tend to act as a barrier to successful 'wetting' of the work pieces by the solder. These oxide films are usually dissolved by some suitable flux in order to expose a clean metallic surface to the alloying action of the solder. Nevertheless some metals, such as aluminium, which have tenacious oxide films are particularly difficult to solder.

17.3.1   In ordinary 'soft' soldering tin–lead base alloys are used. Ideally tin-man's solder should be of eutectic composition (62% Sn–38% Pb) so that it solidifies sharply at 183°C thus reducing to a minimum the time required to produce a joint and also the temperature to which the work piece is exposed. Since tin is extremely expensive relative to lead, commercial solders often contain 50% or more lead. Until the earlier part of the twentieth century domestic plumbing used lead pipes. These were joined or repaired using a tin–lead solder which was 'wiped' around the joint whilst in a part-liquid, part-solid state. In order to achieve this the solder required as wide a liquidus–solidus range as possible. For this reason plumber's solder contained about 67% Pb–33% Sn and solidified between about 265° and 183°C. As the poisonous nature of lead became more apparent lead pipes used in plumbing were replaced by copper and these copper pipes were joined using a lead-free solder (e.g. 98% Sn–2% Ag, molten between 221° and 223°C). Nevertheless the man who installs our copper-pipe system is still referred to as a 'plumber' (Latin: *plumbum*—lead (Pb).

Soft soldering may be accomplished with the aid of some form of soldering 'iron'. In large-scale production, however, a mixture of solder and flux may be preplaced in the joints and the assemblages heated *en masse* by high-frequency electrical currents. In the electronics industry many ingenious methods are used for soldering components to circuit boards. Thus in 'wave' soldering the components are pre-placed in position on the board which is then suspended above a trough of molten solder. A wave is then generated along the surface of the solder so that it touches and 'wets' the contact points to form a joint.

17.3.2 In brazing the soldering iron is replaced by a gas torch. Alternatively brazing solder/flux mixture can be pre-placed and the work pieces heated by high-frequency induction; salt baths; furnace or dip methods. Brazing solders contain between 50 and 66% Cu (balance–Zn). This produces a low-melting point alloy but one which would be extremely brittle were it not for the fact that, during brazing, some zinc is lost so that the resultant joint consists of a tough, ductile α brass instead of a hard, brittle γ structure (Fig. 11.1). Silver solders contain varying quantities of silver in order to reduce the melting range and so produce a brazing alloy with a melting point as low as 690°C.

Welding

17.4 The joining of pieces of metal by first heating them to a temperature above that of recrystallisation and then forging them together is a *pressure*-welding process which the blacksmith has operated for many centuries. A number of mechanical pressure-welding processes are now used, but in the meantime *fusion* welding has gained the ascendancy in the ever-widening range of processes which continue to be introduced and developed.

Fig. 17.3 A classification of welding processes.

Fusion-welding Processes

17.4.1 Gas Welding using the oxy-acetylene torch is now used mainly for jobbing and repair work. By using suitable mixtures of oxygen and acetylene temperatures in the region of 3000°C can be attained so that ordinary metals and alloys are quickly melted. This is necessary in order to overcome the tendency of metallic sheets and plates to conduct heat away from the weld zone so quickly that fusion is impossible.

Metal for the weld joint is, of course, supplied from a welding rod which is normally of a composition similar to that of the work pieces. In addition to metals, some thermoplastics such as rigid PVC are gas-welded.

17.4.1.1 Bronze Welding resembles brazing in that little or no fusion of the work pieces occurs since the bronze filler metal (generally 4.5% Sn; 0.5% P; balance—Cu) melts at a much lower temperature than the work pieces which are usually mild steel, cast iron or nickel alloys. At the same time bronze welding differs from brazing in the application of the filler metal which is in rod form whereas in brazing powdered solder is generally preplaced.

17.4.2 Electric Arc Welding Processes These now constitute by far the most important fusion-welding processes. The earliest electric-welding process employed a carbon electrode from which an arc was struck to the work itself. Weld metal was supplied from a filler rod, the end of which was held in the arc. This method is now virtually obsolete.

17.4.2.1 Metallic Arc Process This is the most widely used of the arc-welding operations. It makes use of a consumable electrode, that is, the electrode is a rod of suitable composition so that when an arc is struck between it and the work piece, the rod melts and provides the necessary weld metal (Fig. 17.4(i)). The electrode is coated with a flux which not

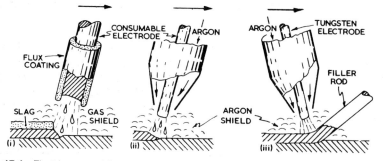

Fig. 17.4 Electric-arc welding processes. (i) The metallic arc—here molten flux provides protection for the weld, helped to some extent by a gas shield provided by burning carbonaceous matter in the flux coating. (ii) The MIG (metallic inert gas) process in which a blanket of argon gas protects the weld. (iii) The TIG (tungsten inert gas) process in which a separate filler rod is used, the tungsten electrode providing the arc but being itself non-consumable.

only provides a protective slag cover for the weld but also assists in stabilising the arc. Although it is used principally for steel, metallic-arc welding is also applicable to many non-ferrous metals and alloys.

17.4.2.2 Submerged-arc Welding   essentially an automatic process used for straight-line joining of metals. It is a variation of the metallic-arc process in which powdered flux is fed into the prepared joint just in advance of the electrode. As the flux melts it envelops the melting end of the electrode and so covers the arc. The weld area is thus protected by a coating of slag so that a smooth clean weld is produced.

17.4.2.3 Gas-shielded Arc Processes   used for welding those metals and alloys in which oxidation is troublesome. In these processes the electrode protrudes from the end of a tube from which an inert gas also issues. In the UK the noble gas argon is generally used whilst in the USA helium, a by-product of the oil-wells, is also available.

The *metallic-inert-gas* (or MIG) process employs a consumable electrode (Fig. 17.4(ii)) as in ordinary metallic-arc welding, whilst in the *tungsten-inert-gas* (or TIG) process a permanent tungsten electrode is used to strike the arc and an external filler rod used to supply the weld metal (Fig. 17.4(iii)). These processes were developed during the Second World War in order to weld alloys containing magnesium, a metal which reacts with most gases, including nitrogen. For this reason relatively expensive inert gases must be used to provide the protective atmosphere. These processes are now used more widely for the welding of other metals and alloys.

The CO_2 *process* was developed more recently and has become very popular for welding steel. It uses an atmosphere of carbon dioxide but since this can oxidise steel at high temperatures, filler rods containing deoxidants such as manganese and silicon are employed.

Although nitrogen tends to be regarded as an unreactive gas it dissolves in steel and gives rise to brittleness. It also reacts with several non-ferrous metals, particularly magnesium, and is therefore of no interest in gas-shielded welding processes.

17.4.2.4 Electro-slag Welding   This was introduced in Russia in 1953 for the joining of heavy steel plates in a vertical *position*. The molten metal is delivered progressively to the vertical gap between the plates rather as in an ingot-casting process. The plates themselves form two sides of the 'mould' whilst two travelling water-cooled copper shoes dam the weld pool until solidification is complete.

17.4.2.5 Plasma-arc Welding   a relatively new process in which a suitable gas such as argon is passed through a constricted electric arc so that it dissociates into positive and negative ions, this ion mixture being termed *plasma*. Outside the arc the ions re-combine and the heat-energy released produces an extremely hot 'electric flame'. Temperatures up to 15 000°C can be produced in this manner. Plasma is used as a high-temperature heat source in cutting, drilling and spraying very refractory materials like tungsten, molybdenum and uranium as well as in welding.

17.4.3 Thermit Welding   used mainly to repair large iron and steel castings. A simple mould is constructed around the parts to be joined, and above this is a crucible containing the Thermit mixture. This thermit mixture consists of powdered ferrous-ferric oxide, Fe_3O_4, and aluminium dust in molecular proportions. The chemical affinity of aluminium for oxygen is greater than the affinity of iron for oxygen hence, on ignition, aluminium reduces the iron oxide:

$$3Fe_3O_4 + 8Al \rightarrow 4Al_2O_3 + 9Fe \quad \Delta_h = -15.4\,MJ/kg$$

The heat of reaction is so intense that molten iron is produced and this is tapped from the crucible into the prepared joint.

17.4.4 Electron-beam Welding   When a stream of fast-moving electrons strikes a suitable target X-rays are generated (20.8.2). At the same time a considerable amount of heat is produced as a result of the conversion of the kinetic energy of the fast-moving electrons. If the electron beam is focused to impinge at a point very high temperatures can be produced. To be really effective the operation must be carried out *in vacuo* otherwise many of the electrons will lose their kinetic energy as a result of collisions with relatively massive molecules of oxygen and nitrogen present in the air space between electron 'gun' and target.

Electron-beam welding is used principally for the refractory metals tungsten, molybdenum and tantalum and also for reactive metals such as beryllium, uranium and zirconium which cannot oxidise if welded *in vacuo*.

17.4.5 Laser Welding   This has been used so far for micro spot-welding and, to some extent, for drilling and cutting. A *laser* (*L*ight *A*mplification by *S*timulated *E*mission of *R*adiation) device consists essentially of a generator of light pulses along with a suitable active substance so that a beam of light of very high intensity is emitted.

Pressure-welding Processes

17.5   The smith has been practising his art for at least 6000 years. In order to weld together two pieces of iron he first heats them to a high temperature and then brings them into contact under pressure by hammering them together on an anvil. Deformation above the recrystallisation temperature (5.8.3) induces recrystallisation which occurs across metal/metal interfaces so that a strong continuous joint is produced.

Most modern pressure-welding processes are similar in principle to smith welding except that the welding temperature is generally localised to the region where joining is to be accomplished.

17.5.1 Electric Resistance-welding Processes   These processes are used to produce either 'spot', 'seam' or 'butt' welds.

17.5.1.1 Spot Welding   In this process the work pieces are overlapped and gripped between heavy metal electrodes (Fig. 17.5(i)). A current sufficient in magnitude to produce local heating of the work pieces to a plastic state is

Fig. 17.5 (i) Spot welding; (ii) projection welding. (SEC = secondary of welding transformer.)

then passed through the electrodes. Since the metal between the electrodes is under considerable pressure a spot weld is produced. This method is particularly suitable for 'tacking' together plates and sheets, often preparatory to producing a more permanent weld.

17.5.1.2 Projection Welding   a modified form of spot welding in which the current flow and hence, the resultant heating effect, are localised by embossing one of the parts to be joined. This method (Fig. 17.5(ii)) is very suitable for joining heavy sections where spot welding would be difficult because of the big heat losses by conduction and consequently heavy currents required.

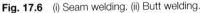

Fig. 17.6   (i) Seam welding. (ii) Butt welding.

17.5.1.3 Seam Welding (Fig. 17.6(i))   resembles spot welding in principle but produces a continuous weld by using wheel electrodes. Plastics can be joined by both spot and seam welding, the electrodes being heated by high-frequency induction since these polymer materials are of course non-conductors of electricity.

17.5.1.4 Butt Welding (Fig. 17.6(ii))   used for joining lengths of wire, rod or tubes. The ends are pressed together and heated at the joint by a current which is passed through the work. Since there will inevitably be a higher electrical resistance at the point of contact of the two work pieces, more heat will be generated there.

17.5.1.5 Flash Welding   somewhat similar to butt welding except that the ends are brought together momentarily in order to strike an arc between them. This melts away any irregularities so that when the ends are then pressed together a sound weld is formed.

17.5.1.6 Induction Welding In this process the work pieces are brought into contact under pressure and a high-frequency induction coil placed around the prospective joint. HF currents induced in the work heats it to welding temperature and the applied pressure causes welding to take place.

17.5.2 Other Pressure-welding Processes Several other pressure-welding processes have been introduced during the last two decades or so.

17.5.2.1 Friction Welding (Fig. 17.7(i)) can be used to join some mate-

Fig. 17.7   (i) Friction welding. (ii) Cold spot-welding.

rials in rod form. The rods are gripped in chucks, one rotating and the other stationary. As the ends are brought together heat is generated by friction between the slipping surfaces, ultimately sufficient to cause welding. Rotation is stopped automatically as welding commences. A modification of this process, used to join plastics, is referred to as spin welding.

17.5.2.2 Cold Welding in the form of lap, seam or butt welding involves making the two surfaces slide relative to each other under great pressure. Oxide films are disrupted and a cold weld forms between the surfaces. Spot welds are made by punching the two work pieces together (Fig. 17.7(ii)).

17.5.2.3 Explosive Welding can be used to 'clad' one sheet of metal with another. The 'flyer' plate, topped with a sheet of high explosive, is suspended at a slight angle to the target plate. On detonation of the explosive the flyer plate strikes the target at speeds of up to 300 m/s and with inter-face pressures of up to 7000 MPa. As a result the material in the region of the impact behaves for a short time as a fluid of quite low viscosity. The flyer plate virtually 'jets' along the surface of the target so that cold welding takes place (Fig. 17.8(i)).

17.5.2.4 Ultrasonic Welding employs ultrasonic vibrations of the order of 20000 Hz to produce slip between two surfaces which are being pressed together. The vibrations lead to disruption of the surface films of oxide so that the surfaces can cold-weld under the pressure used. The sonotrode tip (Fig. 17.8(ii)) clamps the work pieces to an anvil and vibrates laterally.

17.5.2.5 Diffusion Bonding used to join sheet metal. The sheets are held face to face under light pressure in a vacuum chamber. The temperature is

Fig. 17.8 (i) Explosive welding. (ii) Ultrasonic welding.

then raised to the range 0.5 to 0.8 T_m so that bonding occurs in not more than two hours during which time diffusion occurs across the interface so that the surfaces become knitted by a region of solid solution. Steel can be 'clad' with brass in this way.

The Structure and Properties of Welded Joints

17.6 The common welding processes involve the use of high temperatures so that the region containing the weld will be relatively coarse-grained in contrast to the remainder of the work piece which will generally be of a wrought, and consequently, fine-grained structure. If a fusion method of welding has been used the weld metal itself will show a coarse as-cast type of structure. Here the crystals will be large and other features such as coring and segregation of impurities at crystal boundaries may also be present. These features give rise to brittleness which results from weakness at the crystal boundaries. The effect of such faults can be minimised if it is practicable to hammer the weld whilst it is still hot. Not only will this induce fine grain on recrystallisation but it will redistribute more evenly those impurities segregated as a result of coring.

17.6.1 Poor welding technique may result in gas porosity and/or the presence of oxide particles trapped in the weld metal. Gas porosity may arise either from gas dissolved by the molten weld metal or from chemical reactions between such gases and constituents already present. Oxide particles are generally due to reactions between atmospheric oxygen and constituents of the weld metal or those dissolved from the surfaces of the work pieces.

Specific difficulties are associated with the welding of some metals and alloys. Thus, as the carbon content of plain carbon steels increases so do the difficulties in welding them. Carbon tends to oxidise during the welding process and porosity is likely to arise due to the gases released. Thin sections in carbon steel may cool so rapidly as to cause brittleness whilst much thicker sections in oil- or air-hardening steels may become brittle near to the weld as a result of martensite formation. The 13% Cr

Fig. 17.9 (i) Grain growth and martensite formation in and around a weld in 13% Cr stainless steel. (ii) Carbide precipitation in a non-stabilised 18–8 stainless steel. This will lead to 'weld decay'.

air-hardening cutlery steel is particularly difficult to weld satisfactorily since brittleness arises not only from excessive grain growth around the weld but also the formation of martensite in that region (Fig. 17.9(i)).

17.6.2 The 'weld decay' of some 18–8 austenitic steels has been mentioned elsewhere (10.5.1). If they have not been 'proofed' against such an eventuality these steels are normally quenched from 1050°C in order to retain any carbon in solid solution in the austenite. Heating in the range 500–900°C will cause the precipitation of carbide particles and this will lead to corrosion (19.7.1). During a welding process some region either side of the joint will be maintained in this temperature range for long enough for some such precipitation to occur and corrosion will subsequently take place in that region. Many items of plant used in the chemical industries are fabricated by the welding of stainless-steel sheets because highly corrosive conditions are involved. It is then even more important to reduce the tendency towards weld decay. Assuming that the welded structure cannot be solution-treated from above 1050°C—and obviously this will rarely be possible—one of the following means of limiting the precipitation of chromium carbide must be used:

(i)   the use of stainless steel with a carbon content *below* 0.03%;
(ii) the stabilisation of carbides under all conditions by adding niobium or titanium. No dissolved carbon is then available to precipitate as chromium carbide since any carbon present has already been 'tied up' by the strong carbide-forming niobium (or titanium).

17.6.3 Cast irons may be successfully welded if filler rods rich in deoxidants such as manganese and silicon are used. Even so the weld metal may absorb considerable amounts of carbon from the work pieces and since cooling will inevitably be rapid a very brittle white iron structure may be formed (10.6.2). For this reason cast iron is often joined by bronze welding, though subsequent electrolytic corrosion between the bronze weld and the surrounding cast iron must be expected if moisture is present in the environment. Cast iron, being anodic to the bronze weld, would quickly rust.

17.6.4 Aluminium alloys oxidise readily so that during ordinary metallic arc welding they must be protected adequately by a flux. For this reason the bulk of aluminium welding is now carried out with the gas-shielded arc, a process developed to deal with the even more difficult magnesium alloys.

Joining Plastics Materials

17.7 Since thermoplastic polymers can be softened by heat and will harden again on cooling they can be fusion welded by a number of processes. Thermo*setting* polymers, however, cannot be welded but are usually joined using suitable adhesives such as are used for many composite materials (17.9).

17.7.1 Welding Processes for Thermoplastic Polymers These can be divided into two groups: processes where heat is derived from mechanical movement and processes involving external heating.

17.7.1.1 Processes Involving Mechanical Movement These include *ultrasonic welding* (17.5.2.4) which is suitable for mass production industries since weld times of 0.5–1.5 s can be achieved. In the case of plastics materials the parts are held together under pressure and then subjected to ultrasonic vibrations of the order of 20–40 kHz at right angles to the contact area. Applications range from valves and filters used in medical equipment to cassette bodies and vacuum cleaner parts. *Friction welding* has been used for many years and is similar in principle to that used for joining metals (17.5.2.1). Welded acetal floats for domestic cisterns, aerosol bottles and pipe fittings are typical hollow products in addition to solid circular sections.

In *vibration (linear friction) welding* the two surfaces are pressed together using similar pressures to those employed in ultrasonic and friction welding (1–4 MPa). A *linear* vibration of 100–200 Hz is then used to cause fusion and welding in 1–5 s. Large complex linear joints can be made and the process is widely used in the automobile industries for components such as bumpers.

17.7.1.2 Processes Involving External Heating There are a number of techniques which make use of a heating tool. The simplest of these uses a hot flat plate which is clamped between the two surfaces to be joined until they soften. The plate is then withdrawn and the surfaces are brought together under controlled pressure for a specified period. After the fused surfaces have cooled the resultant joint has a strength of at least 90% of the parent material. Brake fluid reservoirs (where strong, liquid-tight joints are required) and lamp assemblies for motor vehicles, plastics pipes for water and sewage are joined in this way. *Hot-gas welding* is similar in principle to oxyacetylene welding of metals (17.4.1), the open flame being replaced by a stream of hot gas. The filler rod is of similar composition to the edges being joined and temperatures of 200–300°C are used. Gases are compressed air, CO_2 and nitrogen—the latter for materials such as polythene which are sensitive to oxidation. This process is used mainly for repair work of PVC, polythene, perspex, nylons and polycarbonates.

Implant welding involves trapping a metal insert between the two halves to be joined and then heating the insert by high-frequency induction so that the plastics material adjacent to the insert fuses to form a joint. The process is used for complex joints in large components such as vehicle bumpers and sailing dinghy hulls. In a similar process, *electrofusion welding* (Fig. 17.10),

Collar with
moulded-in
heater element

Pipe joint

(i)   (ii)

Fig. 17.10   The principles of electrofusion welding of British Gas pipes.

used by British Gas for joining the yellow polythene pipes up to 180 mm diameter, the pipe ends are inserted into a connecting collar which has a heater element moulded in. Resistance heating in this case completes the joint in, typically, 30 minutes.

Joining Ceramic Materials

17.8   Apart from purely mechanical methods such as screwing, shrink fitting and taper fitting ceramics can be bonded to other materials in a number of ways with varying success.

17.8.1 Adhesives   Conventional phenolic and epoxide adhesives are available but their use is restricted by an upper temperature limit of 150°C and a low mechanical stress limit. A typical application is the mounting of ceramic magnets in electric motors. High-temperature adhesives based on organic binding compounds, alkali-metal silicates and phosphates containing a powdered 'filler' material (carbon, alumina, silica, magnesia or zirconia as appropriate) offer service temperatures up to 2200°C. By adjusting the composition of the 'filler' material the coefficient of expansion of the adhesive can be regulated to match that of the work pieces and so minimise stresses generated during temperature changes. Unfortunately joints made this way do not generally reach standards of reliability required in engineering.

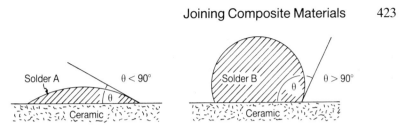

Fig. 17.11 Brazing solder B will not 'wet' the ceramic but brazing solder A will. 'Wetting' implies an angle of contact, θ, of less than 90° between the brazing alloy surface and the ceramic.

17.8.2 Brazing an established method for joining ceramics to ceramics and also ceramics to metals. The main problem is in finding a brazing solder which will 'wet' the ceramic surface (Fig. 17.11).

Generally it is necessary to metallise the ceramic surface either by vapour deposition, 'sputter' ion plating or by coating with a mixture of Mo, Mn and glass powders and then heating in a hydrogen atmosphere at 1500°C. The ceramic can then be brazed by conventional methods, generally using a brazing alloy of 72 Ag–28 Cu (eutectic) composition because of its moderate melting point (779°C) and good ductility to counteract thermal contraction mismatches.

17.8.3 Active Metal Brazing useful in joining oxide or nitride ceramics to steel. A metal which will react with the ceramic surface is added to the brazing solder and so promote 'wetting' of the surface. The best known modified solder is the 72 Ag–28 Cu eutectic to which 5–10% Ti is added mechanically.

17.8.4 Friction Welding (17.5.2.1) A number of ceramics have been successfully welded to metals, e.g. Al_2O_3–Al; SiC–Al and AlN–Al, the latter where aluminium heat sinks need to be attached to AlN components in the electronics industry. Usually the ceramic is held stationary whilst the metal is rotated rapidly in contact with it.

Joining Composite Materials

17.9 Many composites, particularly those with a polymer matrix, are sensitive to high temperatures. They are therefore most conveniently joined, either to each other or to metals, with some form of adhesive. Of the many adhesives available the three groups most suitable are: (i) epoxy; (ii) acrylic; (iii) polyurethane. Each of these groups is available in 'one- or two-part' systems. The 'one-part' adhesives are those which will polymerise and harden after application at ambient temperatures. The polymerisation process can be accelerated in most cases by raising the temperature. The 'two-part' systems, as the term suggests, consists of two separate liquids which, when mixed, begin to co-polymerise until a strong,

Table 17.2 *Some Structural Engineering Adhesives for Composite Materials*

Adhesive group	Strength classification	Most suitable for:	Comments
Polyurethane (one-part)	Very low	Not a true structural adhesive	Simple to use
Polyurethane (two-part)	Very low	Glass, thermoplastics and thermosetting materials	Fast setting possible. Some not suitable for structural work
Acrylic (one-part)	Low/medium	Bonding rubber and plastics materials to composites	Cyanoacrylates ('super glue')—bonds in seconds. Moisture acts as catalyst
Acrylic (pseudo one-part)	Medium	A true structural adhesive—glass, Al, steel, PVC, polyesters, epoxy, zinc	A film of catalyst is spread on one surface, adhesive on the other
Acrylic ((two-part)	Medium	Glass, Al, steel, ABS, PVC, polyesters, nylon, PPE, epoxy, zinc	Sets quickly—some are fortified with epoxy resins
Epoxy (one-part)	Very high	Al, steel, thermosets, epoxy, zinc, polyesters	Requires heat curing
Epoxy (two-part)	Medium/high	Al, steel, epoxy, polyester, glass, zinc, nylon, PU	Slow curing but can be accelerated by heat

NB All of these adhesives will adhere well to paint surfaces but the integrity of the paint may be suspect so that they are only used with catophoretic paints, i.e. paint films deposited electrostatically from baths.

tough, heavily cross-linked solid is the result. Again with some mixtures co-polymerisation is speeded up, and proceeds to give a higher degree of cross linking, by heating.

17.9.1 One or other of these adhesives will bond well with most materials but frequently some surface treatment is required to effect good adhesion. Alternatively additions are necessary to the adhesive to improve its bond with the adherand. Similarly 'rubbery' additions are sometimes made to improve the toughness of an otherwise brittle adhesive. Some characteristics of the main groups of adhesives are given in Table 17.2.

17.9.2 Metal matrix composites (MMCs) can be joined successfully by friction welding processes but fusion welding methods are limited in their application. Thus in the welding of Al-SiC MMCs the use of the TIG process (17.4.2.3) appears to be satisfactory but when electron- or laser-beam welding is attempted the higher temperature attained in the weld pool gives rise to adverse chemical reactions. In particular the SiC fibres are attacked by the molten aluminium of the weld pool:

$$3SiC + 4Al \rightarrow Al_4C_3 + 3Si$$

The aluminium carbide, Al_4C_3, so formed solidifies as an interpenetrating network of thin needle-like plates. Since Al_4C_3 is very brittle such a structure is indeed a materials scientist's nightmare. A weld containing a structure of this type would be brittle in the extreme. Nor does the horror story end there since the weld would corrode and disintegrate in a matter of days in the presence of moisture:

$$Al_4C_3 + 12H_2O \rightarrow 4Al(OH)_3 + 3CH_4$$

Metallic carbides which evolve methane in this way are known as *methanides*.

18

Electrical and Magnetic Properties of Materials

18.1 Inevitably a book of this type is concerned mainly with such mechanical properties of materials as concern the engineer but it is also appropriate to make brief mention of some of the more important electrical and magnetic properties since these are involved in so many important devices in modern engineering practice.

Electrical Conductivity

18.2 It was suggested early in this book that what we refer to as a 'current of electricity' in a metal is in fact a movement of electrons. Thus metals, in which the valency electrons may become de-localised and relatively free to move among the metallic ions which constitute the crystal structure of the metal, are consequently good conductors of electricity. Since electrons will repel each other they tend to move from regions of high concentration to regions of low concentration. Thus Ohm's Law governing the electric current follows the same principle as Fick's Law of Diffusion (7.4.2) governing the movement of solute ions.

Ionic crystals, however, are non-conductors since valency electrons are held rigidly by ions constituting the crystal. Similarly materials such as polymers, in which all chemical bonding is of the covalent type, are also non-conductors since shared valency electrons are unable to move out of their particular molecular orbitals.

18.2.1 Whilst an electric current in metals is due to a movement of electrons, in other materials positive or negative ions may be involved. Thus ions are responsible for conduction through electrolytes.

The Nature of an Electric Current

18.3 The 'electron cloud' theory is a convenient way of explaining some of the properties of metals but in some respects we must regard it as an oversimplification of the true state of affairs. Such shortcomings can be partially explained when quantum restrictions and Pauli's exclusion principle (1.3.3) are applied to the behaviour of electrons.

18.3.1 In a single atom the electrons occupy a number of discrete energy levels and these energy levels become closer together the further they are from the nucleus. In so far as properties like electrical conductivity are concerned it is the outer-shell or valency electrons which interest us. Thus the sodium atom has 11 electrons arranged in 4 shells (i.e. the second shell is divided into two sub-shells, $2s$ and $2p$). Shells $1s$, $2s$ and $2p$ are filled but

Table 18.1 *Electrical Resistivity of Some Materials*

Material	Resistivity $(10^{-8} \Omega.m)$ at 20°C	Temperature coefficient of resistivity $(\times 10^3)$ (0–100°C)
Aluminium	2.69	4.2
Chromium	12.9	2.1
Copper	1.67	4.3
Gold	2.3	3.9
Germanium	46×10^6	—
Iron	9.71	6.5
Lead	20.6	3.4
Magnesium	3.9	4.2
Mercury	95.8	0.9
Nickel	6.84	6.8
Platinum	10.6	3.9
Silicon	23×10^{10}	—
Silver	1.6	4.1
Tin	12.8	4.2
Zinc	5.92	4.2
80 Ni–20 Cr	109	—
18–8 stainless steel	70	—
55 Cu–45 Ni	49	—

the valency sub-shell $3s$ could hold another electron according to Pauli's exclusion principle. The electrons in each sub-shell occupy specific levels (Fig. 18.1(i)). If sodium atoms are now brought into close association, that is a crystal of sodium is considered, they begin to influence one another and the electrons are compelled to seek slightly different energy levels since— again according to Pauli's exclusion principle—not more than two electrons can occupy the same quantum state. Within even a small crystal, containing nevertheless many millions of atoms, the energy *levels* will be replaced by densely filled *bands* (Fig. 18.1(ii)). The sodium atom has only one valency electron so that only half of the energy states in the $3s$ (or valency) band are filled. For this reason the energy required to raise a valency electron to an empty state is negligible. This enables it to move freely within the crystal and so to conduct electricity.

In the case of, say, magnesium the $3s$ band is filled but since it actually

Fig. 18.1 Energy states of electrons in the sodium atom.

overlaps with the $3p$ band (Fig. 18.2) the valency electrons may have their energies increased within the $3p$ band so that magnesium conducts electricity very easily. However, in some materials the adjacent energy bands do not overlap in this way and energy is then required to move electrons across this 'energy gap'. Thus in the diamond allotrope of carbon the valency band is filled by the four outer-shell electrons and a large energy gap exists between this band and the next possible energy band.

18.3.2 In the case of diamond the magnitude of the energy gap is a measure of the amount of energy needed by an electron to break away from the covalent bond and become a free electron. In diamond this energy gap is great (about 8.5×10^{-19} J at 20°C) so that it has a very high electrical resistivity. The other elements of Group IV have similar valency electron structures and crystallise in patterns like that of diamond, that is with four covalent bonds per atom. In silicon and germanium, however, the energy gap between the 'valency band' and the 'conduction band' is much less than in the case of diamond. For silicon it is 1.8×10^{-19} J and for germanium only 1.1×10^{-19} J (at 20°C). Tin, also an element of Group IV, occurs as an allotrope, 'grey tin', stable below 18°C. This allotrope which is non-metallic in character, has a similar structure to that of diamond, silicon and germanium and an energy gap of only 0.13×10^{-19} J. Thus, in each of the elements silicon, germanium and grey tin, the energy gap is so small that they can easily receive enough thermal energy to cause the release of

Fig. 18.2 Energy bands in metals and metalloids which control electrical conductivity.

valency electrons into the conduction band. These elements are known as *semiconductors*.

The Electrical Conductivity of Metals

18.4   The high electrical conductivity of metals is due to the high mobility of the outer-shell or valency electrons which, in turn, is dependent on the ease with which these electrons can pass from the 'valency' band to the 'conduction' band.

Since an electric current is a flow of electrons amongst the positive ions in a metallic crystal any factor which impedes this movement will reduce conductivity. Thus an increase in temperature introduces greater thermal agitation as ions vibrate about their mean positions. This obviously reduces the mean free path which electrons can follow and consequently their mobility, resulting in a decrease in conductivity (Fig. 18.3).

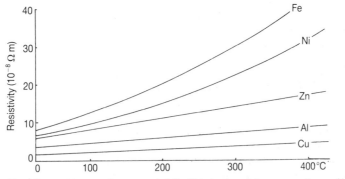

Fig. 18.3   The effects of temperature on resistivity. This is almost linear except near Absolute Zero where the resistivity of some metals falls very rapidly (18.4.2).

Similarly the presence of solute atoms, whether substitutional or interstitial, will increase the possibility of collisions between moving electrons and the ions constituting the crystal structure. Thus alloys are always less conductive than the pure metals from which they are made (Fig. 18.4). The conductivity of copper may be reduced by as much as 40% by the presence of only 0.05% phosphorus (Fig. 18.5).

18.4.1   The presence of lattice faults such as dislocations will also reduce conductivity by increasing the risk of collisions of the mobile electrons with atom cores in the lattice. For example heavily cold-worked copper is some 10% less conductive than annealed copper.

18.4.2 Superconductivity   The electrical resistivity of a metal decreases as its temperature falls for reasons suggested above. However, it was discovered as long ago as 1911 by Kammerlingh Onnes that the electrical resistivity of some metals falls suddenly to zero when they are cooled below

Fig. 18.4 The resistivity of copper–nickel alloys. Alloys have lower conductivities than pure metals.

a certain very low temperature, termed the *transition temperature*, T_c (Fig. 18.6). A number of remarkable experiments have been carried out with these superconductors. One well-known experiment is to cool a metal ring below T_c in a magnetic field at right angles to its plane. If the magnetic field is then altered a current is induced in the ring (this is virtually the principle of the dynamo). This induced current will flow almost indefinitely as long as the temperature of the ring remains below T_c. Such an induced current has been known to flow for several years without appreciable decay.

Most pure-metal superconductors are in the 'transition groups' of the Periodic Classification but some metals, which, by themselves, are not superconductors form intermetallic compounds which are. Thus gold and bismuth as separate pure metals are not superconductive whilst Au_2Bi is. This encouraged further research and by the 1950s a T_c value of 18.1 K had been achieved with the compound Nb_3Sn followed some years later by Nb_3Ge with a T_c of 22.5 K.

A considerable breakthrough came in 1986 when the first of the 'ceramic

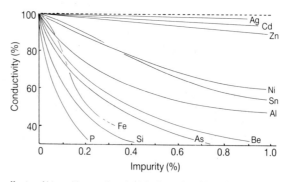

Fig. 18.5 The effects of impurities on the electrical conductivity of copper. Silver and cadmium have little effect and can be used to strengthen over-head conductors (e.g. telephone lines), whilst arsenic and phosphorus must not be present in copper destined for electrical purposes.

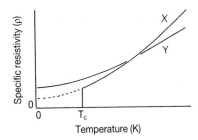

Fig. 18.6 Variations in specific resistivity with temperature for a superconductor (X) and a metal which does not exhibit this phenomenon (Y).

superconductors' was developed. It has a formula $La_{1.8}Sr_{0.2}CuO_4$ and a T_c of 40 K. It was virtually the compound La_2CuO_4 in which La ions had been replaced in some of their lattice sites by Sr ions. In the following year came the most famous of these superconductors to date—$Y.Ba_2Cu_3O_7$ (T_c—93 K). Whilst this temperature, 93 K or $-180°C$, is not one which would be associated with the Costa del Sol—or even the depths of a Siberian winter—$Y.Ba_2Cu_3O_7$ was the first superconductor to break the 'liquid nitrogen barrier' (77 K). The significance of this is that prior to its discovery superconductors would only operate as such when cooled in expensive liquid helium whereas $Y.Ba_2Cu_3O_7$ can be cooled in relatively inexpensive liquid nitrogen. As might be expected the race is now on to produce a superconductor with room-temperature superconductivity, and as I write the lead has passed to the ceramic $Tl_2Ba_2Ca_2Cu_3O_{10}$ with a T_c of 112 K—though I suspect this will have been overtaken before these lines leave the printer!

Theories which attempt to explain superconductivity are generally complex and beyond the scope of this book, but very briefly, some modern theories suggest that with superconductive metals and alloys the 'conduction electrons' become 'paired' when the temperature falls below T_c. This means that an electron with 'spin' in one direction becomes paired with an electron with 'spin' in the opposite direction. As a result of this pairing the electrons produce no resultant field and, since they suffer *no resistance*, can move freely through the lattice.

Unfortunately the value of T_c is not the sole criterion by which the merit of a superconductor can be judged. Under the influence of a suitably strong magnetic field the superconductivity of a material below T_c is destroyed and the material reverts to being a normal conductor. Since magnetism is dependent principally upon the spin of electrons (18.6.1) it is reasonable to suppose that superconductivity is also connected with the movement and spin of electrons, possibly as outlined above.

Superconductivity below T_c is also destroyed when a high current density is employed. Consequently superconductivity disappears:

(i)   if the temperature of the material is raised above T_c;
(ii) if a sufficiently strong magnetic field is applied;
(iii) when a high current density prevails in the conductor.

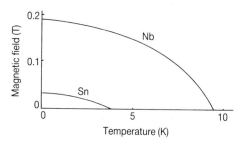

Fig. 18.7   Variations of critical field, H_c, with absolute temperature for a 'hard' superconductor (Nb) and a 'soft' superconductor (Sn).

Critical field strength, H_c, transition temperature T_c, and current density, J_c, are interdependent. The general relationship between field strength, H_c, and temperature, T_c, as this affects superconductors is indicated in Fig. 18.7. Regrettably the value of J_c tends to be very low with most new superconductors and this, coupled with their extreme brittleness, means that much development work is necessary to produce a superconductor of practical use. Given a material with high values of T_c, J_c and H_c and which can *also* be wound to form a solenoid, then superconductors have a great future in both electrical and electronic engineering. Table 18.2 lists properties of some metals, alloys, intermetallic compounds and ceramics which have featured as superconductors since 1911.

Insulators

18.5   Non-conducting materials—or insulators—are those in which bonding is either covalent or ionic. In such materials the valency electrons are

Table 18.2   *Properties of some superconductors*

Material	Critical temperature, T_c (K)	Critical field strength (Tesla)	
Pure metals			
Technetium	11.2	—	
Niobium	9.46	0.194	
Lead	7.18	0.080	
Vanadium	5.30	0.131	at 0 K
Tantalum	4.48	0.083	
Tin (white)	3.72	0.031	
Aluminium	1.19	0.010	
Alloys and intermetallic compounds			
Nb_3Ge	22.4	>360	
Nb_3Sn	18.1	>250	
V_3Ga	16.8	>300	
Pb–Bi eutectic	7.3–8.8	1.5	
Ceramics			
$Tl_2Ba_2Ca_2Cu_3O_{10}$	122	—	
$Y.Ba_2Cu_3O_7$	93	—	
$La_{1.8}Sr_{0.2}CuO_4$	40–50	—	
$(Ba_{0.6}K_{0.4})BiO_3$	30	—	
Fullerene $C_{60}K_3$	30	—	

firmly retained in fixed orbitals and are unable to move freely as they are in metals. Valency bands in insulators are completely filled and there is a wide energy gap between the valency band and the next possible energy band. Under such circumstances a large input of energy would be required in order that an electron could cross the gap. A very small number of electrons may acquire such energy if the material is subjected to a high potential difference and these insulators possess very low but nevertheless measurable conductivities. At high temperatures individual electrons will possess greater energies and so the chances of them 'escaping' is increased. Conductivity will therefore increase. Insulating materials may break down under the pressure of an electrical potential difference which is high enough to raise the energies of large numbers of electrons to the point where they can cross the energy gap and become 'free'. Such a potential difference is generally termed the 'breakdown voltage'.

18.5.1 Intermediate between conductors and insulators are the *semiconductors*, based mainly on the Group IV elements silicon and germanium, which during the latter half of the twentieth century have revolutionised the electronics industry and rendered obsolete the thermionic valve in domestic radio and TV receivers. However, their study is really the field of the electronics engineer and space here must be sacrificed to other important engineering materials topics.

Magnetism

18.6 Iron is the metal generally associated with magnetism but three other 'transition' metals—cobalt, nickel and gadolinium—are also strongly magnetic, or *ferromagnetic*, as it is usually termed. This distinction is necessary because many other elements and materials possess weak magnetic properties. Most metals are *paramagnetic*, that is they are weakly attracted by a strong magnetic field, whilst some metals and all of the non-metals are *diamagnetic*, that is they are repelled by a strong magnetic field.

Only the ferromagnetic metals have characteristics which make them technologically useful as magnets, either alloyed with each other or with other suitable materials.

18.6.1 Ferromagnetism   What we refer to as a 'magnetic field' associated with a ferromagnetic material is produced by the 'spin' of electrons. In fact we can produce an electromagnetic field by passing a stream of electrons through an electrical conductor. In a stable atom the electrons may be regarded as moving particles rotating in orbitals around the nucleus. At the same time each electron behaves as though it were spinning on its own axis and it is this spin which generates the magnetic field, that is, it is in effect a minute magnet.

We have seen that in a stable atom, whether it exists as a single unit or is part of a crystal, not more than two electrons are permitted to occupy the same energy level. These two electrons have spins which are in opposite directions. Consequently in the majority of elements as many electrons will have spins in one direction as in the other and individual fields produced by electrons will cancel out. Hence there will be little, if any, resultant magnetic field associated with the atom. Such elements tend to be either paramagnetic or diamagnetic (Fig. 18.8(i)). Some of the transition ele-

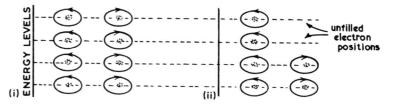

Fig. 18.8   Electron 'spin' as the basis for magnetism.

ments, however, have unfilled *sub*-valency shells so that *unpaired* electrons will be present. As a result, in any atom, more electons will spin in one direction than in the other (Fig. 18.8(ii)) and such an atom will have a resultant magnetic moment. In the elements α-iron, cobalt, nickel and gadolinium the magnetic moments of the atoms are great enough and the ions sufficiently close together that when the atoms become magnetically aligned—during 'magnetisation'— a powerful magnetic field is produced.

18.6.2   Magnetic properties are affected critically by atomic spacing. If atoms are too far apart then forces between them are too weak to resist the effects of thermal agitation which throws spins out of alignment so that individual fields cancel. If atoms are too close together then inter-atomic forces are too great to allow alignment of spins. Conditions appear to be favourable when the atomic radius is between 1.5 and 2.0 times the radius of the shell containing the un-paired electrons. The elements α-iron, cobalt, nickel and gadolinium fulfil these requirements but other transition elements such as titanium, chromium and manganese are only just outside the range. Thus if manganese contains interstitial nitrogen atoms it becomes ferromagnetic due to a modification in the atomic spacing. Similarly interatomic spacing in such ceramics as $Ba\text{-}Fe_{12}O_{19}$ makes them available as magnetic materials. Alloys containing various proportions of the metals iron, chromium, tungsten, cobalt, nickel, aluminium and titanium (Table 18.3) are particularly suitable as permanent magnets.

Contained within the crystals which comprise a ferromagnetic material are small regions known as magnetic *domains*. These are regions, possibly not more than 0.05 mm in diameter, in which atoms are aligned so that their orientations are in the same direction. In the unmagnetised state these domains are randomly orientated so that there is no resultant field, but if these domains are aligned by applying an external magnetic field the material becomes magnetised (Fig. 18.9).

This alignment may be retained when the magnetising field is removed

Table 18.3   Some Permanent-magnet Alloys

Type of material	Composition (%)									Special details	Magnetic properties			Characteristics and uses
	Fe	Co	Al	Ni	Cu	Ti	Nd	Sm	Others		B_{rem} (mT)	H_c (kA/m)	BH_{max} (kJ/m³)	
Carbon steel	99	—	—	—	—	—	—	—	C–1	—	900	4.4	1.5	Long obsolete
'Alnico'	57	12	9	16	5	1	—	—	—	Cast or sintered	725	45	13.5	General purposes
'Columax'**	50	24	8	14	3	—	0.6	—	Si–0.3 Nb–0.6	Full grain orientated	1350	66	60	Widely used in electrical, communication and engineering industries. Anisotropic alloys—can only be produced with a linear axis of magnetisation
'Hycomax 3'**	35	34	7	15	4	5	—	—	—	Cast or sintered	900	128	43	
Rare-earth/cobalt*	—	66	—	—	—	—	—	34	as SmCo₅	Sintered in argon	800	600	127	High performance/low inertia motors for NC machine tools, small stepper motors for watches, camera apertures, etc.
'Neodure'**	76	—	—	—	—	—	16	—	B–8 (as Nd₂Fe₁₄B)	Sintered ceramic	1100	800	215	Magnetic bearings, switches, audiovisual and similar consumer products, disc-file systems, servomotors, etc.
'Feroba 2'**	60 present in BaO(Fe₂O₃)₆									Sintered ceramic	380	160	29	Loudspeakers, door catches, computer track shifters on disc store devices
Magnetic rubber	60 present in BaO(Fe₂O₃)₆									Ceramic in rubber	160	96	4	Magnetic display boards, refrigerator door seals, field magnets for cheap low-performance motors (toys, windscreen washers, etc.)

*Anisotropic materials, properties measured along the preferred axis.

Fig. 18.9 Alignment of domains: (i) unmagnetised; (ii) magnetised.

and in this case the material is said to be *magnetically hard*. In many cases the magnetism is not retained when the magnetising fields is removed and the material is said to be *magnetically soft*. In fact magnetic hardness usually coincides with physical hardness, that is, a hardened steel is often suitable as a permanent magnet, whilst a soft steel is also magnetically soft and therefore unsuitable as a permanent magnet. Possibly the residual strains present in the hardened material oppose *randomisation* of the domains once the material has been magnetised. If a magnetised material is heated it loses its magnetism because the thermal energy supplied provides the necessary activity which permits randomisation of domain alignment.

18.6.3 A permanent magnet is designed to provide a magnetic field of force which does not involve the use of energy from an external source. It must also be able to retain its magnetism for an unlimited period even under the action of limited demagnetising influences. The actual magnetic strength, or *flux*, in a permanent magnet is dependent upon its design and the material from which it is made. The essential characteristics of a permanent-magnetic material after it has been magnetised are:

(i) *coercive force*;
(ii) residual magnetism or *remanence*;
(iii) *energy product value*.

The coercive force is a measure of the material's resistance to demagnetisation by electrical means, whilst remanence relates to the intensity of magnetism remaining in the magnet after the magnetising field has been switched off. The energy product value is virtually the quantity of magnetic energy stored in the magnet following magnetisation. It is derived from the demagnetisation curve RU (Fig. 18.10).

18.6.4 Figure 18.10 shows what is known as a *hysteresis loop* for a magnetically hard material. Here the induced magnetic flux density (measured in *teslas* and denoted by B) is plotted against the magnetising field (expressed in *ampere-turns per metre*, H). Such information is obtained when a ferromagnetic material is placed in a solenoid through which a current can be passed. Increased values of current will produce an increased magnetic field strength H which, in turn, will induce an increasing flux density B in the magnet.

Beginning with a small value of H and increasing it gradually, corres-

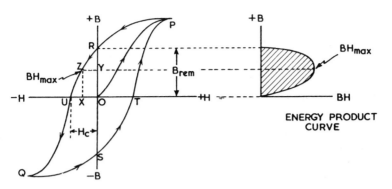

Fig. 18.10 A typical hysteresis curve for a ferromagnetic material.

ponding values of B and H are plotted to give the curve OP. This is the magnetisation curve which ultimately 'levels out' at P where the material reaches magnetic saturation. If the magnetising field is now reduced to zero the corresponding fall in the induced magnetic flux density is indicated by PR. Thus OR represents the value of the magnetic flux density remaining in the material when all external magnetising influence is removed. This is the remanent magnetism or remanence, B_{rem}.

To demagnetise the material and at the same time measure the force required to do this, a current is passed through the solenoid in the reverse direction and gradually increased. This produces the curve RU which is the demagnetisation curve. Intercept OU represents the force required to demagnetise the material completely and is called the coercive force. To complete the hysteresis loop the operations outlined are carried on to 'negative' saturation and then repeated in the reverse sense. The hysteresis loop represents the lagging of the magnetic flux induced behind the magnetising force producing it. The area within the loop is proportional to the quantity of energy lost when the material is magnetised or demagnetised and, for permanent magnet materials, the area of the loop must be as large as possible.

18.6.5 The ultimate criterion of a permanent magnet is the maximum product of B and H obtained from a point Z on the demagnetisation curve, i.e. the position of Z is such that the product of XZ and YZ is a maximum. This value, BH_{max}, corresponds to the maximum energy the magnet can provide in a circuit external to itself.

A simple method of assessing the efficiency of a magnet in terms of its remanence is to determine its 'squareness ratio' (B_{rem}/B_{sat}) where B_{sat} is the saturation value of the applied field and B_{rem} the remanence. A squareness ratio near to one, i.e. $B_{rem}/B_{sat} \simeq 1$, means that any stored pattern of domains is retained in zero field. A high value of H_c ensures that the value B_{rem} will be stable with time and against any stray magnetic fields.

18.6.6 Hardened carbon steels were first used as permanent magnets, later being replaced by alloy steels containing chromium and tungsten. These were followed by 35% cobalt steel in the early days of the 'wireless', until in the 1930s the Fe–Al–Ni–Co range of permanent magnets was developed. Often the choice of a permanent magnet material relates to whether high remanence or a high coercive force is required. Thus 'Columax' has a high remanence whilst 'Hycomax 3' has a high coercive force. More recently, alloys containing 'rare-earth' elements such as neodymium and samarium have provided high values of both remanence and coercive force in a single alloy. In fact the very high flux densities available with the Nd–Fe–B alloys have permitted smaller magnets to be used so that d.c. motors with the magnets in the rotor and windings in the stator are possible—a reversal of previous practice. Some magnetic materials are *anisotropic*, that is they have superior properties along a preferred axis. This effect is obtained by heating the alloy to a high temperature and allowing it to cool in a magnetic field. As thermal agitation is reduced groups of atoms become orientated along the field and this orientation is virtually 'frozen in'.

18.6.7 Whilst magnetically hard materials are required for permanent-magnet applications such as loudspeakers, magnetic chucks, generators, small electric motors, magnetic recording tapes and the like, such components as transformer core and dynamo pole-pieces need to be magnetically soft but of high magnetic permeability. Permeability (μ) is given by the ratio B/H and its value alters with the magnetising field, reaching a maximum and then falling. Since a transformer core is under the influence of alternating current, the core is first magnetised, then demagnetised and then remagnetised with reverse polarity. Such a cycle produces the closed hysteresis loop PRUQST (Fig. 18.10) and the area of this loop represents the amount of energy wasted in overcoming remanent magnetic forces each time the current is reversed (for example 50 times per second). This energy is lost as heat and it is therefore necessary to use materials which give very narrow hysteresis loops and consequently small 'hysteresis loss' (Fig. 18.11(ii)).

18.6.8 Soft pure iron and Fe–4Si have a very low hysteresis and have been used in the form of stampings for transformer cores and the like. Some Fe–Ni alloys are also important as low-hysteresis, *high-permeability* alloys. Thus 74Ni–5Cu–1Mn–bal, Fe has been used in communications engineering, particularly as a shielding material.

As with permanent magnet materials the current tendency is towards the use of magnetically soft ceramic materials known as 'ferrites' the composition of which is based on ferroso-ferric oxide, Fe_3O_4, or more properly, $FeO.Fe_2O_3$. In this formula some of the Fe^{2+} ions in FeO are substituted by either Ni^{2+} and Zn^{2+} or Mn^{2+} and Zn^{2+} giving substances which can be roughly expressed by the formulae, $Fe_2(Ni_xZn_y)O_4$ or $Fe_2(Mn_xZn_y)O_4$ where $x + y = 1$. It should be noted that the composition of either is not strictly stoichiometric (that is, obeying the rules of chemical combinations

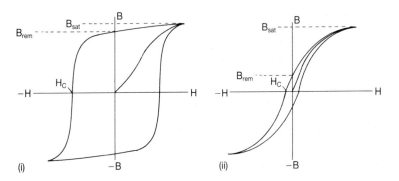

Fig. 18.11 Magnetic hysteresis. (i) This is a curve typical of a magnetically hard material. (ii) Shows a curve for a magnetically soft material. Here both B_{rem} and H_c are very small so that little energy will be lost in overcoming these forces when the material is used for transformer cores.

by fixed proportions). These materials are manufactured as powders and then sintered to give a solid ceramic. At present ferrites dominate the field of electronics. Since their properties can be varied by adjusting composition and also by varying particle size and treatment, they can be tailored to meet specific requirements.

A further group of magnetically soft materials is metallic glasses (13.6.8) the basic composition of which includes transition metals such as Fe, Co, Ni and Cr and non-metals or metalloids like B, Si, P and C. The molten solution is cooled very rapidly (some $10^6\,°C/s$) by squirting on to a cooled rotating wheel (13.6.8) to produce a tape some 50 microns thick. Rapid cooling produces an amorphous (glassy) structure (3.7.5) since the temperature drop is too rapid to permit the onset of crystallisation.

The most important industrial product of this type to date is 'Metglas' 2605/S2 made in the USA. This is an Fe80–B20 metallic glass manufactured as ribbon (Fig. 13.23) for the production of laminated transformer cores. Its lack of crystal boundaries, along with the relative absence of defects, aids the movement of domain walls during the magnetisaton cycle, resulting in a very low hysteresis loss and a high permeability. The hysteresis loss in power-supply transformers using Fe80–B20 'Metglas' is only about 25% of that in those using the conventional Fe–3Si alloy. It has been estimated that the replacement in the UK of transformers based on the conventional alloy by new ones employing Fe80–B20 cores would lead to a fuel saving equivalent to some 1.3 Mt of oil per annum resulting in a reduction of 'greenhouse gas' emissions of 4.5 Mt of CO_2 and 120 kt of SO_2 each year (1.6). Perhaps it is time to introduce the threatened 'Carbon Tax'!

19

The Environmental Stability
of Materials

19.1 The realisation that metallic corrosion is a very costly process is brought home to most of us by the fact that the working life of an expensive 'tin lizzie' is limited by the rusting of its bodywork rather than the wearing of its mechanical parts. Whilst the latter are generally replaceable nothing can be done when the welded shell begins to disintegrate due to creeping rust. A recent EEC survey estimated that a staggering 4.2% of the GNP was the cost to most industrial countries of the effects of corrosion on building structures, ships, automobiles, agricultural equipment and off-shore installations. As far as the EEC is concerned this cost is measured in hundreds of millions of ECUs per annum.

Materials other than metals also possess varying degrees of instability. Rubber 'perishes' particularly when exposed to air and sunlight and some plastics materials become progressively more brittle under the influence of ultra-violet light. Concrete and other ceramic materials are gradually eroded by frost and the chemical effects of our polluted atmosphere. A large group of materials, both metallic and non-metallic, may be damaged to some degree by various forms of radiation.

Most materials will deteriorate more quickly at high temperatures due to attack by atmospheric oxidants such as CO_2 and some sulphur compounds as well as oxygen. Brittle materials of high expansion coefficient such as furnace refractories are liable to spall—or flake away—if the rate of heating or cooling is excessive.

Dry Corrosion or the Oxidation of Metals

19.2 Whilst many metals tend to oxidise to some extent at all temperatures, most engineering metals do not scale appreciably except at high temperatures. When iron is heated strongly in an atmosphere containing oxygen it becomes coated with a film of black scale. A chemical reaction has taken place between atoms of iron and molecules of atmospheric oxygen:

$$2Fe + O_2 \rightarrow 2FeO$$

Although the reaction can be expressed by the above simple equation in fact atoms of iron have been oxidised whilst atoms of oxygen have been reduced. These processes are associated with a transfer of electrons from one atom to the other.

Oxidation $Fe \rightarrow Fe^{++} + 2$ electrons (e^-)

Reduction $O + 2$ electrons $(e^-) \rightarrow O^{--}$

It should be noted that, in the chemical sense, the terms oxidation and reduction have a wider meaning than the combination or separation of a substance with oxygen. Thus a substance is oxidised if its atoms lose electrons whilst it is reduced if its atoms (or groups of atoms) gain electrons. For this reason we say that iron is oxidised if it combines with sulphur, chlorine or any other substance which will accept electrons from the atoms of iron.

19.2.1   In some cases the formation of the oxide film protects the metal from further oxidation yet in many instances oxidation and scaling continue. For this to occur either molecules of oxygen must pass through a very porous film of oxide, or *ions* of either oxygen or the metal must migrate *within* a continuous film.

The chemical composition of a *stable* film of ferrous oxide would be FeO, but the situation indicated in Fig. 19.1, where a piece of iron is being

Fig. 19.1   The mechanism of the scaling of iron at high temperatures. If the mobilities of O^{--} ions and electrons (e^-) are low then the scale film will reduce the rate of oxidation.

strongly heated, results in that part of the oxide film nearest to the iron surface containing an excess of Fe^{++} ions whilst that adjacent to the air contains an excess of O^{--} ions. These O^{--} ions are formed when electrons, released from iron atoms (which then become Fe^{++} ions) pass through the oxide film to oxygen atoms at the surface: $O + 2e^- \rightarrow O^{--}$.

Whilst the film remains thin the force of attraction between O^{--} ions at the oxide surface and Fe^{++} ions at the metal surface will be great so that O^{--} ions will diffuse rapidly to fill vacant 'oxygen sites' at the metal surface. As the oxide film expands away from the metal surface, separation of the Fe^{++} ions and the O^{--} ions increases. Consequently attraction between them decreases and the rate of diffusion and hence growth of the oxide film also decreases.*

Let the thickness of the oxide film be x at a time t, then assuming the concentrations of the ions at the surfaces of the oxide layer remain constant, there will be a difference δc in concentration across the oxide film. In accordance with Fick's Law the rate of diffusion of ions across unit area is $D . dc/dx$, which is proportional to a rate of dx/dt. Therefore:

$$\frac{dx}{dt} \propto x^{-1} \quad \text{or} \quad x^2 = kt$$

where k is related to D, the diffusion constant. This indicates that the film growth follows a parabolic law.

When the tendency of this growth is to increase the density of the oxide film without reducing its mechanical stability then it will obviously improve the protection against further oxidation. If, however, the stresses generated in the film as a result of increased density, lead to fracture and break up of the film, the metal surface will be exposed and further oxidation will be the result.

Oxidation rates also depend upon the porosity of the oxide film. Some metals form ions which are much smaller than the original atoms. Consequently as ions are formed there is a volume reduction of the scale which, as a result, becomes porous and permits easier access of oxygen to the metal surface. Contraction of the oxide film may also cause it to flake off thus exposing a fresh metal surface. In some cases expansion of the film may occur causing it to buckle and become detached.

19.2.2   An oxide film which adheres tightly to the surface of the metal generally offers good protection. Good adhesion is the result of coherence between the film and the metal beneath. In Fig. 19.2(i) there is good

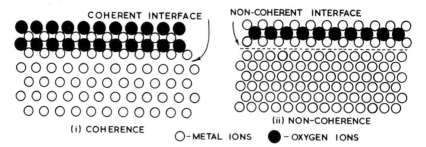

Fig. 19.2 Adhesion between a metal surface and the covering oxide film. In (i) the row of metal ions at the interface 'matches up' with those in the oxide film, but in (ii) ions are 'out of step' at the interface.

---

*With prolonged oxidation of iron at high temperatures such as occurs when large steel ingots are heated for several hours prior to rolling, the oxide layer becomes so thick and the diffusion of O^{--} ions consequently so slow that a layer of Fe_3O_4 ($FeO . Fe_2O_3$) forms on top of the layer of FeO, and this in turn becomes covered with a layer of Fe_2O_3 at the oxide/air interface.

'matching' between the ions in the metal surface and the metallic ions in the oxide film such that the structure is virtually continuous, whilst in Fig. 19.2(ii) matching is absent so that the two surfaces will be non-coherent and there will be little adhesion. The high degree of coherence between aluminium and its oxide leads to the effective protection of this metal especially when the film is artificially thickened by anodising.

19.2.3 It follows from the above that the best way to protect a metal like iron from high-temperature oxidation is to add an element which will coat the surface with a dense and closely adherent oxide layer in which the mobility of both ions and electrons is low. Aluminium forms a dense oxide film with a very high electrical resistance making it ideal in this respect, but it has adverse effects on the mechanical properties of the steel if added in the amounts necessary to provide oxidation protection. The same is true of silicon. This leaves chromium as the next best and most useful metal in this respect. The oxide Cr_2O_3 coats the surface of iron with a dense skin *beneath* the outer film of Fe_2O_3. Chromium is thus an essential constituent of heat-resisting steels (10.5.2).

Some alloys are prone to attack by atmospheres containing sulphurous gases. Since nickel readily forms a sulphide at high temperatures it is particularly liable to be oxidised by gases containing sulphur. Heat-resisting steels used in such conditions must be of the high-chromium type, preferably containing little or no nickel.

Wet Corrosion or Electrochemical Oxidation

19.3 Iron does not rust in a completely dry atmosphere nor will it rust in completely pure, oxygen-free water, but in a moist atmosphere the well-known reddish-brown deposit of ferric hydroxide soon begins to develop. The overall chemical reaction representing rusting can be expressed by a simple chemical equation:

$$4Fe + 6H_2O + 3O_2 \rightarrow 4Fe(OH)_3$$

This general result, however, is achieved in a number of stages and the fundamental principle involved is that atoms of iron in contact with oxygen and water are oxidised, that is they lose electrons and enter solution as ferrous ions (Fe^{++}):

$$Fe \rightarrow Fe^{++} + 2 \text{ electrons } (2e^-)$$

These ferrous ions are ultimately oxidised further to ferric ions (Fe^{+++}) by the removal of another electron:

$$Fe^{++} \rightarrow Fe^{+++} + e^-$$

As iron goes into solution in the form of ions the corresponding electrons are released. These electrons immediately combine with other ions so that overall equilibrium is maintained.

The ease with which a metal can be oxidised in this way depends upon the ease with which valency electrons can be removed from its atoms. Thus metals like calcium, aluminium and zinc hold their valency electrons comparatively loosely and can therefore be oxidised more easily than iron; but the noble metals gold, silver and platinum retain their valency electrons more strongly and are therefore much more difficult to oxidise than iron.

Copper holds on to its valency electrons more strongly than does iron and, under suitable conditions, copper ions will 'steal' valency electrons from atoms of iron. Thus when a pen-knife blade is immersed in copper sulphate solution the blade becomes coated with metallic copper. Copper ions at the surface of the blade have removed electrons from the atoms of iron there so that the resultant ferrous ions have gone into solution thus replacing the copper ions which have been deposited as copper atoms

$$Fe \rightarrow Fe^{++} + 2e^-$$
$$Cu^{++} + 2e^- \rightarrow Cu$$

Copper sulphate solution is sometimes used to coat the surface of steel with a thin layer of copper as an aid to 'marking out', in the tool room.

19.3.1   Thus different metals have different *oxidation potentials* since the amount of energy required to remove their valency electrons varies from metal to metal. The ionisation process builds up an electrical potential called an *electrode potential* and this depends upon the nature of the solution as well as upon the nature of the metal. Electrode potentials are measured with reference to hydrogen (Fig. 19.3) and are useful in assessing

Fig. 19.3 The measurement of electrode potential. Since copper (i) produces a higher electrode potential than does hydrogen it becomes the cathode, whilst hydrogen is the anode. However, hydrogen (ii) produces a higher electrode potential than does iron so in this case hydrogen is the cathode whilst iron is the anode.

Table 19.1 *Electrode Potentials for Some Metals (Standard Molar Solutions of Ions at 25°C)*

	Metal ion	Electrode potential (volts) E_h
(Noble)	Gold (Au^{+++})	+1.50 (Cathodic)
	Platinum (Pt^{++++})	+0.86
	Silver (Ag$^+$)	+0.80
	Copper (Cu$^+$)	+0.52
	(Cu^{++})	+0.34
	Hydrogen (H$^+$)	0.00 (Reference)
	Iron (Fe^{+++})	−0.05
	Lead (Pb^{++})	−0.13
	Tin (Sn^{++})	−0.14
	Nickel (Ni^{++})	−0.25
	Cadmium (Cd^{++})	−0.40
	Iron (Fe^{++})	−0.44
	Chromium (Cr^{+++})	−0.74
	Zinc (Zn^{++})	−0.76
	Aluminium (Al^{+++})	−1.66
	Magnesium (Mg^{++})	−2.37
	Calcium (Ca^{++})	−2.87
	Potassium (K$^+$)	−2.93
(Base)	Lithium (Li$^+$)	−3.05 (Anodic)

the tendencies of the metal to corrode. To do this a standard hydrogen electrode is used as indicated and some form of potentiometer employed to measure the potential difference (in volts) produced between the hydrogen electrode and the metal electrode. Since hydrogen is a gas the electrode used consists of a platinum tube (which is relatively inert) into which hydrogen is passed. Hydrogen dissolves interstitially in platinum and in this way forms an electrode. By using similar methods an *electrochemical series* can be prepared for the metals as indicated in Table 19.1.

19.3.2 Galvanic Action Involving Dissimilar Metals   In Fig. 19.3 the electrode which will supply electrons to the external circuit (when the latter is 'closed') in each case is called the *anode* and the electrode which will receive electrons via the external circuit is called the *cathode*. Thus in Fig. 19.3(i) the hydrogen electrode is the anode whilst in Fig. 19.3(ii) the iron plate is the anode.

At this juncture it is necessary to mention an important point connected with electrochemical nomenclature and one which may otherwise puzzle the reader. The Greek derivation of the term 'anode' implies 'something which is being built up' and in fact the anode was so named in the early days of the study of electricity because it was assumed that positively charged 'particles of electricity' were flowing through the external circuit from the cathode to the anode. It is now realised that what we call an electric current is in fact a flow of *negatively charged particles*—or electrons—in the opposite direction, i.e. from the anode to the cathode via the external circuit. Thus, instead of *receiving* positively charged particles from the external circuit the anode *supplies* negatively charged electrons to the external circuit and is consequently left with positively charged particles (ions) which it immediately releases into solution. We therefore visualise the anode as being negatively charged because it is the source of

Fig. 19.4 The effect of closing the external circuit in the hydrogen-iron cell.

electrons to the external circuit. In the illustrations which follow, electrons will be seen to be flowing in the opposite direction to that which is ascribed to 'current' by long-established convention.

If an external electrical contact is made between the electrodes of the system shown in Fig. 19.3(ii) as indicated in Fig. 19.4, the lower electrode potential of the iron will cause it to lose electrons which will flow away from the anode through the external circuit to the cathode the resultant Fe^{++} ions entering solution. The electrons which arrive at the cathode combine with H^+ ions in the surrounding electrolyte and form molecular hydrogen which subsequently escapes as small bubbles:

$$H^+ + e^- \rightarrow H$$
$$H + H \rightarrow H_2$$

This reaction causes further electrons to flow from the iron anode so that more atoms ionise liberating electrons and entering the electrolyte as Fe^{++} ions. The 'driving force' behind this reaction is the difference in electrode potential between iron and hydrogen. These reactions continue spontaneously, iron dissolving from the anode and hydrogen precipitation at the cathode.

The hydrogen ions (H^+) present in the electrolyte are formed by the ionisation of water molecules:

$$H_2O \rightleftharpoons H^+ + OH^-$$

As H^+ ions are 'neutralised' on receiving electrons to form hydrogen atoms (and subsequently molecules) more water ionises spontaneously. The concentration of hydrogen ions in *pure* water is extremely low—of the order of 10^{-7} gram-atoms/litre—and so the formation of hydrogen atoms and the simultaneous solution of iron would be very slow. In practice some soluble impurities in water lead to a much higher degree of ionisation of the water and so galvanic corrosion takes place at a greater rate. A solution in which free ions are present, and which will therefore allow an electric current to be transmitted in an external circuit, is termed an *electrolyte*. Most organic liquids such as alcohol, turpentine, oils, etc. are non-electrolytes since the presence of covalent bonds throughout their molecules makes ionisation impossible. Galvanic corrosion is not possible in the absence of an electrolyte. As mentioned above pure water is a very weak electrolyte since an extremely small proportion of its molecules are ionised, but when it contains dissolved impurities such as carbon dioxide or sulphur dioxide (prevalent in industrial atmospheres) its degree of ionisation is considerable.

19.3.3   We are now in a position to consider the reactions which will occur when, say, a copper plate and an iron plate are connected externally and are in mutual contact with an electrolyte (Fig. 19.5). Iron is anodic towards

Fig. 19.5   The reactions in an iron-copper galvanic cell.

hydrogen whilst copper is cathodic towards hydrogen. It follows then that iron will be strongly anodic towards copper and so electrons will immediately begin to flow from the iron anode to the copper cathode through the external circuit. Having lost electrons some iron atoms will pass into the electrolyte as Fe^{++} ions. In the meantime the copper cathode becomes negatively charged as it receives electrons from the iron anode, via the external circuit. Positively charged hydrogen ions, H^+, present in the electrolyte will be attracted to the cathode where they will combine with electrons to form hydrogen atoms and, subsequently, molecules:

$$2H^+ + 2e^- \rightarrow 2H \rightarrow H_2$$

thus restoring the balance.

As more and more H^+ ions are neutralised by electrons from the anode in this way to form hydrogen atoms, so more and more water molecules will ionise to restore the balance between molecules and ions. The concentration of hydroxyl ions (OH^-), also formed by the ionisation of water molecules, will build up in the region of the cathode (Fig. 19.6).

Fig. 19.6 Reactions at the cathode.

Although Fe^{++} ions are released into the electrolyte at the anode they do not remain there. There is a mutual attraction between them and the OH^- ions. However, since Fe^{++} ions are much smaller than OH^- ions they are able to migrate rapidly towards the cathode where the concentration of OH^- ions is high. There the two types of ion react:

$$Fe^{++} + 2(OH^-) \rightarrow Fe(OH)_2$$

In addition to the Fe^{++} ions having a greater mobility than OH^- ions, only one Fe^{++} ion is required to neutralise two OH^- ions.

The ferrous hydroxide, $Fe(OH)_2$, which is formed rapidly oxidises to insoluble ferric hydroxide, $Fe(OH)_3$, which collects as the reddish-brown deposit we call 'rust'. Much of this falls from the cathode to the bottom of the cell.

$$4Fe(OH)_2 + 2H_2O + O_2 \rightarrow 4Fe(OH)_3 \downarrow$$
$$\text{'rust'}$$

Dissolved oxygen also plays its part in increasing the rate of corrosion of the anode by reacting with water molecules at the cathode (where electrons are available):

$$O_2 + 2H_2O + 4e^- \rightarrow 4OH^-$$

The rate of formation of hydroxyl ions is thus increased.

19.3.4 Summary of Reactions

The ferrous hydroxide is then gradually oxidised due to the presence of dissolved oxygen:

$$6Fe(OH)_2 + 3H_2O + 1\tfrac{1}{2}O_2 \rightarrow \quad 6Fe(OH)_3$$
$$\text{ferric hydroxide ('rust')}$$

These reactions can be 'combined' to give the following:

$$6Fe + 9H_2O + 4\tfrac{1}{2}O_2 \rightarrow 6Fe(OH)_3$$

This single equation represents chemically the overall reaction but it does not explain the steps by which the result is achieved. In many of the examples which follow the electro-chemistry has been simplified by assuming that Fe^{+++} *ions are formed at the anode* to give a resultant deposit of $Fe(OH)_3$ near to the cathode.

19.3.5 Common Examples of Galvanic Corrosion Involving Dissimilar Materials

Readers who are also practising engineers will by now have realised why it is generally considered bad practice to use dissimilar metals in contact with each other and in mutual contact with an electrolyte albeit the latter is only condensed atmospheric moisture. Nevertheless, in domestic plumbing one often does not need to look far to discover examples of electrochemical malpractice. For example, most cold-water cisterns used to be of galvanised iron and whilst such material was quite adequate when properly used, the life of the zinc coating could only be shortened if a copper inlet pipe was allowed to enter the water adjacent to the side of the tank. Zinc is strongly anodic to copper and since the water was efficiently aerated at the inlet the zinc could only be expected to corrode exposing the iron beneath. Iron too is anodic to copper and so a leaking tank was inevitable. Such leaks usually developed when one was away on holiday, offering a melancholy prospect on one's return. Fortunately PVC cold-water cisterns are now used and are of course immune from any sort of galvanic attack, but if your house was built before about 1970 an inspection of the cistern may be advisable. If it turns out to be of galvanised iron take a closer look at its condition!

Mixing of lengths of iron and copper for carrying water leads to the

Fig. 19.7 An example of bad plumbing.

accelerated corrosion of the iron pipe near to its joint with the copper (Fig. 19.7). Examples of this type of bad practice are often provided by those who should know better. The motor-car industry has a habit of sticking bits of stainless-steel 'trim' on to the mild-steel body-work. This is quite safe so long as the mild steel is completely coated with paint but invariably the paintwork around the fixing hole is damaged during the fixing process and the design is such that water collects in the region of the joint thus constituting a galvanic cell. Since the mild-steel body is anodic to the stainless-steel 'trim' the body-work ultimately rusts away, no doubt to the ultimate satisfaction of the manufacturer who finds such caprices good for trade.

Iron and steel must generally be protected from both atmospheric and aqueous corrosion by coating with some protective film. Painting, electro-plating, tinning and galvanising are processes which can be used to this end. However, when metallic coatings are used the effect of possible film damage must be considered, particularly as to whether the film will be anodic or cathodic to the surface of the iron beneath.

Thus a damaged or discontinuous tin coating can actually accelerate the rusting of the mild steel it is meant to protect. Tin is cathodic to iron and so the galvanic corrosion of iron will take place as indicated in Fig. 19.8. The tin and iron are already in electrical contact with the 'external circuit' and when the scratch is bridged by moisture (containing dissolved oxygen) corrosion will begin. As electrons flow from the iron to the tin, Fe^{+++} ions will enter solution and migrate to the tin cathode where hydroxyl ions will be forming due to the reaction:

$$2H_2O + O_2 + 4e^- \rightarrow 4OH^-$$

Fig. 19.8 The disastrous effect of a defective tin coating on mild steel.

The formation of rust on the cathode then follows the procedure already described, viz.

$$Fe^{+++} + 3OH^- \rightarrow Fe(OH)_3 \downarrow$$

Tin is thus effective as a protective coating only if the coating is perfect. Since it is an expensive metal tin is now only used as a coating in the canning industry and for coating food-processing equipment where its high corrosion-resistance to animal and vegetable substances is utilised, and the fact that any compounds which may be formed are non-toxic.

19.3.6   For engineering purposes either zinc or aluminium are more effective since both are anodic to iron. Zinc is the more widely used of the two since it is easier to apply to mild steel either by galvanising (hot-dipping of the steel into molten zinc); spraying (as in the Schoop process used to coat the Forth Road Bridge); sherardising (a vapour condensation/cementation process); or electroplating.

Consider a damaged zinc coating as indicated in Fig. 19.9. Since zinc is

Fig. 19.9 The 'sacrificial' protection offered by a defective zinc coating. Iron is rendered cathodic so that the zinc goes into solution preferentially.

anodic towards iron it will send electrons through the 'external circuit' to the surface of the iron and as a result Zn^{++} ions enter solution. Hydroxyl ions (OH^-) will form at the surface of the iron as a result of the standard reaction involving oxygen, water and electrons. Zn^{++} ions and OH^- ions will then react:

$$Zn^{++} + 2OH^- \rightarrow Zn(OH)_2 \downarrow$$

Thus, $Zn(OH)_2$ precipitates as a white corrosion product on the surface of the iron cathode.

In this case it is the coating metal which corrodes in preference to the mild steel. This form of corrosion is often termed 'sacrificial', since iron will not rust significantly as long as zinc remains in the vicinity of the initial scratch. It would be wrong to assume, however, that there is any lasting merit in producing a poor-quality porous coating. As with all methods of surface protection the aim should be to produce a perfect coating which will give no opportunity for galvanic action. With a zinc coating, however, the position is less serious if the coating is in fact defective since some measure of temporary sacrificial protection is afforded the mild steel.

Fig. 19.10 The cathodic effect of oxide scale in accelerating the rusting of mild steel sheet.

19.3.7 On the question of the galvanic effect of surface coatings on iron the influence of oxide scale cannot be neglected. Ordinary 'black' sheet will rust more quickly than will clean sheet because oxide scale is cathodic to the metal beneath. Corrosion of the metal will then follow the usual pattern (Fig. 19.10).

19.3.8 Cases involving galvanic corrosion on a microscopic scale are very common. Thus particles of impurity may be either anodic or cathodic to the surrounding metal (Fig. 19.11). In either case pitting of the metal will

Fig. 19.11 The effect of grain-boundary impurities on galvanic corrosion. In (i) the impurity is cathodic to the metal, whilst in (ii) the impurity is anodic to the metal.

result and since impurities are commonly segregated at the crystal boundaries in cast metals then intercrystalline corrosion will take place.

Alloys of duplex structure can be expected to corrode more quickly than single-phase materials of similar composition. Carbon steel in the pearlitic condition rusts far more quickly than it does in the fully hardened state. A steel in the pearlitic condition contains a structure in which the ferrite lamellae are anodic to the cementite lamellae. Consequently the ferrite tends to form Fe^{+++} ions and corrosion proceeds in the usual manner (Fig. 19.12). A steel of similar composition in the quenched state will be wholly martensitic and so galvanic corrosion will be at a minimum because of the uniform nature of the structure. Tempering will increase the tendency towards corrosion as cementite is progressively precipitated.

As a general principle, therefore, very pure metals or uniform solid

Fig. 19.12 The galvanic corrosion of pearlite.

solutions constitute the most corrosion-resistant materials as far as galvanic corrosion is concerned. In *cast* solid solutions, however, excessive coring may lead to galvanic action between different regions of a crystal due to differences in chemical composition.

Pourbaix Diagrams

19.4 The extent to which a metal corrodes depends not only upon its electrode potential relative to other substances in the environment but also upon the chemical nature of the electrolyte and in particular its pH.

The connection between electrode potential, E_h, and the hydrogen ion concentration, pH, was investigated by M. Pourbaix at the Belgian Centre for the Study of Corrosion. He formulated diagrams which connect these values with the electrochemical equilibria of metals in contact with water. Both electrode potential and pH will influence the transfer of electrons in the system so that hydrogen evolution and oxygen absorption are dependent on these factors. These reactions are the basis of electolytic corrosion. Pourbaix diagrams resemble phase-equilibrium diagrams in so far that they do not give any information about the *rate* at which a reaction takes place.

19.4.1 Figure 19.13 illustrates the effects of E_h and pH on the extent to

Fig. 19.13 The E_h/pH diagram for water.

which water will liberate either hydrogen or oxygen by what we usually term 'electrolysis'. Hydrogen evolution is only possible under conditions of E_h and pH which are represented by a point below ab and oxygen can be evolved only if E_h and pH are represented by a point above cd. Therefore there is a *domain* between ab and cd on the E_h/pH diagram where the electrolysis of water is thermodynamically impossible. Thus the electrolysis of water will not take place, regardless of the pH of the electrolyte, unless the electrodes used have a minimum potential difference of more than 1.2 volts, that is the distance apart in terms of E_h of ab and cd.

19.4.2 In Fig. 19.13 the equilibria of hydrogen and oxygen reduction reactions are represented on the E_h/pH diagram. However, equilibria for a metal in conjunction with water can also be represented. Since the corrosion of iron is most important from a technical standpoint this is represented in terms of E_h and pH in Fig. 19.14. The complete Pourbaix

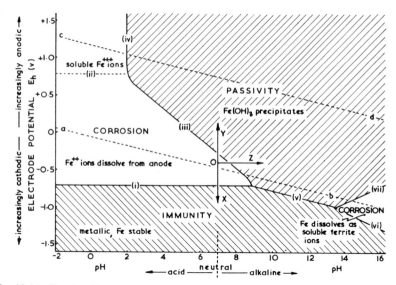

Fig. 19.14   The simplified Pourbaix diagram for the Fe/H$_2$O system.

diagram for iron is complex since a number of equilibria are involved. Fortunately a simplified diagram can be drawn from a study of the following equilibria, a number of which we have already encountered in this chapter:

(i) $Fe \rightleftharpoons Fe^{++} + 2e^-$ Iron atoms form soluble Fe^{++} ions.

(ii) $Fe^{++} \rightleftharpoons Fe^{+++} + e^-$ Fe^{++} ions are oxidised to Fe^{+++} ions.

(iii) $Fe^{++} + 3OH^- \rightleftharpoons Fe(OH)_3 + e^-$ Insoluble ferric hydroxide precipitates.

(iv) $Fe^{+++} + 3H_2O \rightleftharpoons Fe(OH)_3 + 3H^+$ Insoluble ferric hydroxide precipitates.

(v) $Fe + 3H_2O \rightleftharpoons Fe(OH)_3 + 3H^+ + 3e^-$ Insoluble ferric hydroxide precipitates.

(vi) $Fe + 2H_2O \rightleftharpoons FeO_2H^- + 3H^+ + 2e^-$ Ferrite ions go into solution.

(vii) $FeO_2H^- + H_2O \rightleftharpoons Fe(OH)_3 + e^-$ Soluble ferrites precipitate as ferric hydroxide.

If the electrode potential E_h of mild steel is measured in pure water (pH = 7) it will be represented on the diagram (Fig. 19.14) by point O. This indicates that corrosion will take place under such conditions of E_h and pH.

The diagram also suggests three ways in which the extent of corrosion can be reduced. First, the potential can be changed in the negative direction so that it enters the domain of immunity (X). This is the normal cathodic protection such as is offered by zinc, either as a coating or as a separate anode (19.7.4). Secondly, the potential can be changed in a positive direction into the domain of passivity (Y) by applying a suitable external e.m.f. Thirdly, the electrolyte can be made more alkaline (z) so that the metal becomes passive. *Inhibitors* have this effect—amongst others—but need to be carefully controlled since with some metals the passivity domain extends over a very limited pH range and continuous surveillance of the electrolyte composition would be necessary.

Electrochemical Corrosion due to Internal Mechanical Stresses

19.5 Cold-worked materials of *uniform composition* throughout may nevertheless suffer severe electrochemical corrosion due to different energy levels prevailing in different parts of the material. If part of a component is heavily cold-worked it will contain 'locked-up' stresses which are internally balanced and which will produce some elastic strain. The internal energy associated with this state of affairs causes the heavily worked material to corrode more quickly than the less heavily worked material, that is, the heavily worked material is *anodic* to the remainder of the component. This is commonly known as *stress corrosion*.

Consider the cold-formed rim of a motor-car 'wing' (Fig. 19.15(i)). The forming operation will have cold-worked the rim more heavily than the remainder of the wing. If the paintwork is defective in this relatively inaccessible position corrosion can be expected to occur fairly rapidly in the heavily worked and, hence, anodic region. Similarly the cold-forged head of an ordinary nail will be anodic to the remainder of the nail and will rust more quickly as a result (Fig. 19.15(ii)).

It is obvious that heat treatment can be useful in limiting galvanic corrosion of this type. Simple stress-relief annealing processes will be sufficient to remove energy arising from locked-up stress and will thus reduce local corrosion. Such heat treatment is not always possible since cold-working is often a means of stiffening and strengthening a component as in the case of both of those mentioned above.

Fig. 19.15 Examples of electro-chemical corrosion due to local cold-work. (i) The rim of a motor car 'wing'. (ii) The cold-forged head of an iron nail.

19.5.1 Intercrystalline corrosion associated with the presence of impurities at grain boundaries has already been mentioned (19.3.8). However, grain boundaries are also regions of higher energy level, even in very pure metals, possibly due to the irregular arrangement of ions there (Fig. 3.25). Consequently corrosion tends to occur more quickly at grain boundaries. Hard-drawn α-brasses are prone of a phenomenon known as 'season-cracking' in which corrosion follows the crystal boundaries until the component, unable any longer to sustain internal stresses, cracks extensively along these boundaries. The tendency towards this behaviour can be eliminated by a low-temperature anneal which, although it does not cause recrystallisation, results in the removal of locked-up stresses as atoms move small distances into positions nearer to equilibrium.

Electrochemical Action Involving Variations in the Composition of the Electrolyte

19.6 Figure 19.16 illustrates a concentration cell in which both of the electrodes are of exactly the same material but in which the concentration of the electrolyte in contact with each electrode is different.

The general reaction in the cell is represented by:

$$Fe \rightleftharpoons Fe^{++} + 2e^{-}$$

and the direction in which this reaction proceeds is governed, as in many chemical reactions, by the concentration of the reactants. Thus on the left-hand side of the cell a low concentration of Fe^{++} ions will cause the reaction to proceed to the right:

$$Fe \rightarrow Fe^{++} + 2e^{-}$$

so that iron atoms will tend to go into solution as Fe^{++} ions and release

Fig. 19.16 A 'concentration cell'.

electrons into the external circuit. On the right-hand side of the cell the already high concentration of Fe^{++} ions tend to make the reaction move in the other direction:

$$Fe \leftarrow Fe^{++} + 2e^-$$

Thus the left-hand electrode becomes the anode and the right-hand electrode the cathode of the galvanic system as indicated by the direction of electron movement in the external circuit.

19.6.1 Variations in the amount of oxygen dissolved in water will lead to corrosion of this type taking place in many different situations. In the case illustrated in Fig. 19.17 the two steel sheets are of the same composition

Fig. 19.17 The 'crevice corrosion' of steel plate.

and in the same condition but the composition of the electrolyte varies since dissolved oxygen is unlikely to penetrate into the further recesses of the crevice between the plates although water will have been drawn there by capillary action. Consequently the more exposed parts of the sheet become cathodic since electrons will travel to these areas and react with oxygen and water present to form hydroxyl ions:

$$O_2 + 2H_2O + 4e^- \rightarrow 4OH^-$$

Thus, *the region of metal exposed to electrolyte containing less dissolved oxygen will become anodic and go into solution*. This situation is helped by the fact that Fe^{+++} ions have greater mobility than OH^- ions and so rust forms freely outside the crevice. This phenomenon is often termed 'crevice corrosion' and is typified by the fact that corrosion will be more severe in the regions which one might have expected to be the better protected. The general axiom is that *the oxidation/concentration cell accelerates corrosion but it accelerates it where the concentration of oxygen is lower.*

19.6.2   This state of affairs explains why corrosion takes place with increasingly rapidity when once the paintwork of a motor-car has been scratched—or was initially porous as is so often the case (Fig. 19.18).

Fig. 19.18   The sub-paint rusting of a motor-car body—a further manifestation of crevice corrosion.

Water begins to penetrate beneath the paint layer but oxygen does not and so the metal beneath the paint layer becomes anodic and goes into solution thus progressively detaching the paint layer.

Accumulations of otherwise chemically inert dirt on the surface will produce a similar effect in allowing moisture to penetrate to the surface of the metal but at the same time being deficient in oxygen. Hence the reader is advised to wash the under side of his car even more carefully than the upper paintwork, particularly after a frosty spell during which large quantities of salt (a most efficient electrolyte and the car manufacturer's friend) have been thrown on to the roads by the Local Authority. The fact that the salt does *not* pentrate crevices accelerates rather than diminishes the rate of corrosion since Cl^- ions will congregate along with OH^- ions outside the crevice (or dirt patch) and greatly increase the concentration gradient within the electrolyte. Cynics may claim that we get the rotten cars we deserve but the situation is certainly not helped by road salt.

Finally, we will consider the corrosive effect of a simple rain drop on the surface of mild steel (Fig. 19.19). Here the concentration of dissolved

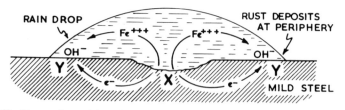

Fig. 19.19   The rusting of iron beneath a rain drop, due mainly to variations in the concentration of dissolved oxygen between centre and rim of the drop.

oxygen in water will obviously be higher at Y than at X where the depth of water is greater. A 'concentration cell' will therefore be set up so that at X iron atoms will go into solution as Fe^{+++} ions, releasing electrons which pass through the external circuit to react with oxygen and water at Y, producing OH^- ions:

$$O_2 + 2H_2O + 4e^- \rightarrow 4OH^-$$

The Fe^{+++} ions released will migrate towards the OH^- ions forming a deposit of $Fe(OH)_3$—'rust'—mainly at the periphery of the rain drop. But for the variation in concentration of oxygen in the rain drop the rusting of mild steel by its presence would be a much slower process since then electrolytic corrosion would depend entirely on the small variation in composition within the mild steel.

The Limitation of Galvanic Corrosion

19.7 Corrosion is a phenomenon which is very expensive to eliminate completely. At best we can limit its action by one of the following methods:

19.7.1 Choosing a Material in Which Galvanic Attack is Unlikely Pure metals and completely uniform solid solutions offer the greatest freedom from galvanic corrosion assuming that steps are taken to prevent 'crevice corrosion' by ensuring that all such fissures are adequately sealed.

Either ferritic or austenitic stainless steels (10.5) offer good resistance to corrosion largely because of the protective film of chromium oxide which adheres to the surface. At the same time, uniformity of the solid solution structure almost eliminates the chance of galvanic corrosion even in the presence of very strong electrolytes. To attain this uniform solid solution structure it is generally necessary to quench stainless steel from 1050°C. This has no effect on the hardness of these very low-carbon steels but has the effect of retaining in solution carbon which would otherwise precipitate as carbides. Slow cooling will allow chromium carbide to precipitate at grain boundaries (Fig. 19.20) thus depleting the region immediately adjacent to the grain boundaries with respect to chromium. These low-chromium regions become anodic to the chromium-rich matrix so that, in the presence of a suitable electrolyte, inter-granular corrosion would occur.

Fig. 19.20 Carbide precipitation and the resultant corrosion in stainless steel.

19.7.2 The Use of Protective Coatings The protection of a surface from galvanic corrosion by sealing off that surface from contact with a potential electrolyte has already been mentioned. Painting, enamelling, various methods of tinning and galvanising—or even coating with a non-reactive grease—all have this fundamental aim in the protection of steel but are seldom completely effective under commercial conditions. The anodising of aluminium in which the natural oxide film is artificially thickened is also a process of this type.

19.7.3 The Avoidance of Potential 'Crevices' Fissures are best sealed or filled in some way to prevent the entrance of thin films of electrolyte which, in the presence of external oxygen, may lead to this form of attack.

19.7.4 The Use of Galvanic Protection In some cases the close proximity of two dissimilar metals in mutual contact with an electrolyte is unavoidable. For example, ships' propellors are commonly of 'manganese bronze' (a high-tensile brass—Table 11.1) which is strongly cathodic towards the steel hull. In the presence of sea water which is a strong electrolyte, the hull could be expected to rust quickly due to the close proximity of the manganese-bronze propellor. This tendency is largely eliminated by fixing some zinc slabs to the hull near to the propellor (Fig. 19.21). Since zinc is

Fig. 19.21 The galvanic protection of a ship's hull.

anodic both to iron and manganese bronze it will corrode sacrificially and can be replaced as required.

Buried iron pipes can be protected in a similar way by burying slabs of zinc or magnesium adjacent to the pipe at suitable intervals. An alternative method (Fig. 19.22) is to employ a small external e.m.f. as shown using the polarity to make the pipe cathodic to its surroundings. Either a battery or a low-voltage d.c. generator (operated by a small windmill) can be used successfully.

Fig. 19.22 Protection of a buried steel pipe by using an impressed e.m.f.

Radiation Damage in Materials

19.8   Whilst the effects of radiation from nuclear sources have come under consideration in recent years we must remember that the effects of other forms of radiation are also significant. Radiation can be classified into two main groups:

(i)   electromagnetic radiations which include radio 'waves'; light; X-rays and γ-rays. These are now commonly considered in terms of small 'energy packets' or *photons*;
(ii)   'radiations' which in fact consist of moving particles. They include fast-moving electrons (originally called β-rays or cathode rays); protons (once known as 'positive rays'); helium nuclei (the original α-rays); and neutrons.

Of these forms of radiation the effects of γ-rays and fast-moving neutrons are of possibly the greatest significance, particularly to the nuclear-power engineer, but in everyday life the effect of ultra-violet light on modern plastics materials cannot be completely neglected.

19.8.1 Radiation Damage in Metals   of interest principally to the nuclear engineer. High-speed neutrons from nuclear fission processes may collide with the nuclei of atoms and if the KE transferred is sufficient the ion may be displaced into an interstitial position leaving behind a vacancy. This will cause limited work hardening since slip will be impeded by the displaced ion.

Of more importance however, is the final resting place of the neutron. If it enters a nearby nucleus the latter may become a radioactive isotope emitting dangerous γ-rays.

19.8.2 Radiation Damage in Polymers   Irradiation of polymer materials with either neutrons or γ-rays may lead to cross-linking, degradation, chain branching or polymerisation. Which of these processes occurs depends upon the nature of the irradiated polymer, the amount of radiation, the temperature and the presence or absence of other substances. Controlled

Fig. 19.23   The scission of a polythene molecule due to irradiation.

irradiation of polythene can be used to effect useful side branching and other desirable properties but uncontrolled irradiation may lead to degradation—or *scission*—even of polythene and is even more likely to occur with many other linear polymers. As a result strength generally falls because of a reduction in van der Waals forces as molecular chains decrease in length.

Radiation in the ordinary visible spectrum (light) is sufficient to cause deterioration in some plastics materials. Museum specimens of Parkesene (12.1) and other early plastics materials currently on display in illuminated cabinets are beginning to crumble to powder.

The Ageing of Thermoplastics and Rubbers

19.9 This often occurs as a result of cross-linking between molecules which contain unsaturated bonds. For example polythene will become progressively more brittle due to the formation of covalent bonds by oxygen atoms which link together adjacent molecules in this way

Fig. 19.24   Cross-linking between polythene molecules.

(Fig. 19.24). Since the covalent cross-link prevents slip between the chain molecules polythene becomes progressively harder and more brittle.

This type of cross-linking with oxygen is likely to occur more readily in the presence of ultra-violet light (strong sunlight) which catalyses the reaction. It can therefore be limited by making the material opaque by adding some carbon black but a more appropriate method is to include an anti-oxidant, usually 0.1 to 0.2% of phenol or an amine.

19.9.1 Poor quality rubber 'perishes' in a similar way due to the formation of oxygen cross-links, particularly at those unsaturated bonds not already used during the vulcanisation process. Ultimately it becomes

Fig. 19.25   The perishing of polybutadiene rubber.

brittle as degradation generally follows the fall in elasticity produced by cross-linking. Again intensity of light and also the presence of ozone will accelerate perishing. The reader may have noticed that continental car owners often cover their tyres to protect them from strong sunlight. Again oxidation can be suppressed by adding carbon black as well as small amounts of aromatic amines or phenol derivatives as antioxidants.

19.9.2   There are of course instances when the controlled disintegration of polymers such as polythene would be desirable. Some years ago we towed our caravan many miles north of the Arctic Circle in Norway following the Arctic Highway, then under construction, until it ran out on us in a wild and lonely place where great glaciers came down to the sea. Sadly pollution is never far away and we were confronted by a shore line littered with thousands of plastics containers presumably carried across from the New World by the Gulf Stream. Processes were developed some years ago by the University of Aston in Birmingham (England) to make such containers 'self destruct' after a reasonable period of time. Unfortunately these processes have not been adopted—possible manufacturers fear that containers may 'destruct' whilst still on the shelves of the supermarket in times when sales are slow. Recycling of plastic bottles would surely be a better solution than throwing them into the sea.

Deterioration of Ceramic Materials

19.10   Most ceramic materials are chemically stable when in contact with ordinary atmospheres, though some natural stones used for building suffer attack by industrial atmospheres containing sulphur dioxide. The latter combines with atmospheric moisture to form sulpurous acid.

19.10.1   Concrete based on Portland cement and used in building foundations is likely to be attacked by sulphate ions (SO_4^{--}) present in the soil, in ground water or in sea water. These ions react first with tri-calcium aluminate in the concrete, followed by a reaction with lime, $Ca(OH)_2$, so exposed, to form a *bulky* reaction product—crystalline $CaSO_4$ which cracks the concrete locally thus exposing new surfaces to continued attack. *Sulphate-resistant concrete* made from cement in which the proportion of tri-calcium aluminate has been reduced to less than 5% and the proportion of tetra-calcium alumino-ferrite (Table 14.4) increased, is therefore used.

19.10.2 Carbonation Shrinkage   Lime, $Ca(OH)_2$, formed during hydration of Portland cement (14.5.2) reacts with any atmospheric CO_2 present:

$$Ca(OH)_2 + CO_2 \rightarrow CaCO_3 + H_2O \qquad (1)$$

This reaction tends to produce a volume *increase* in the cement paste but at the same time CO_2 also attacks and breaks down the hydration products of both calcium silicates and calcium aluminates leading to a volume *decrease* which outstrips the expansion caused in equation (1). Consequently there is an overall 'carbonation shrinkage'. The extent of this is governed by:

(i)   the CO_2 concentration;
(ii) the porosity of the cement paste;
(iii) the moisture content.

The formation of $CaCO_3$ (equation (1)) increases strength simply because $CaCO_3$ is stronger than $Ca(OH)_2$. Other reactions tend to decrease the permeability of the paste. Drying shrinkage which reaches a maximum at 50% relative humidity is particularly affected.

Some firms producing concrete blocks for the building industry treat them in an atmosphere of waste flue gases (rich in CO_2). This accelerates the carbonation shrinkage process so that it is practically complete before the blocks are built into walls. This leaves only the drying-out shrinkage to be completed.

The weathering action of frost is also destructive of many building ceramics. Surface pores which absorb moisture will become disrupted when the moisture freezes and, as a result, expands.

19.10.3   Refractory ceramics used in furnace lining suffer both physical and chemical damage. Since refractory ceramics have a relatively low thermal conductivity they may suffer thermal shock. An appreciable temperature gradient may build up when the surface temperature is suddenly increased and as the hot surface layers expand relative to the cold layers beneath, *spalling*, or flaking of the surface is likely to occur.

Slags may react destructively with some refractories at high temperatures. Sometimes a slag will have a simple solvent action on the refractory but usually such reactions are of an acid/basic nature. Thus silica refractories—which are acid—will be attacked by basic slags containing lime and other metallic oxides.

Since the product of the reaction often has a low melting point fluxing of the furnace lining takes place. The general principle in choosing a refractory material which is to work in contact with a slag is that it should have a similar chemical nature to that of the slag. Thus a basic slag will demand a basic lining of dolomite, magnesite or zircona, whilst an acid slag will require a furnace lining of silica brick or some silica derivative.

19.10.4   Because of the powerful covalent bonds operating so effectively between the atoms in diamonds (3.8.1), they, above all other materials, are commonly regarded as being indestructible. Nevertheless, Averani and Targioni burnt one using a simple 'burning glass' as long ago as 1694. So perhaps the aphorism 'Diamonds are for ever' is not strictly true—sorry girls!

20

Non-destructive Testing

20.1 No material produced under industrial conditions is ever perfect. The incidence of imperfections in any batch of material will depend not only upon the degree of care and attention paid to production by the operator and the extent to which plant has received maintenance, but also upon the type of process. Thus a casting may suffer from one or more of a large number of defects such as incorrect composition; the presence of blow holes, shrinkage cavities, non-metallic inclusions, segregation, surface flaws and cracks; and so on. Many of these defects will not submit to simple visual examination and an apparently acceptable casting may contain a number of internal defects. Rolled, forged or extruded metal, however, is less likely to be defective, assuming that no surface defects are apparent. Had the internal structure been unsatisfactory or its composition incorrect it is likely that the ingot would have cracked during the initial stages of the working process, or that gas-filled cavities would have given rise to blisters near the surface.

20.1.1 Considerable lengths of a wrought section may be produced from a single cast ingot. Assuming that the piped portion of the ingot has been adequately cropped the remainder can be expected to have produced material of reasonably uniform properties and quality. Hence examination by analysis and by mechanical or physical tests of samples taken from the batch will generally be adequate to control quality. The same principles, however, do not apply to components made *individually* by casting, welding and other forms of fabrication involving mainly manual skill. A large number of variable influences often prevails during the manufacture of such a component. Moreover, since it is in finished form ordinary mechanical tests of a destructive nature cannot be used, other than the making of test pieces from the runner or risers of a casting. Hence some form of appropriate non-destructive test must be adopted so that the quality of the finished component can be assessed. The proportion of components thus tested can vary. Whilst for cheap consumer articles the representative proportion which is tested may be very small, in the aircraft industries it will often be 100%. This non-destructive exploration of materials has traditionally been known as *non-destructive testing* (NDT) but, whilst its primary objective is to reveal defects, it can be used in a more sophisticated manner to give an appraisal of the overall quality of the material. It is therefore sometimes referred to as *non-destructive evaluation* (NDE).

The Objects of Non-destructive Testing

20.2   The general aims of NDT may be classified as follows:

20.2.1 Improved Productivity   Defective material may be isolated at an early stage in production so that further manpower is not frustrated in futile processing. Shop capacity is also saved in this way and there is less useless wear and tear on plant. These are obvious advantages of testing at an early stage basic work pieces such as castings and welds. Equally important the knowledge that NDT is being applied often leads to the production of less scrap due to indifferent workmanship. In the case of welding, instances have been reported where a 40% scrap rate fell to less than 5% when it was announced that radiographic examination was to be installed.

20.2.2 Increased Serviceability   Here NDT is used to detect faults which are likely to reduce the service life of the finished component. For example, some defects whether occurring at the surface or below the surface, can initiate fatigue failure (16.5.6). Quench cracks in heat-treated steel components are in this category. The problem of safety must also be considered in this connection. Failure can in some cases lead to disaster particularly in aircraft structures.

Types of Defect

20.3   Defects in metallurgical materials may be inherent in the original casting or they may arise due to faulty technique during subsequent processing. It is convenient to classify defects into two groups:

  (i)   those occurring at the surface. These include *cracks* resulting from unsatisfactory conditions during cooling of the casting; cracks produced during heat-treatment processes; surface *scale* and defects due to the rolling in of *laps* or *scabs* and *seams*;
 (ii)   those occurring below the surface. The more common defects here include *porosity*; *laminations*; *inclusions* of various types; *segregation*; *blowholes*; *internal cracks* and other less obvious faults such as *coarse grain*. Welds may suffer from lack of penetration and unsatis-

factory bonding as well as many of the defects already mentioned such as porosity, inclusions and cracking.

Visual Inspection

20.4   This is undoubtedly the most widely used, yet least costly and very effective method of NDT. Whatever other method of inspection is applied it will invariably be supplemented by visual examination. For example thorough visual inspection of a weld by an experienced operator can reveal the following faults:

(i)   lack of penetration by the weld metal (Fig. 20.1(ii));
(ii)   badly oxidised or rough surfaces;
(iii) the presence of large cracks and their position relative to the weld;
(iv)   surface porosity;
(v)   inadequate fusion of the weld metal as shown by 'undercutting' and concavity of the weld (Fig. 20.1(iii));
(vi)   the use of insufficient weld metal (Fig. 20.1(iv)).

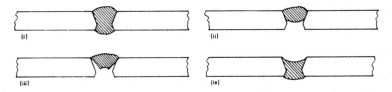

Fig. 20.1   Welding defects which can be assessed by visual inspection. (i) Satisfactory weld. (ii) Lack of penetration by weld metal. (iii) Inadequate fusion. (iv) Insufficient weld metal.

The principal requirement—other than employing an intelligent operator who knows what he is looking for—is to illuminate the component adequately so that it can be examined by the eye or in a few cases by some light-sensitive device such as a photo-electric cell. Visual examination can be assisted by using a small hand-magnifier giving a magnification of about 2.5^x.

20.4.1   Now that many automated processes are in use visual examination by sophisticated techniques is necessary. Thus on-line surface inspection of continuously-cast slabs can be assisted either by high-powered lamps or by laser in conjunction with video camera, CCD camera or photomultiplier, all of which devices intensify the recorded image.

Liquid Penetrants

20.5   Inspection based on the use of liquid penetrants can be employed to detect cracks and other fissure flaws occurring at the surface of a material.

Though most faults of this type could be detected by careful visual inspection the use of the penetrant method provides a more positive aid to detection and makes it less dependent on the human element as well as less time consuming.

Penetrant methods can be used to locate cracks in castings, weld cracks, grinding cracks, pinhole porosity *at the surface* of welds and castings, fatigue cracks, seams and laps in rolled stock, forging laps and many other cavity defects providing they break out at the surface.

20.5.1   The earliest penetrant tests employed oil and whiting. The surface to be examined was coated with a penetrating oil (often paraffin) which was usually heated so that its viscosity was reduced. Moreover the hot oil caused any surface cracks to expand so that oil would be drawn into them. After allowing sufficient time for the oil to penetrate any surface defects the excess was wiped off and the surface thoroughly dried. A thin coating of whiting, either as a dry powder or as a suspension in alcohol (industrial 'methylated spirit') was then applied to the surface and allowed to dry. As the surface cooled any cracks contracted again so that the oil was forced out to stain the coating of whiting.

20.5.2   More sophisticated modifications of this simple test were developed subsequently. In most of these tests a brightly coloured dye is carried in some suitable penetrant, the colour being chosen to produce a high degree of contrast with the coating of 'developer'. In most cases this developer is based on whiting, talc or some such powder suspended in water. Light-alloy castings are examined for surface cracks by first immersing them in a boiling solution containing a brilliant red dye. As the cracks expand the penetrant dye is drawn into them. The castings are then removed from the penetrant tank, rinsed to remove excess penetrant from the surface and then dipped into a hot suspension of whiting in water. On withdrawal of the casting its surface soon dries and as any cracks contract dye is squeezed out into the layer of whiting 'developer' (Fig. 20.2).

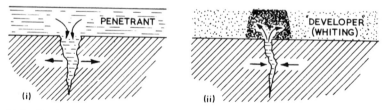

Fig. 20.2   Crack detection by dye penetrant.

20.5.3   Fluorescent penetrants are also used. Most of these substances are oil soluble so that a penetrating oil is used as the vehicle. Dip, brush or spray methods are employed to coat the penetrant on to the test surface. After excess penetrant has been cleaned from the surface the latter is viewed under ultra-violet light (approximately 3.65×10^{-7} m wavelength).

This is outside the visible spectrum and so, assuming that the test surface is inspected in a darkened cubicle, it remains completely dark. The action of ultra-violet light on the fluorescent substance in the penetrant, however, causes it to fluoresce so that any cracks containing the penetrant appear as bright yellow-green lines on a dark background.

Magnetic Methods

20.6 Discontinuities such as cracks, inclusions and cavities in a magnetic material produce distortion in a magnetic field which may be induced in that material. The direction of the magnetic lines of force is altered because the discontinuities have much lower magnetic permeabilities than the main body of the component. Consequently if a magnetic component is carrying a high magnetic flux density any discontinuity will tend to cause 'lines of force' to break out at the surface of the component (Fig. 20.3). If

Fig. 20.3   The effect of cracks and sub-cutaneous cavities on the direction of a magnetic field. Flaws at right angles to the lines of force ((i) and (ii)) will be revealed, whilst those parallel to the lines of force (iii) will not.

some fine magnetic powder is brought into contact with the surface it will collect where the lines of force break out thus indicating the presence of some fault at or near to the surface.

20.6.1   A number of different methods are available for producing the necessary high flux density in the component. The method chosen will depend to some extent on the direction in which the lines of force are required to pass. To obtain maximum sensitivity the lines of force should be perpendicular to the line of the flaw as already indicated in Fig. 20.3. Two ways in which a desired field direction can be produced are indicated in Fig. 20.4.

The magnetic particles can be applied to the surface either by wet or dry techniques. In the 'dry' method finely divided ferromagnetic particles having a high permeability are used and these can be simply shaken on to the surface from a pepper pot. They will align themselves along any lines of force emerging from the surface of the specimen in the region of the flaw. The surplus can be gently blown away. In the 'wet' process a suspension of finely divided magnetic iron oxide in a light petroleum 'fraction' is generally used. This can be applied by spraying or by immersion. Magnetic particles with a fluorescent coating can be used in which case the surface will be observed under suitable ultra-violet light (20.5.3). This method is useful when dealing with rough surfaces.

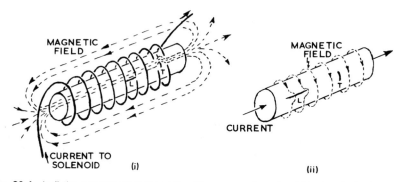

Fig. 20.4   In (i) the component is magnetised by means of a solenoid (or by placing it across the poles of an electromagnet). Here the field will be parallel to the axis so that the transverse crack, T, will be detected whilst the longitudinal crack, L, will not. In (ii) the component is mangetised by passing a current through it so that the induced magnetic field will be perpendicular to the axis of the component. In this case the longitudinal crack, L, will be detected whilst the transverse crack, T, will not.

These methods of fault detection are of course applicable only to ferromagnetic materials but they afford very positive means of locating cracks and crack-like faults so long as they are not more than about 10 mm beneath the surface.

Thermal and Electrical Tests

20.7   A number of non-destructive tests based on thermal properties have been devised from time to time. Many of these rely on the high thermal conductivity of metal and upon the fact that any discontinuity *beneath* the surface will act as a barrier to heat transfer. Thus if the surface of the component is coated with some heat-sensitive indicator and the component then heated at a suitable uniform rate, any build-up of heat will be revealed

Fig. 20.5   The principle of eddy-current testing. The solenoid probe will be near to the surface of the moving test metal. The display unit may be a meter, a digital bar or some form of oscilloscope.

by an appropriate change in the indicator. Areas on the surface where heat has been prevented from escaping inwards, by the presence of some subcutaneous defect, will be indicated by melting (or colour change) of the indicator used. In some of these tests the surface is first 'frosted' by spraying it with a coating of some suitable indicator such as acenaphthene.* The component is then passed at a uniform speed through an induction coil. Eddy currents induced in the *outer skin* of the component generate heat which is prevented from diffusing into the core if any discontinuities of low thermal conductivity exist between the skin and the core. This type of test has been used to examine the degree of continuity between uranium nuclear fuel rods and their protective cladding cans.

20.7.1 Various electrical methods have been developed to test for continuity of structure in conducting materials. Most of these depend upon the measurement of electrical resistance or, alternatively, potential difference across a standard section of material. For example railway rails can be tested by attaching electrodes at each end of a measured length of rail. The potential drop is then measured. The presence of discontinuities such as fissures will cause an increase in electrical resistance, or a decrease in potential drop across the section.

20.7.2 Pulse-video Thermography (PVT) developed in recent years to cope with the NDT of some of the 'new' materials, e.g. lamellar or fibre-reinforced composites. The surface of the test material is heated quickly, generally using a pulsed laser source, and then scanned by an infra-red camera the sensitivity of which is sufficient to detect any hot spots on the surface. These will have been caused by any subcutaneous flaws which were sufficiently large to interrupt the dissipation of heat inwards from the heated surface.

20.7.3 Eddy-current Testing In this method a primary magnetic field is generated by an alternating current passing through a solenoid. This field induces eddy currents in the surface of the metal under test. These flow in concentric paths normal to the direction of the primary field (Fig. 20.5). The magnitude of the eddy currents depends on the frequency of the field, the physical properties of the metal under test and the distance between solenoid and test surface.

The eddy currents in turn generate their own magnetic field which is in opposition to the primary field which is consequently modified so that an impedence change occurs in the solenoid circuit. Cracks or other faults in or near the test surface will deflect the eddy currents which must pass under or round such discontinuities. This causes a variation in the impedence of the solenoid circuit which can be measured and converted into a visual signal.

An advantage of this method of testing is that it is a 'lift off' method, i.e. the solenoid probe does not touch the test surface so that it can be used for

---

* Acenaphthene, $C_{12}H_{10}$, m.pt. 95°C, is derived from the anthracene fraction of coal tar.

testing moving hot metals such as ingot surfaces in a continuous casting process (6.2.1.2). It is unsuitable for steel in the region of the Curie point (769°C) since sudden loss (or gain) of ferromagnetic properties would cause fluctuation of readings.

20.7.4 The A.C. Potential Drop (ACPD) Method   useful in determining the *depth* of surface flaws in a metal. It is based on the principle that a high-frequency electrical current tends to flow close to the surface of a conductor. This is known as the 'skin effect'. Therefore any potential drop measured along the current path is a function of the path length. Since any increase in path length due to the presence of flaws near to the surface will increase the potential drop this gives a measure of the flaw depth.

X-ray Methods of Radiography

20.8   Metals and many other solids are opaque to 'visible' light, that is, electromagnetic radiation of such wavelength as can be 'received' by our organs of sight but are transparent to γ-rays and also to those X-rays of short wavelength. Hence radiography is one of the oldest non-destructive testing methods having been introduced in the 1920s.

Fig. 20.6   The electromagnetic spectrum.

20.8.1   Like light, X-rays travel in straight lines and can therefore be used to produce a reasonably sharp image on a photographic film, the emulsion of which is affected in much the same way by X-rays as it is by visible light. X-rays are absorbed logarithmically:

$$I = I_o e^{-\mu d}$$

where I_o and I are the intensities of the incident and emergent rays respectively, d the thickness and μ the linear coefficient of absorption of radiation for the medium through which the rays are passing. μ is dependent upon the wavelength of the radiation and is lower for that of shorter wavelength. Since X-rays are thus absorbed differentially the density of the photographic image will vary with the thickness of the material through which the rays have passed (Fig. 20.7). Those rays which pass through a region containing a cavity will be less effectively absorbed than those which travel through the entire thickness of material. Conse-

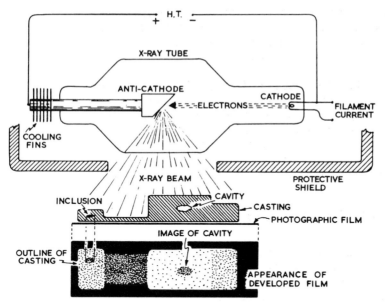

Fig. 20.7   The principles of X-radiography of a casting.

quently the cavity will show as a dark patch on the resultant photographic negative in a similar way to which greater density of visible light produces a darker area on the negative in ordinary monochrome photography.

Although a photographic emulsion is the most sensitive means of detecting both X-rays and γ-rays other means such as Geiger counters and substances which fluoresce under the action of these radiations can also be used. Fluoroscopy is obviously cheaper and quicker than photography but is less sensitive and its use is limited to the less dense metals and alloys.

20.8.2   X-rays are produced when high-velocity electrons strike a metal target. Hence the basic requirements for producing a beam of X-rays are:

(i)   a suitable source of electrons;
(ii)   some means of accelerating these electrons to a high velocity;
(iii) a suitable metal target.

In an X-ray tube the electrons are generated from a heated metal filament by passing a current of electricity through it—electrons from the current become so thermally activated that they 'escape' from the filament. The filament also acts as the cathode of the high-tension circuit (Fig. 20.7) used to accelerate the electrons. The target—or anti-cathode—is of tungsten which will resist melting at the high temperatures involved and is sealed into the other end of the tube. The tube is evacuated since the presence of relatively massive gas molecules would obstruct the passage of electrons. A high potential difference between cathode and target will cause the negatively charged electrons to be accelerated away from the cathode and towards the positively charged target. By the time the electrons reach the target they will have acquired considerable velocity and, hence, kinetic energy and this is absorbed in collision, X-rays being generated. In addition to X-rays, much heat is also liberated and this fact is utilised in electron-beam welding (17.4.4). Here, however, heat is an undesirable by-product and must be conducted away from the target even though the latter is of high melting point tungsten. Internal water cooling of the target or air cooling by a system of radiator fins external to the tube can be used.

20.8.3 X-rays used in metallurgical radiography are 'harder' than those used in medicine. That is, they are of shorter wavelengths which are able better to penetrate metals. At the same time these properties make them more dangerous to human body tissues and the plant producing radiations of this type must be very carefully shielded so that stray radiations do not reach the operators.

The wavelength, λ, of X-rays produced is governed by the accelerating voltage, V, applied between cathode and target:

$$\lambda \propto \frac{1}{V}$$

so that as the potential difference across the tube is increased the wavelength of the X-rays produced becomes shorter.

The absorption of X-rays is a function of the density of the material being penetrated and for this reason inclusions as well as cavities can be detected. At the same time the increased absorption of X-rays by the denser metals means that only thin sections of such metals can be effectively examined. Consequently X-rays are used mainly in the radiography of light-alloy castings and sections. The maximum sensitivity of this method of examination is of the order of 2%, that is a defect about 0.2 mm in diameter may be detected in a sectional thickness of 10 mm.

γ-ray Methods of Radiography

20.9 γ-rays are electromagnetic radiations of very short wavelength (Fig. 20.6) and are able to penetrate considerable thicknesses of many metals. In practice γ-ray methods are used to examine steel from 25 to 250 mm or more in thickness. A further advantage of the use of γ-rays is the very small size of source as compared with that of an X-ray unit. Moreover the cost of the γ-ray source is comparatively low.

20.9.1 Natural γ-ray sources, chiefly radium and radon, have been used for industrial radiography since the late 1920s but their use never became common due to the scarcity and high cost of radium and, in the case of radon, a decay product of radium, its very short half-life. The development of the nuclear reactor, however, has provided new and more effective γ-ray sources. These are generally produced by irradiation of some suitable element in a nuclear reactor. For example when cobalt $^{59}_{27}$ is bombarded with slow-moving neutrons in a nuclear reactor the radioactive isotope $^{60}_{27}Co$ is formed:

$$^{59}_{27}Co + ^{1}_{0}n \rightarrow ^{60}_{27}Co$$

$^{60}_{27}Co$ then undergoes decay:

$$^{60}_{27}Co \rightarrow ^{60}_{28}Ni + ^{0}_{-1}e\ (\beta\text{-particle}) + \gamma\text{-rays}$$

The half-life of $^{60}_{27}Co$ is a convenient 5.3 years.

20.9.2   Radioactive isotopes are also formed as a result of fission but in most cases a mixture of different isotopes is produced and this is generally unsuitable for use as a γ-ray source because, for example, of the rapid decay of some of the components relative to others.

In choosing a suitable radioactive isotope for this type of work the following criteria are important:

(i)   its half-life;
(ii)   the γ-ray energy output;
(iii)   the characteristics of the material to be radiographed.

The half-life of some radioisotopes is but a few days, hours—or even seconds. These will generally be less suitable for this type of work since corrections would need to be applied continually to compensate for their decrease in intensity of radiation. Consequently isotopes with a half-life measured in years or months will be more suitable assuming that its γ-ray emission energy is adequate.

In addition to cobalt $^{60}_{27}Co$ already mentioned, other artificially produced radioisotopes which are useful in this work include $^{192}_{77}Ir$, $^{137}_{55}Cs$ and $^{170}_{69}Tm$ (Table 20.2).

The absorption of γ-rays by matter follows the same general pattern as for X-rays (20.8.1), i.e.:

$$I = I_0 e^{-\mu d}$$

The 'half-value thickness' of a material is sometimes quoted. This is the thickness of the absorber which will reduce the intensity of the beam to one-half of its incident value (Table 20.1).

Table 20.1   *Half-value thicknesses (mm) of some materials*

Material	$^{60}_{27}Co$	$^{192}_{77}Ir$
Lead	12.2	4.8
Steel	22.1	11.2
Aluminium	55.8	30.4
Concrete	68.6	48.2
Water	134.6	81.3

20.9.3 The extreme danger of exposure of the human body to γ-rays and to 'hard' X-rays is well known but whereas the latter are not emitted once the tube has been switched off, a γ-ray source continues to emit as long as the source remains active ($_{27}^{60}$Co has a half-life of 5.3 years). Therefore γ-ray sources must be very carefully stored and shrouded at all times so that operators are protected. A typical storage container is shown in Fig. 20.8(i).

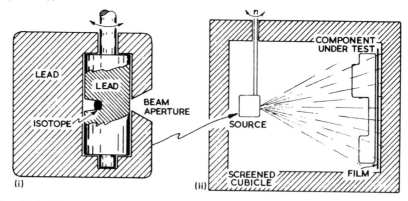

Fig. 20.8 The principles of γ-radiography: (i) a container for storing and shrouding a radioactive source; (ii) outline of a system for γ-radiography.

The general procedure in γ-radiography is similar to that using X-rays in that the component to be radiographed is interposed between γ-ray source and photographic film (Fig. 20.8(ii)). The whole process will be conducted in a radiation-proofed cubicle the source being uncapped by remote control in order that the operator shall be fully protected. A disadvantage in the use of γ-rays is that the wavelength of the radiation (which determines the penetrating power) cannot be varied whereas with X-rays wavelength can be altered to suit the material and also its thickness so that a radiograph of suitable photographic contrast can be obtained.

20.9.4 Table 20.2 shows the main characteristics of the more important γ-ray sources and indicates thicknesses which can be radiographed. These ranges of thickness are only approximate and greater thicknesses can be dealt with provided that exposure times are suitable increased. The lower limit of thickness in each case is that associated with adequate photo-

Table 20.2 *γ-ray Sources Used in Industry*

Isotope	Symbol	Half-life	Relative energy output in terms of γ-radiation	Typical uses
Cobalt–60	$^{60}Co_{27}$	5.3 years	1.3	50–200 mm of steel
Caesium–137	$^{137}Cs_{55}$	33 years	0.35	25–100 mm of steel
Thulium–170	$^{170}Tm_{69}$	127 days	0.0045	2–13 mm of aluminium
Ytterbium–169	$^{169}Yb_{70}$	31 days	0.021	2–13 mm of steel
Iridium–192	$^{192}Ir_{77}$	74 days	0.48	10–90 mm of steel

graphic quality. Lesser thicknesses would produce negatives of lower contrast which may not reveal the defects sought.

The relatively small size of the γ-ray sources and their associated handling equipment make γ-rays useful for radiography on sites inaccessible to bulky X-ray equipment. The initial cost of these γ-ray sources is also much lower than that of X-ray equipment but the results obtained may be inferior to those of X-rays because of the lack of flexibility of wavelength which is obtained with X-rays.

Ultrasonic Testing

20.10 Sonic methods for measuring the depth of the sea and so protecting ships from running aground have been practised for many years. Here a sound is emitted from the bottom of the ship and the echo received and suitably amplified. The interval which elapses between transmitting the sonic impulse and receiving its echo is a measure of the distance the impulse had to travel, that is to the floor of the ocean and back. Sonic methods have also been used to test metals for the presence of flaws. The railway wheel-tapper was for some strange reason always 'good for a laugh' in the old-time music-hall but he was none the less doing a very useful job of NDT. However, ordinary sound waves, audible to the human ear, are of relatively long wavelength and cover a frequency of about 30–16 000 Hz (according to the more optimistic hi-fi experts who probably waste a lot of money producing sounds they cannot hear). Such long wavelengths tend to by-pass small defects in a metallic component and will therefore not reveal them. Consequently ultrasonic frequencies of more than 20 000 Hz are used in this form of testing—generally 0.5 to 15 MHz for metals inspection.

20.10.1 Though here we are concerned principally with the detection of faults these methods are also suitable for measuring wall thicknesses of vessels where other methods are impossible. Thickness of sheet and plate and the thickness of cladding can also be measured in this way.

When an ultrasonic vibration is transmitted from one medium to another some reflection occurs at the interface. Any defect such as an inclusion or an internal cavity will therefore provide a reflecting surface for ultrasonic impulses (Fig. 20.9). The ultrasonic vibrations are produced in a piezo-electric crystal by stimulating it with an electrical impulse. Since ultrasonic vibrations are dissipated very quickly in air there must be close contact

(i)   (ii)

DEFECT

Fig. 20.9   The detection of a fault in plate material by ultrasonic impulses. In (i) the impulse is reflected from the lower surface of the plate whilst in (ii) it is reflected from the defect. Measurement of the time interval between transmission of impulse and reception of echo determines the depth of the defect.

Fig. 20.10 The principles of an ultrasonic testing apparatus. The values of d and t in the test piece are indicated by the values d_1 and t_1 on the CRT.

between the face of the crystal probe and the surface under test and this is most easily achieved by spreading a film of oil between the two. The crystal used is generally either of quartz or barium titanate. In some systems (Fig. 20.10) separate crystals are used as transmitter and receiver and the time interval between transmission of pulse and reception of echo is recorded on a cathode-ray tube (CRT).

20.10.2   In many modern testing devices a single crystal probe is used so that it acts both as transmitter and receiver. Since the crystal is very highly 'damped' a pulse of extremely short duration is emitted. Hence the crystal is dormant and able to receive the ultrasonic echo and so convert it into an electrical signal which is subsequently amplified and displayed on the CRT.

Although ultrasonic testing is closely associated with the exploration of large flat areas of plate materials it is very versatile in other directions. It can be used for testing most materials and in great depths, for example, 10 m or more thickness of steel.

Miscellaneous NDT Methods

20.11   It was once the rather gruesome practice in the East to sacrifice a village virgin into the molten bronze destined for casting large temple bells. If the bell turned out to be a dud it was assumed that the 'virgin' had not been all that she should have been and her father was severely ostracised—or worse—for not exercising adequate surveillance over her.

The test for quality of such a bell was, and is, simple requiring no sophisticated equipment. When struck it should ring, that is vibrate on the required frequencies which will be a combination of its fundamental frequency and certain harmonics and sub-harmonics. Any fault present will cause a damping of vibrations whilst a cracked bell may also vibrate on a different frequency to that intended because its dimensions have in effect been changed. The railway wheel tapper—for some obscure reason always good for a laugh in 'olde tyme' music halls—was using his skill in a similar manner.

20.11.1 NDT methods often have to be devised for special cases and many such methods are very ingenious. For example, in detecting leaks in very long water or oil pipe lines a radioactive method is often used. A quantity of radioactive isotope (in solution as one of its compounds) is injected into the liquid flow at the input end. After sufficient time has passed for the isotope to have travelled about 2.5 km a battery-operated radiation detector/recorder is passed into the line so that it floats along in the liquid stream. Points along the line where the radioactive isotope has leaked out, and is therefore still present in the surrounding sand, are located and recorded by the detector. Small γ-ray sources are sited along the outside of the pipeline at fixed intervals to act as distance calibration markers. The detector/recorder is recovered from the outlet end of the line so that the record can be examined.

Nevertheless the 'big five' among NDT methods are, traditionally, those which involve ultrasonics, X- (and γ-) radiography, magnetic flux leakage, dye-penetration or eddy currents.

Some Applications of NDT

20.12 NDT—or more properly, non-destructive *evaluation*—is now used to examine a vast range of materials and components but is more commonly associated with the following.

20.12.1 Castings Despite improved instrument control during recent decades the incidence of defects in castings is probably greater than in most other manufactured components. This is not surprising when one considers the number of variable factors which prevail during the casting processes. These include melting conditions in so far as they affect gas solution or surface oxidation; casting temperature; rate of pouring; mould temperature and conditions; the design of the mould itself and many other factors.

The principal internal defects likely to arise include:

(i)   gas porosity arising from gas dissolved during melting and precipitated during solidification. Bubbles of entrapped air may also be present if the mould was inadequately vented to allow air to escape as the mould was being filled;

(ii)   shrinkage cavities formed when molten metal is unable to feed into regions where soldification, and consequently shrinkage, is taking place;

(iii) internal inclusions due to entrapped slag and dross.

These internal defects are revealed most effectively by X-radiography particularly in the case of light alloys. For steel castings γ-radiography may be used.

Some of the surface defects in castings include:

(i)   cracks formed by uneven cooling of the casting during or after solidification. Sometimes cracks are initiated by sand cores which are too strong and do not collapse under the pressure of the solidifying metal;

(ii)   cold 'shuts', generally formed when two streams of molten metal fail to unite at an interface because of the presence of surface oxide;

(iii)   surface inclusions of moulding sand, dross or slag.

Extensive hot cracks which consist of several parallel fissures branching to follow grain boundaries may also occur internally during cooling. These, and surface fissures, may show on X-ray film as irregular dark lines. Surface cracks, however, are more easily and cheaply revealed by one of the hot liquid penetrant methods.

The bulk of sheet metal produced now originates from continuous-cast slab (6.2.1.2) and the detection and control of surface imperfections (Fig. 20.11) is most important since, if effective, it makes possible the

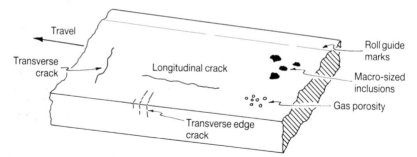

Fig. 20.11   Some defects associated with the surface of continuous-cast slab .

direct rolling of slabs. Most surface cracks originate in the mould due to incorrect cooling rates, stresses due to casting and straightening operations or poor mould design. Inclusions are the result of lack of 'cleanliness' of the molten steel or of inadequate control of the metal level in the pouring trough so that slag enters the mould. Pinholes are due to excessive dissolved gas in the liquid steel. Defects of this type at or near the surface of continuous-cast slabs can be detected by on-line non-contact methods such as optical inspection (20.4.1) or by eddy-current inspection methods (20.7.3).

20.12.2 Welds   The investigation of welded joints has been well developed and most of the orthodox NDT methods are useful.

Radiography, using both X-rays and γ-rays, will detect cavities, inclusions and those surface irregularities which result in a change in thickness of the metal. These defects will also be recorded by ultrasonic testing methods as also will incomplete penetration, lack of fusion and piping.

Cracks and other surface and subsurface imperfections in *magnetic* materials are very reliably detected by using a magnetic method of examination. In non-magnetic materials surface imperfections can be revealed using a fluorescent penetrant method or one of the other penetrant methods, assuming that the defects actually break out to the surface.

20.12.3 Plate and Similar Materials   Such materials may suffer from the presence of inclusions, laminations and other discontinuities elongated in the direction of rolling. These defects are most conveniently detected in cold materials using ultrasonic methods which are very sensitive even when applied to a considerable thickness of material. The on-line inspection of *hot* plate can be carried out by some enhanced optical method or other non-contact system such as eddy-current testing.

20.12.4 Concrete   Concrete structures *in situ* may be tested by some form of pulse method. The pulse may be provided by a simple hammer blow or by some form of electro-acoustic transducer. As with ultrasonic testing methods already described, the time taken for the pulse to travel from one boundary surface to the other—or to be reflected back to the source—is measured. Obstructions or imperfections interfere with the passage of the wave as in the case of ultrasonic testing.

20.12.5 Ceramics, Plastics Materials and Other Non-metallic Materials
Surface cracks are likely to occur in materials such as these and are usually detected by some form of penetrant method. *Very small* surface cracks in ceramics are difficult to detect, yet they are much more serious in these relatively brittle materials. Flaws below the surface of ceramics are also difficult to detect.

20.12.6   Composite materials of the fibre-reinforced and laminated types are now widely tested using ultrasonics. Pulse-video thermography is also useful in detecting impact damage which has occurred *below* the surface and is therefore invisible to simple visual inspection. Since composites are relatively transparent to X-rays, radiographic methods employing low-voltage X-ray sources are used. If the presence of faults which break the surface is suspected, then the composite is first treated with a penetrant containing a substance which is opaque to X-rays. Faults will then show more positively when the material is X-rayed.

Appendix I

Error Functions

y	erf(y)	y	erf(y)	y	erf(y)
0.000	0.000	0.510	0.529	1.00	0.843
0.050	0.0564	0.520	0.538	1.01	0.847
0.060	0.0676	0.530	0.546	1.02	0.851
0.070	0.0789	0.540	0.555	1.03	0.855
0.080	0.0901	0.550	0.563	1.04	0.859
0.090	0.101	0.560	0.572	1.05	0.862
0.100	0.112	0.570	0.580	1.06	0.866
0.110	0.124	0.580	0.588	1.07	0.870
0.120	0.135	0.590	0.596	1.08	0.873
0.130	0.146	0.600	0.604	1.09	0.877
0.140	0.157	0.610	0.612	1.10	0.880
0.150	0.168	0.620	0.619	1.11	0.884
0.160	0.179	0.630	0.627	1.12	0.887
0.170	0.190	0.640	0.635	1.13	0.890
0.180	0.201	0.650	0.642	1.14	0.893
0.190	0.212	0.660	0.649	1.15	0.896
0.200	0.223	0.670	0.657	1.16	0.899
0.210	0.234	0.680	0.664	1.17	0.902
0.220	0.244	0.690	0.671	1.18	0.905
0.230	0.255	0.700	0.678	1.19	0.908
0.240	0.266	0.710	0.685	1.20	0.910
0.250	0.276	0.720	0.691	1.21	0.913
0.260	0.287	0.730	0.698	1.22	0.916
0.270	0.297	0.740	0.705	1.23	0.918
0.280	0.308	0.750	0.711	1.24	0.921
0.290	0.318	0.760	0.718	1.25	0.923
0.300	0.329	0.770	0.724	1.26	0.925
0.310	0.339	0.780	0.730	1.27	0.928
0.320	0.349	0.790	0.736	1.28	0.930
0.330	0.359	0.800	0.742	1.29	0.932
0.340	0.369	0.810	0.748	1.30	0.934
0.350	0.379	0.820	0.754	1.31	0.936
0.360	0.389	0.830	0.760	1.32	0.938
0.370	0.399	0.840	0.765	1.33	0.940
0.380	0.409	0.850	0.771	1.34	0.942
0.390	0.419	0.860	0.776	1.35	0.944
0.400	0.428	0.870	0.781	1.36	0.946
0.410	0.438	0.880	0.787	1.40	0.952
0.420	0.447	0.890	0.792	1.50	0.966
0.430	0.457	0.900	0.797	2.00	0.995
0.440	0.466	0.910	0.802	∞	1.000
0.450	0.475	0.920	0.807		

y	erf(y)	y	erf(y)	y	erf(y)
0.460	0.485	0.930	0.812		
0.470	0.494	0.940	0.816		
0.480	0.503	0.950	0.821		
0.490	0.512	0.960	0.825		
0.500	0.520	0.970	0.830		
		0.980	0.834		
		0.990	0.839		

Appendix II
Some Properties of the Elements

Element	Symbol	Atomic number (Z)	Relative atomic mass	Melting point (°C)	Boiling point (°C)	Relative density	Crystal structures
Hydrogen	H	1	1.0080	−259.2	−252.5	0.0899 (g)	Hexagonal
Helium	He	2	4.0026	−272.2	−268.9	0.178 (g)	—
Lithium	Li	3	6.941	180.5	1329	0.534	CPH $\xrightarrow{-196°C}$ BCC
Beryllium	Be	4	9.01218	1284	(2400)	1.85	CPH $\xrightarrow{1200°C}$ BCC
Boron	B	5	10.811	(2300)	(2550)	2.3	Rhombohedral $\xrightarrow{1200°C}$ simple rhombohedral
Carbon	C	6	12.01115	(5000) Gr.	(5000)	2.25 (graphite) 3.51 (diamond)	α-graphite (hexagonal); β-graphite (rhombohedral); diamond (cubic)
Nitrogen	N	7	14.0067	−210.0	−195.8	1.2506 (g)	Hexagonal; complex cubic
Oxygen	O	8	15.9994	−218.8	−183.0	1.429 (g)	∝ (orthorhombic)⇌β (rhombohedral)⇌γ (cubic)
Fluorine	F	9	18.9984	−218.0	−188.0	1.69 (g)	—
Neon	Ne	10	20.179	−248.7	−245.9	0.9002 (g)	
Sodium	Na	11	22.9898	97.8	883	0.97	CPH $\xrightarrow{-222°C}$ BCC
Magnesium	Mg	12	24.305	650	1103	1.74	CPH
Aluminium	Al	13	26.9815	660	2480	2.699	FCC
Silicon	Si	14	28.086	1412	3240	2.4	Diamond structure (cubic)
Phosphorus	P	15	30.9738	44.1	280 (yellow)	1.82	Orthorhombic; complex cubic
Sulphur	S	16	32.06	112.8	444.5 (rhombic)	2.07	Orthorhombic; monoclinic; rhombohedral

Element	Symbol	At. no.	At. weight	m.p. (°C)	b.p. (°C)	Density	Crystal structure
Chlorine	Cl	17	35.453	−101.0	−34.1	3.214 (g)	—
Argon	Ar	18	39.948	−189.4	−185.8	1.78 (g)	—
Potassium	K	19	39.102	63.6	775	0.86	BCC
Calcium	Ca	20	40.88	843	1350	1.55	FCC $\overset{448°C}{\rightleftharpoons}$ BCC
Scandium	Sc	21	44.9559	1538	(2870)	2.5	CPH $\overset{1334°C}{\rightleftharpoons}$ BCC
Titanium	Ti	22	47.90	1667	3285	4.54	CPH $\overset{822.5°C}{\rightleftharpoons}$ BCC
Vanadium	V	23	50.9414	1920	3380	6.0	BCC
Chromium	Cr	24	51.996	1900	2690	7.19	BCC
Manganese	Mn	25	54.9380	1244	2020	7.43	Complex cubic $\overset{727°C}{\rightleftharpoons}$ complex cubic $\overset{1095°C}{\rightleftharpoons}$ FCC $\overset{1133°C}{\rightleftharpoons}$ BCC
Iron	Fe	26	55.847	1536	(3070)	7.87	BCC $\overset{910°C}{\rightleftharpoons}$ FCC $\overset{1400°C}{\rightleftharpoons}$ BCC
Cobalt	Co	27	58.9332	1492	(2900)	8.90	CPH $\overset{400°C}{\rightleftharpoons}$ FCC
Nickel	Ni	28	58.71	1453	(3000)	8.90	FCC
Copper	Cu	29	63.546	1083	2575	8.96	FCC
Zinc	Zn	30	65.37	419.5	907	7.133	CPH
Gallium	Ga	31	69.72	29.8	2420	5.91	Orthorhombic
Germanium	Ge	32	72.59	937	2700	5.36	Diamond structure (cubic)
Arsenic	As	33	74.9216	(817)(36 at.)	616	5.73	Rhombohedral
Selenium	Se	34	78.96	220.5	685	4.81	Hexagonal; monoclinic
Bromine	Br	35	79.904	−7.3	58	3.12	—
Krypton	Kr	36	83.80	−157	−152	3.708 (g)	—
Rubidium	Rb	37	85.4678	38.8	680	1.53	BCC
Strontium	Sr	38	87.62	770	1350	2.6	FCC $\overset{215°C}{\rightleftharpoons}$ CPH $\overset{605°C}{\rightleftharpoons}$ BCC
Yttrium	Y	39	88.9059	1500	3030	5.51	CPH $\overset{1490°C}{\rightleftharpoons}$ BCC
Zirconium	Zr	40	91.22	1860	4400	6.5	CPH $\overset{865°C}{\rightleftharpoons}$ BCC
Niobium (Columbium)	Nb (Cb)	41	92.9064	2468	4735	8.57	BCC
Molybdenum	Mo	42	95.94	2620	4650	10.2	BCC
Technetium	Tc	43	(99)	2720	5030	11.5	CPH
Ruthenium	Ru	44	101.07	2310	4120	12.2	CPH
Rhodium	Rh	45	102.9055	1960	(3760)	12.44	FCC
Palladium	Pd	46	106.4	1552	2940	12.0	FCC
Silver	Ag	47	107.868	960.8	2210	10.49	FCC
Cadmium	Cd	48	112.40	321	765	8.65	CPH
Indium	In	49	114.82	156.4	2075	7.31	Face-centred tetragonal
Tin	Sn	50	118.69	231.9	2623	7.298	Gray tin (cubic) $\overset{13°C}{\rightleftharpoons}$ White tin (tetragonal)
Antimony	Sb	51	121.75	630.5	1610	6.62	Rhombohedral

Element	Symbol	Atomic number (Z)	Relative atomic mass	Melting point (°C)	Boiling point (°C)	Relative density	Crystal structures
Tellurium	Te	52	127.60	449.8	990	6.24	Hexagonal
Iodine	I	53	126.9045	113.6	183	4.93	—
Xenon	Xe	54	131.30	−112	−108	5.851 (g)	—
Caesium	Cs	55	132.9055	29.7	700	1.9	BCC
Barium	Ba	56	137.34	710	(1700)	3.5	BCC
Lanthanum	La	57	138.9055	920	(3420)	6.16	CPH $\rightleftharpoons$330°C FCC $\rightleftharpoons$864°C BCC
Cerium	Ce	58	140.12	804	3500	6.77	FCC $\rightleftharpoons$−178°C CPH $\rightleftharpoons$−10°C FCC $\rightleftharpoons$725°C BCC
Praseodymium	Pr	59	140.9077	932	(3020)	6.77	CPH $\rightleftharpoons$792°C BCC
Neodymium	Nd	60	144.24	1024	(3060)	7.01	CPH $\rightleftharpoons$862°C BCC
Promethium	Pm	61	(147)	—	—	—	—
Samarium	Sm	62	150.4	1052	(1900)	7.54	Rhombohedral $\rightleftharpoons$917°C BCC
Europium	Eu	63	151.96	826	1490	5.17	BCC
Gadolinium	Gd	64	157.25	1350	(3000)	7.87	CPH $\rightleftharpoons$1264°C BCC
Terbium	Tb	65	158.9254	1356	(2500)	8.25	CPH $\rightleftharpoons$near m.pt. BCC
Dysprosium	Dy	66	162.50	1500	(2630)	8.56	CPH $\rightleftharpoons$near m.pt. BCC
Holmium	Ho	67	164.9303	1500	(2700)	8.80	CPH $\rightleftharpoons$near m.pt.°C BCC
Erbium	Er	68	167.26	1530	(2600)	9.06	CPH $\rightleftharpoons$1370°C BCC
Thulium	Tm	69	168.9342	1525	(2400)	9.32	CPH $\rightleftharpoons$near m.pt. BCC
Ytterbium	Yb	70	173.04	824	(1800)	6.96	FCC $\rightleftharpoons$798°C BCC
Lutecium	Lu	71	174.97	1700	(3500)	9.85	CPH $\rightleftharpoons$near m.pt. BCC
Hafnium	Hf	72	178.49	2220	4450	11.4	CPH $\rightleftharpoons$1755°C BCC
Tantalum	Ta	73	180.9479	3010	5240	16.6	BCC
Tungsten	W	74	183.85	3380	(5500)	19.3	BCC
Rhenium	Re	75	186.2	3150	(5700)	20.0	CPH
Osmium	Os	76	190.2	3045	> 5300	22.5	CPH
Iridium	Ir	77	192.22	2443	4400	22.5	FCC

Element	Symbol		Atomic mass			Density	Crystal structure
Platinum	Pt	78	195.09	1769	3880	21.45	FCC
Gold	Au	79	196.9665	1063	(2950)	19.32	FCC
Mercury	Hg	80	200.59	−38.9	357	13.55	Rhombohedral; body-centred tetragonal
Thallium	Tl	81	204.37	303	1460	11.85	CPH $\xrightarrow{234°C}$ BCC
Lead	Pb	82	207.2	327.4	1740	11.34	FCC
Bismuth	Bi	83	208.9806	271	1680	9.80	Rhombohedral
Polonium	Po	84	(209)	246	965	(9.2)	Simple cubic $\xrightarrow{75°C}$ rhombohedral
Astatine	At	85	(210)	(300)	—	—	—
Radon	Rn	86	(222)	−71	−61.8	—	—
Francium	Fr	87	(223)	(24)	(750)	—	—
Radium	Ra	88	226.0254*	960	(1140)	5.0	—
Actinium	Ac	89	(227)	(1050)	2927	10.07	FCC
Thorium	Th	90	232.0381*	1750	(4850)	11.5	FCC $\xrightarrow{1400°C}$ BCC
Protactinium	Pa	91	231.0359*	(1425)	(4410)	15.4	Body-centred tetragonal
Uranium	U	92	238.029*	1130	3930	18.7	Orthorhombic $\xrightleftharpoons{660°C}$ tetragonal $\xrightleftharpoons{775°C}$ BCC
Neptunium	Np	93	237.0482*	637	3380	19.5	Orthorhombic $\xrightleftharpoons{278°C}$ tetragonal $\xrightleftharpoons{540°C}$ BCC
Plutonium	Pu	94	(244)	640	3454	(19.0)	Simple monoclinic $\xrightleftharpoons{122°C}$ BC monoclinic $\xrightleftharpoons{206°C}$ FC orthorhombic $\xrightleftharpoons{319°C}$ FCC $\xrightleftharpoons{451°C}$
Americium	Am	95	(243)	(995)	2600	11.7	Double CPH $\xrightleftharpoons{451°C}$ BC tetragonal $\xrightleftharpoons{476°C}$ BCC
Curium	Cm	96	(247)	—	—	—	—
Berkelium	Bk	97	(249)	—	—	—	—
Californium	Cf	98	(251)	—	—	—	—
Einsteinium	Es	99	(254)	—	—	—	—
Fermium	Fm	100	(257)	—	—	—	—
Mendelevium	Md	101	(256)	—	—	—	—
Nobelium	No	102	(254)	—	—	—	—
Lawrencium	Lw	103	(257)	—	—	—	—
		104					
		105					
		106					

(g) denotes density of the gas in g/dm^3.

*Relative atomic mass of the most commonly available long-lived isotope.

( ) values shown in parentheses are uncertain.

INDEX